PLANT PATHOGENS

Detection and Management
for Sustainable Agriculture

PLANT PATHOGENS

*Detection and Management
for Sustainable Agriculture*

Edited by
Pradeep Kumar
Ajay K. Tiwari
Madhu Kamle
Zafar Abbas
Priyanka Singh

Apple Academic Press Inc.
4164 Lakeshore Road
Burlington ON L7L 1A4, Canada

Apple Academic Press Inc.
1265 Goldenrod Circle NE
Palm Bay, Florida 32905, USA

© 2020 by Apple Academic Press, Inc.

First issued in paperback 2021

Exclusive worldwide distribution by CRC Press, a member of Taylor & Francis Group
No claim to original U.S. Government works

ISBN 13: 978-1-77463-463-9 (pbk)
ISBN 13: 978-1-77188-788-5 (hbk)

Library and Archives Canada Cataloguing in Publication

Title: Plant pathogens : detection and management for sustainable agriculture / edited by Pradeep Kumar [and four others].

Other titles: Plant pathogens (Oakville, Ont.)

Names: Kumar, Pradeep (Professor of biotechnology), editor.

Series: Innovations in plant science for better health.

Description: Includes bibliographical references and index.

Identifiers: Canadiana (print) 20190137231 | Canadiana (ebook) 20190137274 | ISBN 9781771887885 (hardcover) | ISBN 9780429057212 (ebook)

Subjects: LCSH: Phytopathogenic microorganisms. | LCSH: Phytopathogenic microorganisms—Detection. | LCSH: Phytopathogenic microorganisms—Control. | LCSH: Sustainable agriculture.

Classification: LCC SB731 .P63 2019 | DDC 632/.3—dc23

Library of Congress Cataloging-in-Publication Data

Names: Kumar, Pradeep (Professor of biotechnology), editor. | Tiwari, A. K. (Ajay K.), editor. | Kamle, Madhu, editor. | Abbas, Zafar (Professor of botany), editor. | Singh, Priyanka, 1975- editor.

Title: Plant pathogens : detection and management for sustainable agriculture / edited by Pradeep Kumar, Ajay K. Tiwari, Madhu Kamle, Zafar Abbas, Priyanka Singh.

Description: 1st edition. | Palm Bay, Florida : Apple Academic Press, 2019. | Includes bibliographical references and index. | Summary: "Plant Pathogens: Detection and Management for Sustainable Agriculture addresses the most critical issues in the management of emerging diseases throughout the world. Experts in plant pathology from internationally renowned institutes share their research and examine key literature on vital issues in pathogen disease diagnosis and management. They look at both traditional pathology as well as new and advanced biotechnological and molecular diagnosis approaches. This book is divided into four parts, covering viral and fungal disease detection and management, nematode diseases and management, bio-control, and biotechnological approaches and impact of climate change. The authors look at the challenges of crop protection against diseases caused by plant pathogens for the most economically important crops, including fruits, vegetables, and cereals. The establishment and management of plant diseases using conventional and eco-friendly methods are discussed with an emphasis on the use of beneficial microbes and modern biotechnological approaches. Plant Pathogens: Detection and Management for Sustainable Agriculture focuses on expert disease diagnosis and integrated management practices with molecular diagnostic techniques to achieve disease free-plants from a wide array of pathogens. The volume will be a valuable source of information for those involved with and studying plant pathology and crop disease management"-- Provided by publisher.

Identifiers: LCCN 2019026294 (print) | LCCN 2019026295 (ebook) | ISBN 9781771887885 (hardcover) | ISBN 9780429057212 (ebook)

Subjects: LCSH: Phytopathogenic microorganisms--Research.

Classification: LCC SB732.5 .P53 2019 (print) | LCC SB732.5 (ebook) | DDC 632/.3072--dc23

LC record available at https://lccn.loc.gov/2019026294

LC ebook record available at https://lccn.loc.gov/2019026295

About the Editors

Pradeep Kumar, PhD

Pradeep Kumar, PhD, is currently working as an Assistant Professor in the Department of Forestry, North Eastern Regional Institute of Science and Technology (NERIST) Deemed to be University-MHRD, Government of India, Nirjuli, Arunachal Pradesh, India. Before he joined NERIST, he worked as an international research professor/assistant professor in the Department of Biotechnology, Yeungnam University, South Korea. He was Postdoctorate Researcher in the Department of Biotechnology Engineering, Ben-Gurion University of the Negev, Israel, and was awarded a PBC-outstanding Post-Doc Fellowship for more than three years. His areas of research and expertise are wide, including microbial biotechnology, plant pathology, bacterial genetics, insect-pest biocontrol, gene expression, cry genes, and molecular biology. He has been honored with an international travel grant from Ben-Gurion University of Negev, Israel, to attend an international conference. He was awarded with prestigious Early Career Research Award from Department of Science & Technology, Government of India. He is the recipient of a best paper presentation and the Narasimhan Award by the Indian Phytopathological Society, India. He has presented several oral and poster presentations at various national and international conferences. He has published three book and 50 research and review articles in peer-reviewed journals and several book chapters with international publishers, including Springer, CABI, Bentham, and Apple Academic Press. He is serving as an associate editor for PLOS ONE, *BMC-Complementary and Alternative Medicine,* a guest editor for *Evidence Based Complementary and Alternative Medicine,* and he also provides his services to over 30 journals as editor, editorial board member, technical editor, and peer reviewer.

Ajay Kumar Tiwari, PhD

Ajay K. Tiwari, PhD, is a Scientific Officer at the UP Council of Sugarcane Research, Shahjahnapur, UP, India. He has published 70 research articles, nine review articles in national and international journals, several book chapters in edited books, and has also authored several edited books published by Springer, Taylor & Francis, and others. He has submitted more than 150 nucleotide sequences of plant pathogens to GenBank. He is a regular reviewer of many international journals as well as an editorial board member. He is Managing Editor of the journal *Sugar Tech* and Chief Editor of the journal *Agrica*. He has received several young researcher awards and was nominated for the Narshiman Award by the Indian Phytopathological Society, India. Dr. Tiwari is also the recipient of many international travel awards given by governmental agencies in India and other countries. He has visited Belgium, Brazil, China, Italy, Germany, and Thailand for conferences and workshops and has delivered several invited talks at Oman University for PhD students on phytoplasma disease diagnosis and management. Dr. Tiwari has been involved in the research on molecular characterization and management of agricultural plant pathogens for the last nine years. Currently, he is working on molecular characterization of sugarcane phytoplasmas and their secondary spread in nature. He is a regular member of several professional organizations, including the British Society of Plant Pathology, Indian Phytopathological Society, Sugarcane Technologists Association of India, International Society of Sugarcane Technologists, Society of Sugarcane Research and Promotion, and others. Dr. Tiwari earned his PhD from CCS University, Meerut, Uttar Pradesh, India.

Madhu Kamle

Madhu Kamle is currently working as an Assistant Professor in Department of Forestry, North Eastern Regional Institute of Science & Technology, Nirjuli, Arunachal Pradesh, India. Her area of research is plant biotechnology, plant–microbe interaction, microbial genomics, and plant disease diagnosis. She did her PhD in plant biotechnology at ICAR-CISH, Lucknow, and Bundelkhand University, Jhansi, India. She had been awarded a prestigious PBC Outstanding Post-doctoral Fellowship (2014–2016) from the Council of Higher Education, Israel, and is a recipient of a post-doc fellowship from the Jacob Blaustein Institute of Desert Research, Ben Gurion University, Israel (2013–2014). She has also worked as an International Research Professor in the School of Biotechnology, Yeungnam University, Gyeongsan, Republic of Korea.

She has 10 years of research experience and has published 25 research papers in peer-reviewed journals, 10 book chapters, and one edited book from Springer Nature, Switzerland. She is working as an editor for Science

Alert journals and Taylor & Francis journals, and as a reviewer for various Springer, Taylor & Francis, Frontiers, and PLOS journals. She is a life member of the Nano-Molecular Society and member of the American Society of Microbiology.

Zafar Abbas, PhD

Zafar Abbas, PhD, is a Senior Associate Professor and Chairman in the P. G. Department of Botany at G. F. College, M. J. P. Rohilkhand University in Shahjahanpur, Uttar Pradesh, India. He has 40 years of research experience in plant and crop physiology, with a specialization in plant nutrition. He has attended several national and international seminars and conferences. At present, eight PhD students have completed their doctorate degrees under his supervision. Dr. Abbas is a life member and member of editorial boards of several Indian and foreign journals and societies, and he has authored a book and has published over 30 articles.

Priyanka Singh, PhD

Priyanka Singh, PhD, is a Scientific Officer of the Uttar Pradesh Council of Sugarcane Research, (UPCSR) Shahjahanpur, India. She has worked at the Indian Institute of Sugarcane Research, Lucknow, India, for nine years. She has 19 years of research experience with a specialization in organophosphorus chemistry and in the area of cane quality/postharvest management of sugar losses with the help of chemicals as well as ecofriendly compounds. She has extensive experience using electrolyzed water to preserve cane quality and is responsible for the first time that it was reported that electrolyzed water has immense potential to be used in the sugar industry to preserve postharvest sucrose losses. She has synthesized and characterized 37 new organophosphorus compounds belonging to the chalcone series, of which two important chemicals (chalcone thiosemicarbazone and chalcone dithiocarbazate) were found to be highly fungitoxic to the sugarcane parasitic fungi *Colletotrichum falcatum, Fusarium oxysporum,* and *Curvularia pallescens.* She has worked on the extraction of volatile constituents from higher plants and their biological activity against agricultural pests. She has also worked on the management of postharvest formation of nonsugar and polysaccharide compounds in sugarcane and the effect of bioproducts on growth, yield, and quality of sugarcane and soil health. She has also worked on indicators of postharvest losses in sugarcane and reported that mannitol is one of the most important indicators.

Presently she is working on varietal spectrum of sugarcane for the selection of elite sugarcane varieties so as to recommend the proper varietal balance of sugarcane varieties in Uttar Pradesh, as well as working on modulating the activities of sucrose metabolizing enzymes through bioactive silicon (orthosilicic acid) for increased cane and sugar productivity, which will benefit farmers as well as the sugar industry in remarkable way. She is also managing and working on a project on "Assessment of postharvest quality deterioration in promising sugarcane varieties under sub-tropical condition" which is expected to reduce postharvest losses and will increase sugar recovery. She is carrying a project on "Varietal screening for jaggery" production at UPCSR, Shahjahanpur, for the recommendation of elite sugarcane varieties for commercial production of jaggery.

Dr. Singh is a research advisor for a dissertation on genetic diversity in sugarcane and a training advisor for MSc students. She has organized several short-term training programs on techniques in microbiology, biotechnology, and molecular biology. She received an "Award of Excellence" from Sinai University, Al Arish, Egypt, in 2008. She is also serving as Managing Editor for the journal *Sugar Tech* and as Executive Editor for journal *Agarica*. In addition, she is a reviewer for several international journals. She is one of the editors of the Proceeding of International Conference IS-2011. A prolific author, Dr. Singh has authored a book on innovative healthy recipes with jaggery, edited three books on postharvest management of sugarcane, written several annual reports, written on 100 years of sugarcane research, and published three book chapters and more than 50 research papers in various national and international journals and proceedings. She has attended several national and international conferences and workshops in China, Egypt, Thailand, and India, and has coordinated technical as well as plenary sessions in India, China, and Egypt.

Dr. Singh completed her PhD on "Efficacy of organophophorus derivatives against fungal pathogens of sugarcane" in 2000 from DDU, Gorakhpur University (Uttar Pradesh, India). She was awarded a postdoctoral fellowship from DST, New Delhi in the years 2006 and 2010.

Contents

Contributors

Elvis Asare-Bediako
Department of Crop Science, University of Cape Coast, Cape Coast, Ghana.
E-mail: easare-bediako@ucc.edu.gh

Bhavin S. Bhatt
Shree Ramkrishna Institute of Computer Education and Applied Sciences, Surat.
E-mail: bhavin18@gmail.com

Mohammad Danish
Department of Botany, Aligarh Muslim University, Aligarh 202002, Uttar Pradesh, India.
E-mail: danish.botanica@gmail.com

Utpal Dey
Division of Crop Production, ICAR Research Complex for NEH Region, Umiam 793103,
Meghalaya, India. E-mail: utpaldey86@gmail.com

Quazi Mohd. Imranul Haq
Deparment of Biological Sciences and Chemistry, College of Arts and Sciences,
University of Nizwa, Nizwa, Sultanate of Oman

Hisamuddin
Department of Botany, Aligarh Muslim University, Aligarh, 202002, Uttar Pradesh, India

Touseef Hussain
Division of Plant Pathology, ICAR—Indian Agricultural Research Institute, Pusa,
New Delhi 110012, India. E-mail: hussaintouseef@yahoo.co.in

G. P. Jagtap
Department of Plant Pathology, College of Horticulture, VNMKV, Parbhani, Maharashtra, India

Vinayaka S. Kanivebagilu
Department of Botany, Kumadvathi First Grade College, Shimoga Road, Shikaripura 577427,
Shimoga, Karnataka, India. E-mail: ks.vinayaka@gmail.com

Samiullah Khan
Department of Botany, Aligarh Muslim University, Aligarh, 202002, Uttar Pradesh, India

Shveta Malhotra
Arya Mahila PG College, Shahjahanpur 242001, Uttar Pradesh, India.
E-mail: shvetamudit@gmail.com

Archana R. Mesta
Department of Botany, Kumadvathi First Grade College, Shimoga Road, Shikaripura 577427,
Shimoga, Karnataka, India

S. K. Mishra
Punjab Agricultural University, Regional Station, Faridkot 151203, Punjab, India

A. K. Misra
Department of Agricultural Meteorology, B.A. College of Agriculture,
Anand Agricultural University, Anand 388110, Gujarat, India.
E-mail: ashueinstein@gmail.com

Aamir Raina
Department of Botany, Aligarh Muslim University, Aligarh, 202002, Uttar Pradesh, India

Richa Salwan
Department of Veterinary Microbiology, CSK-Himachal Pradesh Agricultural University,
Palampur 176062, India

Ravulapenta Sathish
Senior Research Fellow (Entomology), National Institute of Plant Health Management (NIPHM),
Hyderabad 500030, India

Vivek Sharma
Department of Plant Pathology, CSK-Himachal Pradesh Agricultural University,
Palampur 176062, India. E-mail: ankvivek@gmail.com

K. Shukla
Department of Botany, D.D.U. Gorakhpur University, Gorakhpur, Uttar Pradesh, India

Achuit K. Singh
Crop Improvement Division, ICAR Indian Institute of Vegetable Research, Varanasi,
Uttar Pradesh, India. E-mail: achuits@gmail.com

Naresh Pratap Singh
Department of Biotechnology, Sardar Vallabhbhai Patel University of Agriculture and Technology,
Meerut 250110, Uttar Pradesh, India. E-mail: naresh.singh55@yahoo.com

Deepa Srivastava
Department of Botany, D.D.U. Gorakhpur University, Gorakhpur, Uttar Pradesh, India

B. S. Sunanda
Assistant Scientific Officer (Nematology) and Centre In-charge, AICRP (Nematode).
E-mail: patilsunanda722@gmail.com

Diganggana Talukdar
Department of Plant Pathology and Microbiology, College of Horticulture,
Central Agricultural University, Sikkim, India

M. K. Tripathi
College of Agriculture, Rajmata Vijayaraje Scindia Krishi Vishwa Vidyalaya, Gwalior 474002,
Madhya Pradesh, India

Vaishali
Department of Biotechnology, Sardar Vallabhbhai Patel University of Agriculture and Technology,
Meerut 250110, Uttar Pradesh, India

Vibha
Department of Plant Physiology, Jawaharlal Nehru Krishi Vishwa Vidyalaya, Jabalpur 482004, India.
E-mail: vibhapandey93@gmail.com

S. B. Yadav
Department of Agricultural Meteorology, B.A. College of Agriculture,
Anand Agricultural University, Anand 388110, Gujarat, India

S. M. Yahaya
Department of Biology, Kano University of Science and Technology, Wudil P.M.B. 3244, Nigeria.
E-mail: sanimyahya@gmail.com

Abbreviations

AMP	antimicrobial peptide
ANN	artificial neural network
BA	betaine aldehyde
BADH	betaine aldehyde dehydrogenase
BCA	biocontrol agent
CBD	cannabidol
CCN	cereal cyst nematode
CMV	cucumber mosaic virus
CYMV	Catharanthus yellow mosaic virus
DP	Dhan Pant
dsRNA	double-stranded ribonucleic acid
ELISA	enzyme-linked immunosorbent assay
EPSPS	5-enol-pyruvyl shikimate-3-phosphate synthase
FOL	*Fusarium oxysporum* f. sp. lycopersici
FYM	farm yard manure
GUS	β-glucuronidase
HAD	helicase-dependent amplification
IR	inverted repeat
LSC	large single copy
NDH	NADH-dehydrogenase
NDR	Narendra
NRPS	nonribosomal peptide synthetase
OLCD	okra leaf curl disease
OMD	okra mosaic disease
PCR	polymerase chain reaction
PEG	polyethylene glycol
PGPR	plant growth promoting rhizobacteria
pHBA	p-hydroxybenzoic acid
PPN	plant parasitic nematodes
PSB	phosphorus solubilizing bacteria
PSF	phosphorus solublizing fungi
PSMs	phosphate solubilizing microorganisms
PYVV	potato yellow vein virus

RBS	ribosomal-binding site
RFLP	restriction fragment length polymorphism
RKN	root-knot nematodes
RNAi	RNA interference
RPA	recombinase polymerase amplification
SCP	soluble crude protein
siRNA	small interfering RNA
SSC	small single copy
TCP	tri-calcium phosphate
TIBA	tissue blot immunoassay
tsp	total soluble protein
TSWV	tomato spotted wilt virus
UTR	untranslated region
VOC	volatile organic compound

Preface

Plant Pathogens: Detection and Management for Sustainable Agriculture addresses one of the most critical issues for the management of emerging diseases throughout the world. Plant diseases caused by fungi, bacteria, viruses, etc., collectively represent a significant burden to crop production and a threat to global food security and agriculture sustainability. The agricultural productivity must increase with the global population increase and the climate change scenario. The diagnosis of plant diseases can be difficult at the early stages of disease on individual crops as well as at the early stages of an epidemic. However, for many diseases, symptoms do appear during early stages, and thus based on diagnosis, the applicable management approaches, including cultural, chemical, biological, have been considered worldwide for sustainable productivity. Accurate estimates of disease incidence, disease severity, and the negative effects of diseases on the quality and quantity of agricultural produce are important for field crops, horticulture, plant breeding, and for improving the fungicide efficacy as well as for the basic and applied research. Therefore, it is significant to have expert disease diagnosis and integrated management practices advanced with molecular diagnostic techniques to obtain disease-free plants from a wide array of pathogens.

This book volume consists of 14 chapters and basically provides expert knowledge on new approaches, updated techniques, and useful information on crop diseases caused by various pathogenic agents and also on their management. This book is divided into four parts: Part I includes viral and fungal disease management and contains seven chapters; Part II consists of nematode diseases and management and contains two chapters; Part III consists of biocontrol and contains two chapters; and Part IV consists of biotechnological approaches and the impact of climate change and contains three chapters.

In this book, expert researchers share their research knowledge and key literature on vital issues covering the pathogen disease diagnosis and management addressing with traditional pathology as well as biotechnological approaches with advanced molecular diagnosis approaches. We are extremely delighted and grateful to all the authors for their expert contributions in the form of chapters, making this volume edition possible.

We are extremely grateful to the staff of Apple Academic Press and others concerned with CRC Press and Taylor & Francis Group for their untiring effort and immense support throughout. This book presents intense information on crop disease diagnosis and management for sustainable agriculture and would be extremely helpful for wide array of researchers, scientists, and academicians. We also hope that it will be useful to all concerned.

—**Pradeep Kumar**
Ajay K. Tiwari
Madhu Kamle
Zafar Abbas
Priyanka Singh

PART I
Viral and Fungal Disease and Management

CHAPTER 1

Viral Diseases of Okra in Ghana and Their Management

ELVIS ASARE-BEDIAKO*

Department of Crop Science, University of Cape Coast, Cape Coast, Ghana

*Corresponding author. E-mail: easare-bediako@ucc.edu.gh

ABSTRACT

Viral diseases are major biotic factors that affect productivity of okra (*Abelmoschus esculentus* L. Moench) worldwide. Okra mosaic disease (OMD) caused by *Okra mosaic virus* (OkMV; genus *Tymovirus*; family *Tymoviridae)* and Okra leaf curl disease (OLCD) caused by a complex of begomoviruses: *Cotton leaf curl Gezira virus* (CLCuGV]), *Okra yellow crinkle virus* (OYCrV), *Hollyhock leaf crumple virus* (HoLCrV), and *Okra leaf curl virus* (OLCV) are the major viral diseases of okra in West Africa including Ghana. OLCD and OMD are commonly observed among okra crops in Ghana, with disease incidence of up to 100% depending on the okra cultivar and stage of growth. Management of OLCD and OMD involves the use of both synthetic and phytopesticides against the *Bemisia tabaci* and *Podagrica* spp. vectors, respectively, as well as the use of compost and fertilizers to ensure healthy growth of plant and to improve the tolerance of plants against viral infection. Resistance and tolerant okra genotypes have so been identified and their integration with phytopesticides and judicious use of chemical pesticides is recommended for effective management of these viral diseases.

1.1 BACKGROUND

Okra (*Abelmoschus esculentus* L. Moench) is a member of the family *Malvaceae* and a native to West and Central Africa but is now widely grown throughout the tropics (Kochhar, 1986; Schippers, 2000). The world production of common okra as fresh vegetable is estimated at 1.7 million tons year^{-1} (Schippers, 2000; Asare-Bediako et al., 2014). Ghana is the eighth largest producer of okra in the world (FAOSTAT, 2014).

Okra crop is the third most important vegetable in Ghana after pepper and tomato, with production of 80,000 tons estimated at $51, 189,000 USD (FAOSTAT, 2011; 2013).

It can be grown anywhere in Ghana but the major producing centers are Brong Ahafo, Ashanti, Northern, Volta, Greater Accra, and Central regions (NARP, 1993). Okra production provides livelihood, employment, and income to rural smallholder farmers and retailers in urban centers. Okra is an important fruit vegetable crop in Ghana, and a source of energy for human consumption (Babatunde, 2007). The crop is a rich source of protein, fat, carbohydrate, fiber, thiamine, riboflavin, nicotinamide, and ascorbic acid (Hamon, 1988; Schippers, 2000; Babatunde, 2007). It also contains significant amount of potassium, magnesium, calcium, and iron (Hamon and Charrier, 1997). Okra is a multipurpose fruit vegetable due to its diverse uses of the fruits (pods), fresh leaves, buds, flowers, stems, and seeds (Mihretu et al., 2014). Immature okra fruits and fresh leaves are usually consumed as vegetables while the dried fruits are ground into powder and used in stews and soups (Siemonsma, 1982a). Okra seeds can be used as substitutes or additives in feed preparation (Purseglove, 1974), in the preparation of okra seed meal (Martin and Roberts, 1990), in the confectionery industry (Adetuyi et al., 2011), and in blood plasma replacement or blood volume expander.

In spite of the significant contribution (75%) of the West and Central African region including Ghana to okra production in Africa, average productivity in the region (2.5 t ha^{-1}) is far below that of East (6.2 t ha^{-1}) and North Africa (8.2 t ha^{-1}) (FAOSTAT, 2008). In Ghana, yield potential of up to 3.0 t ha^{-1} has been reported for Okra (MoFA, 2007), depending on the cultivar, harvesting frequency, and period for harvesting (Cudjoe et al., 2005) but current average yield is 2.1 t ha^{-1} (FAOSTAT, 2014). The wide yield gap of okra in Ghana could be attributed to several production constraints including biotic and abiotic factors. Insect pests and plant viruses are important biotic factors causing severe constraints on the productivity of okra in Ghana (Obeng-Ofori and Sackey, 2003; Asare-Bediako et al., 2014a). Viral diseases are major constraints to okra production worldwide (Ndunguru and Rajabu, 2004; Asare-Bediako et al., 2014a, b). The productivity of okra is affected by at least 19 plant viruses. Of these, *Okra mosaic virus* (OkMV; genus *Tymovirus*; family *Tymoviridae*), *Bhendi yellow vein mosaic virus* (BYVMV, genus *Begomovirus*), *Cotton leaf curl Gezira virus* (CLCuGV, genus *Begomovirus*), and Okra leaf curl virus (OLCuV; genus *Begomovirus*) are the most common and well-studied (Brunt et al., 1990; Swanson and Harrison, 1993; Tiendrebego et al., 2010; Sayed et al., 2014). Okra mosaic disease (OMD) and okra leaf curl disease (OLCD) are the common viral diseases affecting okra production in Ghana (Siemonsma, 1991; Norman, 1992; Asare-Bediako et al., 2014a). *Okra yellow vein mosaic virus* (OYVMV) is a major limiting factor to okra production in India (Sayed et al., 2014).

1.2 OKRA MOSAIC DISEASE

OMD caused by OkMV is the most common viral disease of okra in West Africa including Ghana. The virus contains a single-stranded positive-sense RNA (approximately 6.2 kb) with isometric particles of 28 nm in diameter with icosahedral symmetry and 32 morphological units (Koenig and Givord, 1974; Givord and Hirth, 1973). OkMV is transmitted in a nonpersistent manner by flea beetles (*Podagrica* species) (Brunt et al., 1990, 1996). The virus can also be mechanically transmitted (Koenig and Givord, 1974). It has a wide host range, and infects 105 plant species and varieties in 23 dicotyledonous families (Givord and Hirth, 1973) including both crop and weed species.

Symptoms of OkMV infection are prevalent in okra fields in Ghana. Field surveys conducted by Asare-Bediako et al. (2014c) at the Komenda–Edina–Eguafo–Abirem (KEEA) municipality of the Central region showed mean disease incidences ranging from 78% to 83%. A recent field survey carried out by Agyarko (2016) revealed mean OMD incidences of 67.6%, 76.2%, and 75% at the coastal savannah, forest, and transition agroecological zones, respectively, of the Central region of Ghana. The corresponding mean symptom severity scores at these three zones were 1.60, 2.04, and 1.9, respectively, indicating mild infection (Agyarko, 2016). Field trial involving 20 okra genotypes showed disease incidence of up to 100% depending on the cultivars and the growth stage (Asare-Bediako et al., 2017). Double antibody sandwich ELISA (DAS-ELISA) detected OkMV in all the 20 okra genotypes. Prevalence of OMD in okra fields has also been reported in Ivory Coast (Givord et al., 1972; Fauquet and Thouvenel, 1987) and Nigeria (Koenig and Givord, 1974; Alegbejo, 2001; Fajinmi and Fajinmi, 2010). Common symptoms associated with OkMV infection of okra include mosaic, vein chlorosis, and vein-banding and stunted growth (Koenig and Givord, 1974; Brunt et al., 1990; Swanson and Harrison, 1993). OMD has been reported to cause yield losses of up to 100% in okra crops (Atiri, 1984; Alegbejo, 2001).

1.3   OKRA LEAF CURL DISEASE

Okra leaf curl disease (OLCD) is a major constraint on okra production in West Africa. In Africa, the disease is associated with a number of bego-moviruses of the family *Geminiviridae*, which are transmitted by *Bemisia tabaci* Genn. (Brown and Bird, 1992; Brown and Czosnek, 2002; Brown, 2007, 2010). These begomoviruses include: *Cotton leaf curl Gezira virus* (CLCuGV; [Tiendrebego et al., 2010]), *Okra yellow crinkle virus* (OYCrV; [Shih et al., 2007]), and *Hollyhock leaf crumple virus* (HoLCrV; [Bigarré et al., 2001; Idris et al., 2002]), and *Okra leaf curl virus* (OLCuV; [Brunt et al., 1990; Swanson and Harrison, 1993]). The disease causes leaf wrinkle, curl, vein distortion, leaf yellowing, stunted growth, and reduced yields (Askira, 2012). OLCD has been reported to cause yield losses of up to 100% depending on the date of planting, cultivar, and locality (Fauquet and Thouvenel, 1987; Brown and Bird, 1992; Basu, 1995). The average economic losses due to OLCD have been estimated between 11,100 USD and 1950 USD for 1 ha of crop, depending on the okra variety (Tiendre-bego et al., 2010a). Report by Asare-Bediako et al. (2014a) revealed high incidences and severities of both OLCD in all the communities surveyed in the KEEA municipality of the Central region of Ghana. Furthermore, survey conducted by Agyarko (2016) revealed the highest mean incidence and symptom severity of OLCD at the coastal savannah zone, followed by the forest zone and then the transition zone of the Central region. Field experiment involving 20 okra accessions conducted by Agyarko (2016) at the coastal savannah zone of the Central region showed disease incidence of up to 100% depending on the cultivar. Field trial conducted by Oppong-Sekyere in the forest zone of the Ashanti region of Ghana involving 25 okra accessions showed high incidence of OLCD. OLCD has also been reported in other African countries including Burkina Faso (Tiendrebego et al., 2010), Cameroon (Leke, 2010), Ivory Coast (N'Guessan et al., 1992), Mali (Konet al., 2009), Nigeria (Atiri and Ibidapo, 1989; Alegbejo, 1997;

Askira, 2012), Niger (Shih et al., 2009), Nigeria (Askira, 2012), and Sudan (Idris and Brown. 2002).

1.4   MANAGEMENT OF VIRAL DISEASES

Effective management of viral diseases is quite pertinent in order to improve yields of okra. Various strategies are employed in the management of plant virus diseases and these are mainly directed at preventing virus infection by eradicating the source of infection to prevent the virus from reaching the crop, reducing the spread of the disease by managing its vector, using virus-free planting material, and planting resistant varieties (Naiduand Hughes, 2003). Karim (2016) stated that although most farmers practice strict monitoring or calendar spraying with chemical insecticides to control insects that vector these viruses, they still observe severe yellowing on plants and probably because viruses responsible for the yellowing are not mainly insect-transmitted.

1.4.1   THE USE OF RESISTANT VARIETIES

Study conducted by Asare-Bediako et al. (2014c) revealed that the majority of farmers acquire their planting materials (okra seeds) from their own farm (uncertified source) and this practice contributes to the spread of diseases. The planting of resistant cultivars has therefore been universally considered the most effective method to control diseases caused by viruses in okra. In screening 21 okra genotypes against OkMV infection, Asare-Bediako et al. (2017) showed that nine genotypes GH2052, GH2063, GH2026, GH3760, GH5302, GH5332, GH5793, GH6105, and UCCC6 exhibited mild symptoms of OMD, and were less susceptible to flea beetle infestation and associated leaf damage during both major and minor cropping seasons in Ghana. Asare-Bediako et al. (2016) also identified okra genotypes GH3760, GH2052, GH5332, UCC6, GH5302, GH5793, and GH2063 showed mild symptoms of OLCD, when they screened 21 okra genotypes against OLCV infection under natural conditions.

In assessing the performance of 25 okra accessions against viral infection, Oppong-Sekyere (2011) identified accessions Atuogya-tenten, GH3736 Fetri, Atuogya-tiatia, Atuogya-Asante, and GH4376 Atuogya to show high tolerance to OLCD/OMD and pests. He also reported that accessions KNUST/SL1/07Nkrumahene, DA/08/02Dikaba, DA/08/03Sheo mana, DA/08/004Agbodro, DA/08/02Asontem, DA/08/02Sheo mana, DA/08/001Wun mana, and GH 5787Asontem did not show any signs of viral infestation, and hence can be said to exhibit field resistance against viral infection. In screening *A. esculentus* and *A. callei* cultivars against OLCD and OMD, under field conditions in Nigeria, Udengwu, and Dibua (2014) identified *A. callei* cultivars EbiOgwu, Ojoogwu, Tongolo, VLO, Oruufie, and Ogolo to be resistant to these two viral diseases. They discussed the potential of incorporating these resistant genes from *A. callei* cultivars into the susceptible *A. esculentus* cultivars.

1.5   MANAGEMENT OF THE VECTOR WITH INSECTICIDES AND PHYTOPESTICIDES

According to Naidu and Hughes (2003) and Bhagati and Goswami (1992), management of viral disease can be directed at controlling the vector that transmits the virus. Positive correlations between incidences of viral diseases and the vectors that transmit them have been reported in several host plant–vector–viruspathosystems (Bhagati and Goswami, 1992). In Ghana, positive association between the populations of *B. tabaci* and the severity of OLCD has been reported by Asare-Bediako et al. (2014b). They demonstrated that phytopesticides can significantly reduce the *B. tabaci* vector population and reduce incidence and severity of OLCD, leading to improved fruit yield. Aqueous neem leaf and garlic extracts were found to be more effective than that of mahogany, bougainvillea, chili pepper, and pawpaw leaves in reducing the populations of whitefly and decreasing incidence and severity of OLCD (Asare-Bediako et al., 2014b). These botanicals have been shown to possess virus inhibition, and insectrepellent/ anti-feedant properties (Schmutterer, 1990; Revkin, 2000; Nevala, 2000; Asare-Bediako et al., 2014b). In assessing the effectiveness of different plant extracts against *Podagrica* spp. infestation and OMD, Asare-Bediako et al. (2014a) indicated that the phytopesticides exhibited moderate to high level of efficacy in decreasing the insect populations and the incidence and severity of OMD. Botanicals or plant leaf extracts have been used in the control of OMV in which karamja extract treated plants had minimal virus incidence, maximum plant height, flower production, fruit forma-tion, and highest yield as reported by Bhyan et al. (2007). Obeng-Ofori and Sackey (2003) also reported Actellic (synthetic pesticides), neem seed extract (botanical), and *Bacillus thurigiensis* (Bt bacteria) are very effective in reducing the population and damage caused by the major insect pests of okra including flea beetles and whiteflies, thereby improving yield and quality of okra fruits.

1.5.1 ALTERING THE PLANTING DATE

It has been reported that the ecology of the flea beetle and whitefly vectors should be ascertained so as to alter the planting date of okra such that the period of vector abundance coincides with the growth stage when plants are old enough to tolerate effects of viral infection on the field (Fajinmi and Fajinmi, 2010a). Agyarko (2016) observed significantly higher infestations of whitefly and higher final severity of OLCD in the dry season than in the wet season. This suggests that dry season plantings of okra can result in higher vector (whitefly) infestations and severe viral infections than wet season plantings. Similarly, Asare-Bediako et al. (2017) observed that the overall mean severity of OMD recorded at 10 weeks after planting in the minor cropping season was significantly higher than that of the major cropping season.

1.5.2 THE USE OF PHYSICAL BARRIER TO CONTROL THE FLEA BEETLE VECTORS

Fajinmi and Fajinmi (2010b) reported that when a netting barrier is erected around okra plants till 21 days after emergence, it excludes the plants from infestations by flea beetles which vector OkMV and hence decreases incidence of OMD.

1.5.3 THE USE OF SOIL AMENDMENT

Application of compost also reduced incidence and severity of OMD, and improved the yield and quality of okra fruits though it did not significantly influence the population of flea beetles that infested the okra plants (Agyei et al., 2017). According to Badejo and Togun (1998), compost application ensures release of nutrients in balanced proportions that prevent excessive gaseous and leaching losses to ensure synchrony between nutrient supply and crop uptake. This in turn ensures healthy growth of okra plants which improves the resistance and/or tolerance of the plants to viruses and pests attacks, thereby improving yield and quality of produce (Agyei et al., 2017).

1.5.4 INTEGRATED PEST AND DISEASE MANAGEMENT APPROACH

Classical integrated pest management (IPM) employs a systematic combination of practices (hygienic, cultural, agronomic), the use of natural control mechanisms employing natural enemies and plant extract, in addition to the judicious use of chemical pesticides to achieve economic management of pest levels above an economic threshold. This has been adopted as the national crop protection policy for Ghana (Kyofa-Boamah et al., 2005). IPM against viral diseases and pests of okra involves planting of resistant varieties, rogueing out of diseased plants, rotation with non-hosts crops, avoiding smoking when handling or working in okra fields, controlling insect vectors with recommended insecticides and/or use of insecticidal soaps before disease spread (MoFA, 2013). In Ghana, crude neem seed extracts and *Bacillus thurigiensis* (Bt) can be used effectively by farmers as a component of IPM in okra (Obeng-Ofori and Sackey, 2003).

1.6 CONCLUSIONS

OMD and OLCD have been demonstrated as the major viral diseases affecting okra production in Ghana. These diseases are prevalent in Ghana with disease incidences of up to 100% reported depending on the cultivar. Incidence and severity of OLCD are highest at the coastal savannah zone, followed by forest while the transition agroecological zone had the lowest. Field screening of okra genotypes against viral infection showed genotypes with tolerance and resistance to both OMD and OLCD. Other management strategies adopted so far include the use of physical barrier (wire netting), both synthetic pesticides and phytopesticides against insect vectors and the use of soil amendment (compost/fertilizers). Integration of various methods to ensure effective management of these diseases is therefore recommended.

KEYWORDS

- okra
- viruses
- okra mosaic disease
- viruses
- management strategies
- pesticides

REFERENCES

Adetuyi, F. O.; Osagie, A. U.; Adekunle, A. T. Nutrient, Antinutrient, Mineral and Zinc Bioavailability of Okra *Abelmoschus esculentus* (L) Moench Variety. *Am. J. Food Nutr.* **2011**, *1* (2), 49–54.

Agyei K. F.; Asare-Bediako, E.; Amissah, R.; Daniel Okae-Anti, D. Influence of Compost on Incidence and Severity of Okra Mosaic Disease and Fruit Yield and Quality of Two Okra (*Abelmoschus esculentus* L. Moench) Cultivars. *Int. J. Plant Soil Sci.* **2017**, *16* (1), 1–14.

Alegbejo, M. D. In *Evaluation of Okra Genotype for Resistance to Okra Mosaicvirus*, 15th Annual Conference of the Horticultural Society of Nigeria, National Horticultural Research Institute: Ibadan, 1997; p 60.

Agyarko, F. Studies of Okra Mosaic and Okra Leaf Curl Disease in the Central Region of Ghana. M.Phil. Thesis, Department of Crop Science, School of Agriculture, University of Cape Coast, Ghana, 2016, p 150.

Asare-Bediako, E.; Agyarko, F.; Verbeek, M.; Taah, K. J.; Asare, A. T.; Frimpong, K.; Agyei, K. F.; Sarfo; Eghan, M. J.; Combey, R. Variation in the Susceptibility of Okra (*Abelmoschus esculentus* L. Moench) Genotypes to Okra Mosaic Virus and Podagrica Species under Field Conditions. *J. Plant Breed. Crop Sci.* **2017**, *9* (6), 79–89.

Asare-Bediako, E.; Agyarko, F.; Kingsley, J.; Taah, K. J.; Aaron Asare, T. A.; Agyei Frimpong, K.; Sarfo, J. Phenotypic and Serological Screening of Okra Genotypes for Resistance Against Okra Mosaic Disease. RUFORUM Working Document Series (ISSN 1607-9345) No. 14, 2016, pp 571–580.

Asare-Bediako, E.; Addo-Quaye, A. A.; Bi-Kusi, A. Comparative Efficacy of Phytopesti-cides in the Management of *Podagrica* spp and Mosaic Disease on Okra (*Abelmoschus-esculentus* L). *Am. J. Exp. Agric.* **2014a**, *4* (8), 879–889.

Asare-Bediako, E.; Addo-Quaye, A. A.; Bi-Kusi, A. Comparative Efficacy of Plant Extracts in Managing Whitefly (*Bemisia Tabaci* Gen) and Leaf Curl Disease in Okra (*Abelmoschus esculentus* L). *Am. J. Agric. Sci. Technol.* **2014b**, *2* (1), 31–41.

Asare-Bediako, E.; Van der Puije, G. C.; Taah, K. J.; Abole, E. A.; Baidoo, A. Prevalence of Okra Mosaic and Leaf Curl Diseases and *Podagrica* spp. Damage of okra (*Albelmoschus esculentus*) Plants. *Int. J. Curr. Res. Acad. Rev.* **2014c**, *2* (6), 260–271.

Askira, A. B. A Survey on the Incidence of *Okra leaf curl virus* on Okra in Lake Alau Area of Borno State, Nigeria. *Int. J. Agric.* **2012**, *4* (1), 1–6.

Atiri, G. I.; Ibidapo, B. Effect of Combined and Single Infections of Mosaic and Leaf Curl Virus on Okra Growth and Yield. *J. Agric. Sci.* **1989**, *112*, 413–418.

Badejo, M. A.; Togun, A. O. *Strategies and Tactics of Sustainable Agriculture in the Tropics.* College Press Ltd., 1998.

Basu, A. N. Bemisia Tabaci (Gen.) Crop Pest and Principal Whitefly Vector of Plant Viruses. West View Press: Boulder, San Francisco, Oxford, 1995, p 183.

Bhagathi, V. K.; Goswani, B. K. Incidence of Yellow Vein Mosaic Disease on Okra in Relation to Whitefly Population and Different Sowing Time. *Ind. J. Virol.* **1992**, *8*, 37–39.

Bhyan, B. S.; Alam, M. M.; Ali, M. S. Effect of Plant Extracts on Okra Mosaic Virus Incidence and Yield Related Parameters of Okra. *Asian J. Agric. Res.* **2007**, *1* (3), 112–118.

Brown, J. K.; Bird, J. Whitefly Transmitted Geminiviruses and Associated Disorders in the Americas and the Caribbean Basin. *Plant Dis.* **1992**, *76*, 220–226.

Brunt, A.; Crabtree, K.; Gibbs, A. J. *Viruses of Tropical Plants.* CAB International: Wallingford: UK, 1990.

Cudjoe, A. R.; Kyofa-Boamah, M.; Nkansah, G. O.; Braun, M.; Owusu, S.; Adams, E.; Monney, E.; Attasi, R.; Owusu, P.; Sarpong, S. Commercial Okra Production in Ghana— Good Agricultural Practices/Code of Practice and IPM Strategies. In *Handbook of Crop Protection Recommendations in Ghana, Ministry of Food and Agriculture, Accra*; Kyofa-Boamah, M., Blay, E., Braun, M., Kuehn, A., Eds; 2005; pp 75–92.

Fajinmi, A. A.; Fajinmi, O. B. Epidemiology of *Okra mosaic virus* on Okra under Tropical Conditions. *Int. J. Vegetable Sci.* **2010a**, *16* (3), 287–296.

Fajinmi, A. A.; Fajinmi, O. B. Incidence of Okra Mosaic Virus at Different Growth Stages of Okra Plants (Abelmoschus esculentus L. Moench) under Tropical Condition. *J. Gen. Mol. Virol.* **2010b**, *2* (1), 028–031.

FAOSTAT. Food and Agricultural Organization of the United Nations. Online and Multilingual Database, FAO, Rome, Italy. http://faostat.fao.org/foastat/

Givord, L; Hirth, L. Identification, Purification and Some Properties of a Mosaic Virus of Okra (Hibiscus esculentus). *Ann. Appl. Biol.* **1973**, *74*, 359–370.

Hamon, S.; Charrier, A. Les Gombos. In: *L'amelioration des plantestropicales*; Charrier, A., Jacquot, M., Hamon, S., Nicholas, D., Eds; CIRAD/ORSTOM: Montpellier, France, 1997; pp 313–333.

Hamon, S. Evolutionary Organization of Its Kind *Abelmoschus* (okra). Co-adaptation and Evolution of Two Species Grown in West Africa, *A. esculentus* and *A. caillei.* Paris, ORSTOM, DTP Works and Documents, 1988, p 191.

Idris, A. M.; Brown, J. K. Molecular Analysis of *Cotton Leaf Curl Virus*-Sudan Reveals an Evolutionary History of Recombination. *Virus Genes* **2002**, *24*, 249–256.

Kochhar, S. L. Tropical Crops. A Text Book of Economic Botany. Macmillan Publishers Ltd.: London and Basingstoke; Macmillan Indian Ltd, 1986, pp 467.

Kon, T.; Rojas, M. R.; Abdourhame, I. K.; Gibertson, R. L. Roles and Interactions of Begomoviruses and Satellite DNAs Associated with Okra Leaf Curl Disease in Mali, West Africa. *J. Gen Virol.* **2009**, *90* (4), 1001–1013.

Konate, G.; Barrow, N.; Fargette, D.; Swanson, M. M.; Harrison, B. D. Occurrence of Whitefly-transmitted Geminiviruses in Crops in Burkina-Faso, and their Serological Detection and Differentiation. *Ann. Appl. Biol.* **1995**, *126*, 121–129.

Kyofa-Boamah, M.; Blay, E.; Braun, M.; Kuehn, A. Commercial Okra Production in Ghana-Good Agricultural Practices/Code of Practice and IPM Strategies: A. *Handbook of Crop Protection*; Recommendations in Ghana, Ministry of Food and Agriculture: Accra, 1995; pp 75–92.

Leke, W. N. Molecular Epidemiology of Begomoviruses that Infect Vegetable Crops in Southwestern Cameroon. Swedish University of Agricultural Sciences, Uppsala, 2010. Available at: http://pub.epsilon.slu.se/id/eprint/2338 (accessed April 22, 2016).

Mihretu, Y.; Wayessa, G.; Adugna, D. Multivariate Analysis among Okra (*Abelmoschus esculentus* (L.) Moench) Collection in South Western Ethiopia. *J. Plant Sci.* **2014**, *9* (2), 43–50.

Ministry of Food and Agriculture (MoFA) Okra Production. Horticultural Development Unit, MoFA, Accra, 2013.

Naidu, R. A.; Hughes, J. D. A. Methods for the Detection of Plant Virus Diseases. Plant Virol. Sub Saharan Africa, **2003**, 233–253.

National Agricultural Research Project (NARP) Horticultural Crops. Vol. 3, Accra: NARP, CSIR, 1993.

Ndunguru, J.; Rajabu, A. C. Effect of *Okra Mosaic Virus* Disease on the Above-ground Morphological Yield Components of Okra in Tanzania. *Scientia Horticulturae* **2004**, *99* (3), 225–235.

Nevala, A. E. In the Southwest, a New Plant Reserve Protects the Mother of All Chilies. National Wildlife, 2000, pp 14.

N'Guessan, K. P.; Fargette, D.; Fauquet, C.; Thouvenel, J. C. Aspects of the Epidemiology of Okra Leaf Curl Virus in Cote d'Ivoire. *Trop Pest Manage.* **1992**, *38*, 122–126.

Norman, J. C. *Tropical Vegetable Crops*. Arthur H. Stockwell Ltd.: Devon, UK, 1992; p 252.

Obeng-Ofori, D.; Sackey, J. Field Evaluation of Non-synthetic Insecticides for the Management of Insect Pests of Okra Abelmoschus esculentus (L.) Moench in Ghana. *Ethiopian J. Sci.* **2003**, *26*, 145–150.

Oppong-Sekyere, D. *Assessment of Genetic Diversity in a Collection of Ghanaian Okra Germplasm (Abelmoschus Spp. L) Using Morphological Markers*. M.Sc. Thesis, Kwame Nkrumah University of Science and Technology, Kumasi, Ghana, 2011; p 96.

Purseglove, J. W *Tropical Crops: Dicotyledons*. Longman Group, EUA: London, 1974; pp 17–30.

Revkin, A. C. Need Elephant Repellent? Try This Hot Pepper Brew. The New York Times, 2000.

Sayed, S. S.; Rana, D.; Krishna, G.; Reddy, P. S.; Bhattacharya, P. S. Association of Begomovirus with Okra (*Abelmoschus esculentus* L.) Leaf Curl Virus Disease in Southern India. *SAJ Biotechnol.* **2014**, *1* (1), 102.

Schippers R. R. African Indigenous Vegetables: An Overview of the Cultivated Species, 2000.

Schmutterer, H. Properties and Potential of Natural Pesticides from the Neem Tree, *Azadirachta indica*. Ann. Rev. Entomol. **1990**, *35*, 271–297.

Siemonsma, J. S. *Abelmoschus*: A Taxonomical and Cytogenetical Overview. Int. Crop Network Ser. 5. Int. Board Plant Genet. Resources, Rome, Italy, 1991, 52–68.

Siemonsma, Y. *La culture du gombo (Abelmoschus spp.) Itgume fruit tropical avec Reference Spéciale de la Cote d'Ivoire.* Thesis, University of Wageningen, The Netherlands, 1982.

Sinnadurai, S. Vegetable Production in Ghana. *Acta Horticulturae.* (ISHS) **1973,** *33,* 25–28.

Swanson, M. M.; Harrison, B. D. Serological Relationships and Epitope Profiles of Isolates of Okra Leaf Curl Gemini Virus from Africa and the Middle East. *Biochimie.* **1993,** *75* (8), 707–711.

Tiendrebego, F.; Lefeuvre, P.; Hoareau, M.; Villemot, J.; Konate, G.; Traore, A. S.; Baro, N.; Traore, V. S.; Reynaud, B.; Traore, O.; Lett, J. -M. Molecular Diversity of Cotton Leaf Curl Gezira Virus Isolates and Their Satellite DNAs Associated with Okra Leaf Curl Disease in Burkina Faso. *Virol. J.* **2010,** *7* (1), 48–49.

Tindall, H. *Vegetables in the Tropics.* London: Macmillan Press Ltd, 1986.

Udengwu, O. S.; Dibua, U. E. Screening of Abelmoschus esculentus and Abelmoschu scallei Cultivars for Resistance Against Okra Leaf Curl and Okra Mosaic Viral Diseases, Under Field Conditions in South Eastern Nigeria. *African J. Biotechnol.* **2014,** *13* (48), 4419–4429.

CHAPTER 2

Current and Prospective Approaches for Plant Virus Diseases Detection

TOUSEEF HUSSAIN[1,*] and QUAZI MOHD. IMRANUL HAQ[2]

[1]*Department of Botany, Plant Pathology Section, Aligarh Muslim University, Aligarh-202002, India*

[2]*Deparment of Biological Sciences and Chemistry, College of Arts and Sciences, University of Nizwa, Nizwa, Sultanate of Oman*

Corresponding author. E-mail: hussaintouseef@yahoo.co.in

ABSTRACT

Due to the globalization of trade through the Free Trade Agreement and rapid climate change patterns, promotes the transfer of virus from one country to another and its hosts and vectors, therefore, the diagnosis of viral diseases is getting more important now a days. The lack of general reliability of the methods of visual identification and the variability in the characteristic expression within the host plant, it is very difficult to detect virus infections in the plants. For effective management practices, it is necessary to reduce the spread of diseases, to monitor the health of the plants and detect pathogens in the initial stage. Because the symptoms of viral diseases are not different with great diversity and are confused with abiotic stresses, symptomatic diagnosis may not be appropriate. DNA-based (PCR, RCA) and serological methods (ELISA) now provide essential tools for accurate plant disease diagnosis, in addition to the traditional visual scouting for symptoms. From the last three decades, different forms of (ELISAs), has been developed based on serological principle, that have been widely used. Although serological and PCR-based methods are the most available and effective to confirm the diagnosis of the disease, volatile, provide immediate results and can be used to detect infections in asymptomatic stages. We explain how these tools will help plant disease management and complement serological and DNA-based methods.

2.1 INTRODUCTION

Plant viruses diseases are one of the major threats all around the world, which cause loss of billion dollars per year by destroying various economically important crops.[43] Plant virus symptoms occur on stems, leaves, flowers, or fruits and vary from mild to severe damage, slow growth, and rarely death of the plant.[1]

Important aspects like agronomic, economic, and social impact are frequently influenced by a wide range of viruses infecting plants. Early stage detection of viruses which causes infection is crucial to reduce economic losses. For diagnostic purpose, biological indexing and serological enzyme-linked immunosorbent assay (ELISA) are most extensively and most widely used methods. Furthermore, modern molecular techniques have revolutionized plant virus detection and identification.

During early research, detection and identification of plant on symptomatology of infected plants was not reliable because the symptoms differed depending on the cultivar, growth stage, and virus strain, etc. Biological assays are still most commonly and frequently used diagnostic methods for many plant viruses until now because the methods are simple and easy.[21,39] Previously, virus diagnosis was mainly done by virology specialist having many years of experience.

Rolling circle amplification (RCA) is reliable, convenient, and cheaper than polymerase chain reaction (PCR) for the diagnosis of plant viruses with small single-stranded circular DNA including geminiviruses. In future, this shortcut will extensively speed up the genomics of gemini, circo, and nanoviruses.[58]

2.2 HISTORY

Globally, horticultural and agricultural crops are infected by plant viruses and are a major threat to them. Methods for detection and identification of viruses play a crucial role in virus disease control. Diagnostic techniques for plant viruses are mainly divided in two major categories: biological properties and intrinsic properties of the virus itself. Detection methods depend on ELISA, immunoblotting, and coat protein includes agglutination tests. Viral nucleic acid-based techniques like prospective, dot-blot hybridization are more accurate than other diagnostic methods. This diagnostic method for plant viruses provides increased sensitivity, more flexibility, and specificity for quick diagnosis of plant virus diseases.

Early molecular hybridization technologies were rapidly supplanted by more powerful nucleic acids amplification methods based on the PCR. Although molecular methods are highly discriminatory, allowing strain typing, routine testing has been hampered by problems in reproducibility. Continuous efforts have been made to overcome these barriers. Improved systems to prepare plant or insect samples have been developed. Efforts have also been directed at increasing the sensitivity and specificity of detection, which can be limited by the high content of enzyme inhibitors in plant materials. Nested and multiplex PCR offer high sensitivity and the possibility to detect several targets in one assay, respectively. There are many other technologies, which allow the amplification of nucleic acids in an isothermal reaction (nucleic acid sequence-based amplification [NASBA] or reverse transcription loop-mediated isothermal amplification [RT-LAMP] procedures). High-throughput testing has been achieved by real-time polymerase chain reaction (RT-PCR), in which the automation of PCR combined with fluorimetry. RT-PCR simultaneously permits detection and quantification of targets gene. In the near future, nucleic acid arrays and biosensors assisted by nanotechnology could revolutionize the methodology for diagnosis of plant viruses.

New technologies are slow, which requires more knowledge of what makes a better routine diagnostic methods to begin, which also requires rate of uptake of understanding. This can be achieved by keeping in mind the two most successfully used plant geminivirus detection methods: RT-PCR and ELISA. The publication based on ELISA method for the diagnosis of plum pox virus (PPV; genus Potyvirus, family Potyviridae) and Arabis mosaic virus (ArMV; genus Nepovirus, family Secoviridae) by Clark and Adams[55] was a major hike in virus diagnostics.

2.3 PRINCIPLES OF SEROLOGY

Serological tests are important for the final confirmation of an unknown plant virus and also to study the virus species and strains relationship. The most important advantage of this serology method is based on the specificity between viral antigen (Ag) and antibody (Ab).

Study of serums, especially their reactions and properties, is called serology. Detection of plant viruses with direct immunoblotting is done in this technique. The antiserum is used to detect plant viruses. Antiserums are produced by injecting an Ag into rabbit or goat. Any protein or substance that the animal's immune system recognizes as foreign is known as antigen. The polyclonal Ab's are also used which is produced by purifying a particular Ag and injecting the purified virus (the Ag, the virus particle) into rabbit. All the plant virus detection techniques that use antisera are known as serological techniques.

Virus particles and associated proteins have many epitopes with different amino acid sequences and have the ability of inducing the specific Ab's production (Fig. 2.1). The virus particles, their protein capsid, and the different types of virus induced proteins can function as Ag's.[39,56,57,59,60]

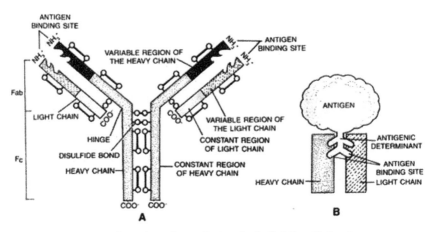

A, structure of an antibody molecule. B, Antigen binding site.

FIGURE 2.1   Diagrammatic representation of (A) structure of an immunoglobulin G (IgG) molecule. Fab, F(ab)2, and Fc represent fragments obtained by enzyme cleavage of IgG, (B) antigen binding site.

Source: biologydiscussion.com

The significance of the serological methods for the study of plant viruses and plant viral diseases was demonstrated for the first time by the pioneer work of Purdy-Beale.[42]

2.4 DIAGNOSIS

Viruses cannot be isolated and grown on cell-free media as fungi or bacteria, as they are obligate parasites, but they have to be maintained on susceptible host plants by artificial inoculation at regular intervals, under controlled conditions.[40] Many different types of viruses may exhibit same symptoms and the disease phenotype can give only less information for disease diagnosis purpose. Many specific and reliable techniques used for virus identification mainly depend on the different aspects of the viruses such as pathogenicity, transmissibility, architecture of virus particles, presence of virus-specific skeleton in infected cells, properties of the protein coat.

Most extensively, Ab-based technique used for diagnostic of viruses is known as ELISA. In this method, the plant tissue extract being tested for virus detection incubated in the well of ELISA plate. Then, Ab is added which binds to the specific Ag (the virus particle). Secondary Ab linked to an enzyme is supplemented which gets attached to the previous Ab already existed in the well. After washing, a colorless substrate is supplemented which produces colored product (yellow) after reaction that indicates the existence of virus and the intensity of the color is used to calculate the virus concentration.

After the development of ELISA, new methods have been focused. Nucleic acid spot hybridizations (NASH) have been generally used for some plant viruses and viroids. PCR technique published in the early 1990s is most extensively used method for plant virus diagnostics.[61]

2.5 CLASSIFICATION OF VIRUSES

According to Baltimore Classification System, based on the way in which a virus produces messenger RNA (mRNA) during infection, the viruses have been classified into seven major groups:

i)   Group I: double-stranded DNA viruses
ii) Group II: single-stranded DNA viruses
iii) Group III: double-stranded RNA viruses
iv) Group IV: positive-sense single-stranded RNA viruses
v)   Group V: negative-sense single-stranded RNA viruses
vi) Group VI: reverse transcribing Diploid single-stranded RNA viruses
vii) Group VII: reverse transcribing circular double-stranded DNA viruses

Within each of these groups, many different characteristics are used to classify the viruses into families, genera, and species. Typically, combinations of characters are used and some of the most important are: particle morphology, genome properties, biological properties, and serological properties.

2.5.1 PARTICLE MORPHOLOGY

Amongst plant viruses, the most frequently encountered shapes are:

2.5.1.1 ISOMETRIC

Apparently spherical and (depending on the species) from about 18nm in diameter upwards. The example here shows *Tobacco necrosis virus*, genus *Necrovirus* with particles 26 nm in diameter.

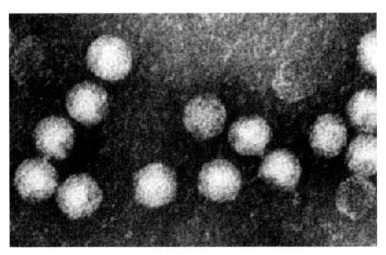

2.5.1.2 ROD-SHAPED

About 20–25 nm in diameter and from about 100 to 300 nm long. These appear rigid and often have a clear central canal (depending on the staining method used). Some viruses have two or more different lengths of particle and these contain different genome components. The example here shows *Tobacco mosaic virus*, genus *Tobamovirus* with particles 300 nm long.

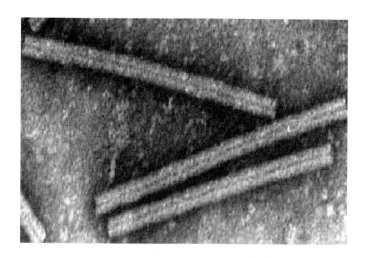

2.5.1.3   *FILAMENTOUS*

Usually about 12 nm in diameter and more flexuous than the rod-shaped particles. They can be up to 1000 nm long, or even longer in some instances. Some viruses have two or more different lengths of particle and these contain different genome components. The example here shows *Potato virus Y*, genus *Potyvirus* with particles 740 nm long.

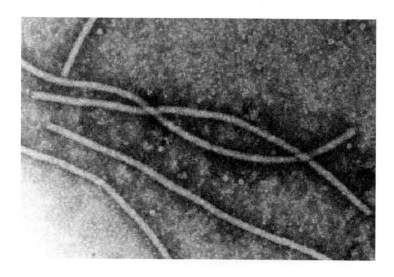

2.5.1.4   GEMINATE

Twinned isometric particles about 30 × 18 nm. These particles are diagnostic for viruses in the family *Geminiviridae* which are widespread in many crops especially in tropical regions. The example here shows *Maize streak virus*, genus *Mastrevirus*.

2.5.1.5   BACILLIFORM

Short round-ended rods. These come in various forms up to about 30 nm wide and 300 nm long. The example here shows *Cocoa swollen shoot virus*, genus *Badnavirus* with particles 28 × 130 nm.

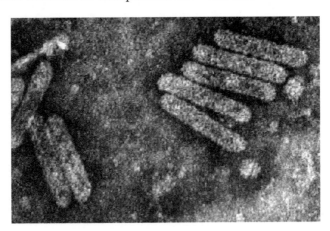

2.6 DETECTION OF PLANT VIRUSES BASED ON BIOLOGICAL PROPERTIES

2.6.1 SYMPTOMATOLOGY

The plant virus symptoms are mainly used to describe a disease and also for the removal of infected plants from the field to control the diseases. Visual observation of the symptoms is comparatively simple, whereas several factors such as time of infection, host plant variety, virus strain, and environment can affect the symptoms.[31] However, some viruses cause asymptomatic infection. It is mandatory that visual observation for symptoms is done in addition with other confirmatory tests to make sure of the meticulous diagnosis.[3]

2.7 PLANT VIRUS TRANSMISSION TESTS

This test is very important for virus investigation, identity, vector transmission, and mechanical transmission to susceptible host plants.[21] Mechanical transmission is possible with minimum facilities and the symptoms produced allow both the detection and identification of large number of plant viruses.[18] For other plant, mechanically transmissions are not possible as well as the viruses of small fruit can be identified through graft/vector transmission.[16,37,38]

2.8 PHYSICAL PROPERTIES OF PLANT VIRUSES

Physical properties of the plant viruses (such as dilution end point, longevity in vitro, and thermal inactivation point) used to measure infectivity of the virus, even though these physical properties are not consistent and not recommended for plant virus diagnostic purpose.[15]

2.9 BIOCHEMICAL TECHNIQUES

Virus infection causes many changes in biochemical and physiological activity of the host plants. Infection caused by the viruses could be detected by variability pattern in isozyme of infected and healthy leaves, determined by sodium dodecyl sulfate polyacrylamide gel electrophoresis technique.

2.10   DETECTION PLATFORMS

Few methods for virus detection are extensively used to find out the products of various amplicons.

Common gel electrophoresis techniques are widely applicable method to check the application; however, this is time consuming.

2.11   ISOTHERMAL METHODS

A simple method to get isothermal DNA amplification is to separate the double-stranded DNA strands of the template in a non-thermal way, such as, PCR, helicase-dependent amplification (HDA),[63] and recombinase polymerase amplification (RPA)[62] are two examples of this approach. At a single temperature, the HDA can be performed, but a brief incubation at 95–96°C before adding HDA enzymes increasing the sensitivity has been shown. The main benefit of RPA is short time for reaction to take place usually <30 min.

Based on transcription, this NASBA method is used for isothermal amplification of RNA. This technique has been employed for the finding out of various plant pathogens.[66]

2.12   MICROSCOPIC ANALYSIS

Electron microscopy is very useful to know the virus particles particle morphology for virus detection.[2,33] Rod-shaped and filamentous potexviruses and tobamoviruses can be differentiated than other viruses. Less concentration of viruses in plant sap extract are difficult to see unless its concentrated prior observation. Many plant viruses induce distinctive intracellular inclusions and their detection by EM can give a rapid, easy, and cheap method in terms of money to confirm viral infections.[12]

2.13   DETECTION AND IDENTIFICATION METHODS USING VIRAL COAT PROTEIN AGGLUTINATION TESTS

Precipitin tests based on the visible precipitate formation, when appropriate quantity of virus particles and specific antibodies, are in touch with each other.[47] Precipitin/microprecipition tests are routinely done by some researchers, whereas double diffusion and agglutination tests are more commonly used. The Ouchterlony double diffusion method is mainly used to differentiate alleged, but separate virus strains. In this test, on an inert carrier particle, Ab is coated and the reaction between positive Ag–Ab results in aggregation which can be seen on both surface by naked eyes or under microscope. Compared with other precipitin tests, it is more sensitive and can be performed with low concentration of reactants.[20,48]

These detection tests can be performed with less facility and, therefore, are routinely used in many research institutions with limited facilities and have an availability of antiserum.[39]

2.14 IMMUNOELECTRON MICROSCOPY

The important advantage of immunoelectron microscopy is being able to be applied to tissue homogenates and of requiring very low quantities of virus and antiserum. Relationships can be studied in different ways: (1) by the differential trapping of virus particles on electron microscope grids coated with antisera to different viruses; (2) by endpoint dilution of an antiserum which effectively coats (decorates) the virus particles; (3) by the observation of a clumping reaction at different serum concentrations. Pretreatment of grids with protein A of *Staphylococcus aureus* may enhance virus adsorption[44] but meaningful results occur only under certain conditions. It may be useful to follow trapping of virus particles on serum-coated grids with decoration,[32] particularly as trapping may be susceptible to bias due to nonspecific adsorption of virus particles to the grids.[29] To distinguish from nonspecific decoration, decoration of particles by antisera to distantly related viruses may be difficult however, especially in crude sap.

2.15 ELISA

After the development of ELISA technique, serological application was totally revolutionized and subsequent modification led to the double-antibody sandwich technique which allowed the use of small amounts of reagents and test tissue, and was easily applicable to processing large numbers of samples. This technique uses to diagnose proteins. Antibodies are coated to the wells within ELISA plate and the plant sap extract is compiled to the well. It will bind to the antibodies if there is a virus coated/ present on the surface. A secondary antibody is added. This secondary Ab permits for indirect detection of the virus particles as it is attached with enzyme, due to its sensitivity and adaptability. It is used in many situations, mainly for testing various samples in a very short time.

Different types of ELISA have been developed[47] which belong to two main categories: direct and indirect ELISA method. In direct ELISA method, the Ab's are coated to the well of microtiter plate, which captures the virus present in the sample. This captured virus is then detected by adding with an Ab–enzyme conjugate followed by addition reagents for color development. The detecting and capturing Ab's detecting antibodies can be from the same or from other sources, as the virus is sandwiched between two Ab's; therefore, it is also known as double-antibody sandwich (DAS) ELISA.

An alternative to the DAS method is indirect ELISA. In the second step, virus antibody (primary antibody) is added. Then the primary Ab is detected with antispecific antibodies conjugated with an enzyme and finally reagents are added for the color development. The secondary antibody binds directly to the primary antibody. It has some disadvantages such as virus particles for sites on the plate, competition between plant sap, and high background reactions.[39]

Another broadly used ELISA is triple-antibody sandwich-ELISA (TAS-ELISA). This is very similar to DAS-ELISA, except that one more extra step is done prior to addition of secondary antibody enzyme conjugate. The third type called protein A-sandwich (PAS) ELISA, in this type of ELISA, prior to the addition of primer Ab, the microtiter wells are coated with protein A. Indirect ELISA is more suitable and economical for virus detection in many situations including disease surveys and quarantine programs.

Serological methods are grouped into solid and liquid phase tests. The latex agglutination reaction (LAR) and passive hemagglutination (PHA) tests are the examples of liquid-phase tests or the precipitin test and ELISA. Gel-based assay (double immune-diffusion gel assay, DIGA) has been also developed.

Other important considerations for serological detection and diagnosis are the type of epitopes recognized and the quality of antisera.

Various techniques are developed for the diagnosis of plant viruses, each with its own advantages and disadvantages. Some of them are given below with their brief introduction and a pertinent example.

2.15.1 TISSUE BLOT IMMUNOASSAY

Tissue blot immunoassay, like ELISA, uses antibodies prepared against plant viruses. The plant tissue extract is blotted on the nitrocellulose membrane and the virus if finally detected by using labelled probes. This method is simple, rapid, sensitive, and inexpensive. Various virus diagnostic kits are also available.

2.15.2 QUARTZ CRYSTAL MICROBALANCE IMMUNOSENSORS

This is a unique technique for detection of plant viruses using virus-specific antibodies. Voltage is applied across the disk, making the disk warp slightly via a piezoelectric effect. Adsorption of virus particles to the crystal surface changes its resonance oscillation frequency in a concentration-dependent manner. It is therefore qualitative and quantitative. The designers of the technique claim that it is as sensitive but more rapid than ELISA, and economical. As little as 1 ng of particles of cymbidium mosaic virus and odontoglossum ring spot virus were detected in crude sap, extracts depicted first use of quartz crystal microbalance for plant viruses.[14]

2.15.3 FLUORESCENCE RT–PCR USING TAQMAN TECHNOLOGY

In this technique, two specific primers and a third primer labelled with fluorescence anneals between them. Once the labeled flanking primers extend, the specify primer results in the released of and fluorescence takes place. In this method, to detect the reaction products, post-reaction processing is not required. For the detection of *Potato spindle tuber viroid*, a Taqman assay was 1000 times more sensitive than a chemiluminescent assay.[7] By this method, the thrips vector of Tomato spotted wilt virus was perfectly screened for the presence of virus. To detect two orchid viruses, multiplex fluorescence PCR was used simultaneously.[8]

2.15.4 COMPETITIVE FLUORESCENCE PCR (CF–PCR)

This is a modified version of the above method used to segregate between multiple virus infections and virus strains, simultaneously. In this method, several sets of primers attached with many types of fluorescent marker are added. Potatoes infected with mixed Potato virus Y strains were detected and identified using this technique.[49]

2.15.5 BIOELECTRIC RECOGNITION ASSAY (BERA)

Some important roles such as biosecurity, food safety, environmental monitoring, homeland security, and medical diagnostics are played by this.

The BERA is a method which detects the electric response of culture cells and possesses its physiology. In previous studies[23,24,25,26] the possible appliance of the technique for ultra-rapid and ultra-cheap tests for the diagnosis of human and plant viruses were developed.

A further manipulation of the technique, called the "6th sensor generation" uses 5th generation sensors which have engineered cells expressing target antibodies on their nitrocellulose membrane (NCM).[27,28,34,35] The combination of sensitivity, simplicity, and reliability makes BERA suitable for large screening and monitoring of environment. It is used for plant viruses, such as the *Cucumber Green Mottle Mosaic Virus* (CGMMV) and the *Tobacco Rattle Virus* (TRV), using suitable plant cells as the sensing elements, in order to establish an efficient system for the plant virus detection. Artificial neural networks (ANN) were applied with various architecture. Next, ANN was described and more specifically multilayer perceptions which are one of the widely used and most popular feed forward classification models.

2.16 CONCLUSION

Various plant virus detection techniques as mentioned above is now available. The use of various plant detection methods results in increased sensitivity, specificity, and explores a large number application of the diagnostics to eradicate the effects of many of the threatening virus diseases.[70]

The positive result in both nucleic acid-based and serological-based plant virus detection techniques does not imply the existence of plant viruses.[67,68,69] Solving this problem requires rendering diagnostic kits from a same source to scientists around the word, so that differences found with specificity and quality of different reagents in assay kits are reduced. It is also necessary to aggregate, accredit, and expand durable diagnostic.

Plant viruses are becoming more prevalent and threats of novel viral epidemics, as seen in the last 10 years. Therefore, it is mandatory that the viruses' movements all over the world must be checked and quarantined in required places where necessary. However, presently, it is costly, having technical difficulties of constructing, crafting, and utilizing microarrays. As chips become easily available and as scale economies are realized, hopefully, costs will reduce.

Depending on the target virus sequence information, from early 2000s, several methods such as reverse transcription-loop-mediated isothermal amplification and RT-PCR have been established that are able to differentiate closely related viruses, rapid, and sensitive. Recent techniques such as RT-PCR can be used to measure the pathogen and to find out virus population dynamics and metagenomic analyses by next-generation sequencing. Conventional diagnostic techniques have only given information which depends on target pathogens. A number of other options for the researchers to detect and investigate the plant pathogens have culminated with the advancement in the technologies.

Recent technologies, for example, next-generation sequencing and microarrays also hold valuable for use in diagnosing disease in rice. In spite of continuous research and development of new methods, only few technologies get adopted for continuous use in research laboratories. In plant virus diagnostics (e.g., Geminivirus), current developments can be divided into three main categories: (1) multiplex methods (e.g., LBA), (2) methods favorable to the discovery of viruses (e.g., NGS), and (3) methods that can be featured in the field (e.g., LAMP). It was unthinkable 10 years ago that real-time PCR would become first choice, in comparison to ELISA.

The evolution of plant viruses through genetic pressure and drift to infect particular crops has enabled them to overcome host–plant defences (reviewed by Garcia-Arenal and Fraile, 2008). Viral diagnostics is one of the most valuable tools for plant disease management. Virologists have to move ahead more from traditional use of biological indicator hosts to molecular diagnostics and sequence data to establish relationships, groups, genera, including families among the ever growing list of new viruses[53] or viruses on new hosts in new locations.[51,52]

Unlike other plant pathogens, plant viruses are mainly liable to fallacious diagnosis if done wholly on symptoms.[2] Nucleic acid-based tests provide a great opportunity for specific, sensitive, and quick diagnostic of plant viruses. However, in the cases where both nucleic acid-based methods and serology give indistinguishable information through specificity and detection sensitivity, in that case, serology is the favored method of diagnostic purpose.

KEYWORDS

- **ELISA**
- **geminiviruses**
- **PCR**
- **antigen**
- **diseases**

REFERENCES

1. Agrios, G. N. *Plant Diseases Caused by Viruses*: *Plant Pathology*. Elsevier Academic Press: San Diego, California, 2005; pp 723–824.
2. Baker, K. K.; Ramsdell, D. C.; Gillett, J. M. Electron Microscopy: Current Applications to Plant Virology. *Plant Dis.* **1985**, *69*, 85–90.
3. Bock, K. R. The Identification and Partial Characterization of Plant Viruses in the Tropics. *Trop. Pest Management* **1982**, *28*, 399–411.
4. Bawden, F. C. The Relationship Between the Serological Reactions and the Infectivity of Potato Virus X. *J. Expel. Pathol.* **1935**, *16*, 435–43.
5. Bawden, F. C. *Plant Viruses and Virus Diseases*. Chronica Botanica Co.: Waltham, Massachusetts, 1950; p 294.
6. Boyd, W. C. *Fundamentals of Immunology*, 2nd ed. Interscience Publishers, Inc.: New York, 1947; p 503.
7. Boonham, N. Development of a Real-time RT–PCR Assay for the Detection of Potato Spindle Tuber Viroid. *J. Virol. Methods* **2004**, *116*, 139–146.
8. Boonham, N. The Detection of Tomato Spotted Wilt Virus (TSWV) in Individual Thrips Using Real Time Fluorescent RT–PCR (Taqman™). *J. Virol. Methods* **2000**, *101*, 37–48.
9. Carpenter, P. L. *Immunology and Serology*. W. B. Saunders Company: Philadelphia, Pennsylvania; London, England, 1956; p 351.
10. Chen, H.; Wei, T.; Shikamoto, Y.; Shimizu, T.; Saotome, A.; Mizuno, H. Aggregation Ability of Virus-specific Antibodies is Correlated with Their Capacity to Neutralize Rice Dwarf Virus. *JARQ* **2012**, *46*, 65–71.
11. Dvorak, M. J. The Effect of Mosaic on Globulin of Potato. *Infectious Dis.* **1927**, *41*, 215–221.

12. Edwardson, J. R.; Christie, R. G.; Purcifull, D. E.; Petersen, M. A. Inclusions in Diagnosing Plant Virus Diseases. In *Diagnosis of Plant Virus Diseases*; Matthews, R. E. F. CRC Press: Boca Raton, Florida, USA, 1993; pp 101–128.

13. Eggins, B. *Biosensors—An Introduction*. Wiley and Teubner: Chichester, 1996.

14. Eun, A. J.-C.; Huang, L.; Chew, F.-T.; Li, S. F.-Y.; Wong, S. M. Detection of Two Orchid Viruses Using Quartz Crystal Microbalance (QCM) Immunosensors. *J. Virol. Methods* **2002**, *99*, 71–79.

15. Francki, R. I. B. Limited Value of the Thermal Inactivation Point, Longevity In Vitro and Dilution End Point as Criteria for the Characterization, Identification, and Classification of Plant Viruses. *Intervirology* **1980**, *13*, 91–98.

16. Fridlund, P. R. Glasshouse Indexing for Fruit Tree Viruses. *Acta Horticulturae* **1980**, *94*, 153–158.

17. Hibino, H.; Usugi, T.; Oraura, T.; Tsuchizaki, T.; Shohara, K.; Iwasaki, M.Rice Grassy Stunt Virus: a Plant Hopper-borne Circular Filament. *Phytopathology* **1985**, *75*, 894–899.

18. Horvath, J. New Artificial Hosts and Non Hosts of Plant Viruses and Their Role in the Identification and Separation of Viruses. XVIII. Concluding Remarks. *Acta Phytopathologica Hungarica* **1983**, *18*, 121–161.

19. Hamilton, R. I.; Edwardson, J. R.; Francki, R. I. B.; Hsu, H. T.; Hull, R.; Koenig, R.; Milne, R. G. Guidelines for the Identification and Characterization of Plant Viruses. *J. Gen. Virol.* **1981**, *54*, 223–241.

20. Hughes, J.d'A.; Ollennu, L. A. The Viro Bacterial Agglutination Test as a Rapid Means of Detecting Cocoa Swollen Shoot Virus. *Ann. Appl. Biol.* **1993**, *122*, 299–310.

21. Jones, A. T. Experimental Transmission of Viruses in Diagnosis. In *Diagnosis of Plant Virus Diseases*; Matthews, R. E. F, Ed.; CRC Press: Boca Raton, Florida, USA, 1993; pp 49–72.

22. Kabat, E. A.; Mayer, M. M. *Experimental Immunochemistry*, 1st ed. Charles C Thomas: Springfield, Illinois, 1948; p 567.

23. Kintzios, S.; Pistola, E.; Panagiotopoulos, P.; Bomsel, M.; Alexandropoulos, N.; Bem, F.; Biselisand, I.; Levin, R. Bioelectric Recognition Assay (BERA). *Biosensors Bioelectronics* **2001a**, *16*, 325–336.

24. Kintzios, S.; Pistola, E.; Konstas, J.; Bem, F.; Matakiadis, T.; Alexandropoulos, N.; Biselis, I.; Levin, R. Application of the Bioelectric Recognition Assay (BERA) for the Detection of Human and Plant Viruses: Definition of operational parameters. *Biosensors Bioelectronics* **2001b**, *16*, 467–480.

25. Kintzios, S.; Makrygianni, E.; Pistola, E.; Panagiotopoulos, P.; Economou, G. Effect of Amino Acids and Amino Acid Analogues on the In Vitro Expression of Glyphosate Tolerance in Johnson Grass (*Sorghum halepense* L. pers.). *J. Food Agric. Environ.* **2003**, *3*, 180–184.

26. Kintzios, S.; Bem, F.; Mangana, O.; Nomikou, K.; Markoulatos, P.; Alexandropoulos, P.; Fasseas, C.; Arakelyan, V.; Petrou, A. L.; Soukouli, K.; Moschopoulou, G.; Yialouris, C.; Simonian, A. Study on the Mechanism of Bioelectric Recognition Assay: Evidence for Immobilized Cell Membrane Interactions With Viral Fragments. *Biosens. Bioelectron.* **2004**, *20*, 907–916.

27. Kintzios, S.; Goldstein, J.; Perdikaris, A.; Moschopoulou, G.; Marinopoulou, I.; Mangana, O.; Nomikou, K.; Papanastasiou, I.; Petrou, A. L.; Arakelyan, V.; Economou,

A.; Simonian, A. *The BERA Diagnostic System: An All-purpose Cell Biosensor for the 21st Century*. 5th Biodetection Conference, Baltimore, MD, USA, 2005.

28. Kintzios, S.; Marinopoulou, I.; Moschopoulou, G.; Mangana, O.; Nomikou, K.; Endo, K.;Papanastasiou, I.; Simonian, A. Construction of a Novel, Multi-analyte Biosensor System for Assaying Cell Division. *Biosens. Bioelectron.* **2006**, *21* (7), 1365–1373.

29. Lesemann, D. E.; Bozartrt, R. F.; Koenig, R. The Trapping of Tymovirus Particles on Electron Microscope Grids by Adsorption and Serological Binding. *J. Gen. Virol.* **1980**, *48*, 257–262.

30. Luria, E. S. *General Virology*. John Wiley and Sons: New York and Chapman and Hall, Ltd.: London, England, 1953, p 427.

31. Matthews, R. E. F. Host Plant Responses to Virus Infection. In *Comprehensive Virology, Vol. 16, Virus–Host Interaction, Viral Invasion, Persistence, and Diagnosis*; Fraenkel-Conrat, H.; Wagner, R. R. Plenum Press: New York, USA, 1980; pp 297–359.

32. Milne, R. G.; Luisoni, E. Rapid Immune Electron Microscopy of Virus Preparations. *Methods Virol.* **1977**, *6*, 265–281.

33. Milne, R. G. Electron Microscopy of In Vitro Preparations. In *Diagnosis of Plant Virus Diseases*; Matthews, R. E. F, Ed.; CRC Press: Boca Raton, Florida, USA, 1993; pp 215–251.

34. Moschopoulou, G.; Kintzios, S. Membrane Engineered Bioelectric Recognition Cell Sensors for the Detection of Sub Nanomolar Concentrations of Superoxide: A Novel Biosensor Principle. International Conference on Instrumental Methods of Analysis (IMA) 2005, Crete, Greece.

35. Moschopoulou, G.; Kintzios, S. Application of "Membrane-Engineering" to Bioelectric Recognition Cell Sensors for the Detection of Picomole Concentrations of Superoxide Radical: A Novel Biosensor Principle. *Anal. Chim. Acta* **2006**, *573–574*, 90–96.

36. Matsumoto, T.; Somazawa, K. Immunological Studies of Mosaic Diseases. III. Further Studies on the Distribution of Antigenic Substance of Tobacco Mosaic in Different Parts of Host Plants. *J. Soc. Trop. Agr.* **1933**, *5*, 37–43.

37. Martelli, G. P. Leafroll. Graft-transmissible Diseases of Grapevines. Handbook for Detection and Diagnosis, Martelli, G. P. ICVG/FAO, Rome, Italy, 1993; pp 37–44.

38. Nemeth, G. Virus, Mycoplasma, and Rickettsia Diseases of Fruit Trees. Akademiai Kiado: Budapest, Hungary, 1986.

39. Naidu, R. A.; Hughes J. d'A. Methods for the Detection of Plant Virus Diseases. *In Plant Virology in Sub-saharan Africa*; Hughes, J.d'A; and Odu, B. O., Eds.; Proceedings of a Conference Organized by IITA, International Institute of Tropical Agriculture, Nigeria, 2001; pp 233–260.

40. Narayanasamy, P. *Microbial Plant Pathogens-Detection and Disease Diagnosis Viral and Viroid Pathogens*, Vol. 3. Springer Dordrecht Heidelberg: London, 2001; p 343.

41. Pan, S. Q.; Ye, X. S.; Kuć, J. A Technique for Detection of Chitinase S-1,3-Glucanase and Protein Patterns After a Single Separation Using Polyacrylamide Gel Electrophoresis or Isoelectron Focusing. *Phytopathology* **1991**, *81*, 970–974.

42. Purdy-Beale, H. A. Immunological Reactions with Tobacco Mosaic Virus. *Proc. Soc. Exptl. Biol. Med.* **1928**, *25*, 702–703.

43. Thresh, J. M. Crop Viruses and Virus Diseases: A Global Prospective.In *Virus Diseases and Crop Biosecurity*; Cooper, J. I.; Kuhne, T.; Polischuk, V. P., Eds. Springer: Dordrecht, The Netherlands, 2006; p 148.

44. Shukla, D. D.; Gough, K. U. The Use of Protein A, From *Staphylococcus aureus*, in Immune Electronmicroscopy for Detecting Plant Virus Particles. *J. Gen. Virol.* **1979**, *45*, 533–536.

45. Smith, K. M. *Recent Advances in the Study of Plant Viruses*, 2nd ed. J. and A. Churchill, Ltd.: *London, England*, 1951, p 300

46. Timian, R. G.; Peterson, C. E.; Hooker, W. J. *Am. Potato J.* **1955**, *32*, 411–417.

47. Van Regenmortel, M. H. V.; Dubs, M. C. Serological Procedures. In *Diagnosis of Plant Virus Diseases*; Matthews, R. E. F. CRC Press: Boca Raton, Florida, USA, 1993; pp 159–214.

48. Walkey, D. G. A.; Lyons, N. F.; Taylor, J. D. An Evaluation of a Virobacterial Agglutination Test for the Detection of Plant Viruses. *Plant Pathol.* **1992**, *41*, 462–471.

49. Walsh, K.; North, J.; Barker, I.; Boonham, N. Detection of Different Strains of Potato Virus Y and Their Mixed Infections Using Competitive Fluorescent RT–PCR. *J. Virol. Methods* **2001**, *91*, 167–173.

50. Wilson, G. S.; Mile, A. A. *Topley and Wilson's Principles of Bacteriologyand Immunity*, 4th ed. Edward Arnold, Ltd.: London, England, 1995; p 2331.

51. Alfaro-Fernandez, A.; Cebrian, M. C.; Cordoba-Selles, C.; Herrera-Vasquez, J. A.; Jorda, C. First Report of the US1 strain of *Pepino mosaic virus* in Tomato in the Canary Islands, Spain. *Plant Dis.* **2008**, *92*, 1590–1590.

52. Garcia-Arenal, F.; Fraile, A. Questions and Concepts in Plant Virus Evolution: AHistorical Perspective. In *Plant Virus Evolution*; Roossinck, M. J., Ed.; Springer-Verlag: Berlin, 2008; pp 1–14.

53. Jordan, R. L.; Guaragna, M. A.; Van Buren, T.; Putnam, M. L. First Report of a New Potyvirus, *Tricyrtis virus Y*, and *Lily virus X*, a Potexvirus, in *Tricyrtis formosana* in the United States. *Plant Dis.* **2008**, *92*, 648–648.

54. Koike, S. T.; Kuo, Y.-W.; Rojas, M. R.; Gilbertson, R. L. First Report of *Impatiens Necrotic Spot Virus* Infecting Lettuce in California. *Plant Dis.* **2008**, *92*, 1248–1248.

55. Clark, M. F.; Adams, A. N. Characteristics of the Microplate Method Of Enzyme-linked Immunosorbent Assay for the Detection of Plant Viruses. *J. Virol. Methods* **1977**, *34* (3), 475–483.

56. Astier, S.; Albouy, J.; Maury, Y.; Robaglia, C.; Lecoq, H. Principles of Plant Virology: Genome, Pathogenicity, Virus Ecology. Science Publishers, 2007, ISBN: 1578083168, New Hampshire, USA.

57. Hiebert, E.; Purcifull, D. E.; Christie, R. G. Purification and Immunological Analyses of Plant Viral Inclusion Bodies. In *Methods in Virology*, Vol. VII.; Maramorsch, K.; Koprowski, H., Eds., Academic Press: London, 1984; 225–280, ISBN 0-12-470270-4.

58. Haible, D.; Kober, S.; Jeske, H. Rollling Circle Amplification Revolutionizes Diagnosis and Genomics of Geminiviruses. *J. Virol. Methods* **2006**, *135*, 9–16.

59. Lima, J. A. A.; Sittolin, I. M.; Lima, R. C. A. Diagnose e Estratégias de Controle de Doenças Ocasionadas Por Vírus. In *Feijão Caupi: Avanços Tecnológicos*; Freire Filho, F. R.; Lima, J. A. A.; Silva, P. H. S; Ribeiro, V. Q., Eds. Embrapa Informação Tecnológica: Brasília, Brazil, 2005; pp 404–459.

60. Purcifull, D. E.; Hiebert, E.; Petersen, M.; Webb, S. Virus Detection–Serology. In *Encyclopedia of Plant Pathology*; Maloy, O. C.; Murray, T. D., Eds; John Wiley & Sons, Inc., 2001; pp 1100–1109. ISBN: 0-471-29817-4.

61. Vunsh, R.; Rosner, A.; Stein, A. The Use of the Polymerase Chain Reaction (PCR) for the Detection of Bean Yellow Mosaic Virus in Gladiolus. *Ann. Appl. Biol.* **1990,** *117,* 561–569.

62. Piepenburg, O.; Williams, C. H.; Stemple, D. L.; Armes, N. A. DNA Detection Using Recombination Proteins. *PLoS Biol.* **2006,** *4,* e204.

63. Vincent, M.; Xu, Y.; Kong, H. Helicase-dependent Isothermal DNA Amplification. *EMBO Rep.* **2004,** *5,* 795–800.

64. Derrick, K. S. Immuno-specific Grids for Electron Microscopy of Plant Viruses. *Phytopathology* **1972,** *62,* 753–754.

65. Compton, J. Nucleic Acid Sequence-based Amplification. *Nature* **1991,** *350,* 91–92.

66. Klerks, M. M.; Leone, G. O.; Verbeek, M.; van den Heuvel, J. F.; Schoen, C. D. Development of a Multiplex Amplidet RNA for the Simultaneous Detection of Potato Leafroll Virus and Potato Virus Y in Potato Tubers. *J. Virol. Methods* **2001,** *93,* 115–125.

67. Konaté, G.; Neya, B. J. Rapid Detection of Cowpea Aphid-borne Mosaic Virus in Cowpea Seeds. *Ann. Appl. Biol.* **1996,** *129,* 261–266.

68. Konaté, G.; Barro, N. Dissemination and Detection of Peanut Clump Virus in Groundnut Seed. *Ann. Appl. Biol.* **1993,** *123,* 623–627.

69. Johansen, E.; Edwards, M. C.; Hampton, R. O. Seed Transmission of Viruses: Current Perspectives. *Annual Rev. Phytopathol.* **1994,** *32,* 363–386.

70. Martin, R. R.; James, D.; Le'vesque, C. A. Impacts of Molecular Diagnostic Technologies on Plant Disease Management. *Annual Rev. Phytopathol.* **2000,** *38,* 207–239.

CHAPTER 3

Virus Afflictions of Anticancerous Medicinal Plant *Catharanthus roseus* (L.) G.Don

DEEPA SRIVASTAVA* and K. SHUKLA

Department of Botany, D.D.U. Gorakhpur University, Gorakhpur, Uttar Pradesh, India

Corresponding author. E-mail: drdeepasrivastav@gmail.com

ABSTRACT

Catharanthus roseus (L.) G. Don Madagascar periwinkle is one of the most extensively investigated medicinal plants. Pharmacological studies revealed that *C. roseus* contains more than 70 different types of alkaloids (indole alklaloids) and chemotherapeutic agents. Its high commercial value is due to the presence of anticancerous compound vincristine and vinblastine. Due to the presence of these compounds, viral infections in this plant are comparatively less. However, when it occurs, it affects the plant morphologically, physiologically, nutritionally, at a cellular level, as well as the quality and quantity of pharmaceutical interest compound found in plants in its healthy stage. The question arises that if *C. roseus* has an antiviral agent in it, then why some viruses are able to invade it. Why it is unable to defend itself when it is infected? What is the difference between the viruses that afflict *C. roseus* and the viruses against which *C. roseus* can work? Are there some common factors in those viruses that infect *C. roseus*? It is a potential antiviral agent, then why so many viruses are reported in it? The answer to these questions may give a new dimension to fight against viral infections in plants as well as in animals. The aim of this chapter is to summarize the progress made in the detection of viruses in *C. roseus* so that comparative analysis of viruses that invade and infect *C. roseus* can be made for future studies. We here report 54 viruses distributed worldwide in *C. roseus*, their name, natural host, symptoms, and their mode of transmission in alphabetical order. The particle size, shape, and other information about the viruses are reported in tabular form. These information may be useful for further studies. Although *C. roseus* is most extensively worked medicinal plants but most of the work has been done on its phytochemical properties. This study may induce some interest toward its virological study as its viral infection has its own significance.

3.1 INTRODUCTION

Catharanthus roseus (L.) G. Don, Madagascar periwinkle is one of the most extensively investigated medicinal plants that belongs to family Apocynaceae, which produces a class of secondary metabolites termed terpenoid indole alkaloids.[1] Pharmacological studies have revealed that *C. roseus* contains more than 70 different types of alkaloids (indole alklaloids) and chemotherapeutic agents.[2,3] These are treated for anticancer,[4-6] antidiabetic, and antihypertensive remedies.[7-10] Two important *Catharanthus* alkaloids, namely vinblastine and vincristine, have been developed into cancer chemotherapy agents since the 1960s and also marketed as vinblastine sulfate (Velbe®) and vincristine sulfate (Oncovin®).[11] In-vitro studies have shown that this plant produces a large number of alkaloids upon elicitation.[12] The enormity of work conducted on this medicinal plant is so large that since the 1950s more than 2500 publications have come in, ironically, only handful of data are available with regard to its viral infections. Due to very little emphasis on the diseases of the plant, very limited records of virus infections are available. This study compiles the available records on virus diseases on this plant. Espinha and Gaspar reported cucumber mosaic virus (CMV) infection in *C. roseus*, showing mild mosaic, chlorosis, and plant distortion.[13] Tomato spotted wilt virus (TSWV) has also been reported in *C. roseus* with black spots, systemic mosaic, leaf deformation, and browning of larger leaves at the bottom part of the plant.[14] Samad et al. reported the natural infection of *C. roseus* with an isolate of CMV in India.[15] Complete DNA sequences of two begomoviruses infecting this plant from Pakistan were determined. The sequence of one begomovirus (clone KN4) shows the highest level of nucleotide sequence identity (86.5%) to chili leaf curl India virus, and then (84.4% identity) to papaya leaf curl virus, and thus represents a new species, for which the name "Catharanthus yellow mosaic virus" (CYMV) was proposed.[16] Seabra et al. described the occurrence of an uncharacterized potyvirus causing mosaic and leaf malformation in this species.[17] In Sao Paulo State, Brazil, Maciel et al. reported a new potyvirus in this plant, exhibiting mosaic symptoms followed by leaf malformation and flower variegation named Catharanthus mosaic virus on the basis of biological, immunological, and molecular data.[18] There are many such available information. As a medicinal plant, tremendous research efforts have been given to study the bioactive compounds of *C. roseus* compared to its phytopathological aspect. However, this chapter will focus on viral infections reported in *C. roseus*.

3.2 REVIEW OF LITERATURE ON VIRUSES DESCRIBED ON *CATHARANTHUS ROSEUS*

Worldwide Distribution of *Catharanthus* viruses has been given in Table 3.1. A survey of the available literature reveals that identification and classification of viruses are mainly based on symptomatology, host range, transmission, biological properties, physical properties, particle morphology, and serological relationships. *C. roseus* has been reported as a susceptible host of many distinct viruses such as pepper ringspot virus, Narcissus mosaic virus, poplar mosaic virus, CYMV, tobacco streak virus, alfalfa mosaic virus, tobacco ringspot virus, Zantedeschia mild mosaic virus,[19] Carnation mottle virus,[9] potato yellow vein virus (PYVV),[20] and TSWV.[21] Natural occurrence of CMV on *C. roseus* has also been recorded.[15] Table 3.2 includes the particle size and shape of the viruses occurring on *C. roseus*.

3.3 DESCRIPTION OF VIRUSES REPORTED ON *CATHARANTHUS ROSEUS*

3.3.1 ABELIA LATENT TYMOVIRUS

First reported in *Abelia grandiflora*; from Maryland, USA.[22]

Symptoms: *C. roseus* is not a natural host of this virus but when experimentally infected mostly show necrotic local lesions, mottles.[23–25]

TABLE 3.1 Worldwide Distribution of *Catharanthus* Viruses.

Viruses and disease associated	Country(s)	References
Abelia latent tymovirus	Spreads in the North American region, the USA (eastern)	[25]
Alfalfa mosaic alfamovirus	Probably distributed worldwide	[24,156,157]
Apple mosaic ilarvirus	Probably distributed worldwide	[158]
Bean Pod mottle comovirus	Spreads in the North American region	[36,159]
Beet Curly top Hybrigemini virus	Spreads in the African region, the Eurasian region, the Mediterranean region, the south and central American region, Argentina, Bolinia, Brazil, Canada, Costa, Rica, Cyprus, Egypt, India, Iran, Italy, Mexico, Puerto Rico, Spain, Turkey, and the USA	[38,44]
Belladonna mottle tymovirus	Spreads in Bulgaria and Germany	[44,162]
Cacao yellow mosaic tymovirus	Spreads in Sierra Leone	[44,161]
Carnation mottle carom virus	Probably distributed worldwide	[9,162,163]
Cassava green mottle nepovirus	Spreads in Eurasian region, Australia, China, New Zealand, Turkey, and the former USSR	[44,48]
Cherry leaf roll nepovirus	Spreads in the USA and Australia	[44,164]
Citrus leaf rugose ilarvirus	Spreads in Argentina	[44,54]
Citrus ringspot virus	Spreads in the North American region	[44,61]
Clover yellow mosaic potexvirus	Spreads in Canada and USA	[44,64]
Cowpea severe mosaic comovirus	Spreads in the North American region and south and central American region	[44,165,166]
Cucumber mosaic cucumovirus	Probably distributed worldwide	[44,167–169]
Dogwood mosaic nepovirus	Spreads in the USA	[44,75]
Dulcamara mottle tymovirus	Spreads in the United Kingdom	[44,77]
Elm mottle ilarvirus	Probably distributed worldwide	[44,81]

TABLE 3.1 *(Continued)*

Viruses and disease associated	Country(s)	References
Erysimum latent tymovirus	Spreads in Germany	[44,81]
Foxtail mosaic potexvirus	Spreads in the USA	[44,170]
Humulus japonicus ilarvirus	Spreads in China	[44,85]
Lilac ring mottle ilarvirus	Found, but with no evidence of spread, in Canada, the Netherlands, and the United Kingdom	[44,171]
Nandina mosaic potexvirus	Spreads in the USA (in California)	[44,87]
Narcissus mosaic potexvirus	Probably distributed worldwide, spreads in particularly the Netherlands and the United Kingdom	[44,172]
Pea seed-borne mosaic potyvirus	Spreads in the Central Asian region, the Eastern Asian region, and the North American region, Australia, and the United Kingdom	[44,97]
Peach enation nepovirus	Spreads in Japan	[44,98]
Peanut stunt cucumovirus	Spreads in France, Japan, Korea, Morocco, Poland, Spain, and the USA	[173]
Pepper ringspot tobravirus	Spreads in the South and Central American region, Brazil	[44,103]
Plum American line pattern ilarvirus	Spreads in the North American region, Canada, and the USA	[44,171]
Poplar mosaic carlaravirus	Spreads in the Netherlands and the United Kingdom	[44,174]
Potato black ringspot nepovirus	Spreads in the South and Central American region, Peru	[44,175]
Potato T trichovirus	Spreads in the South and Central American region, Bolivia, and Peru	[44,113]
Prune dwarf ilarvirus	Probably distributed worldwide	[44,176]
Prunus necrotic ringspot ilarvirus	Probably distributed worldwide	[44,177]
Scrophularia mottle tymovirus	Spreads in Germany	[44,120]
Tobacco rattle tobravirus	Spreads in the Eurasian region, the North American region, and the South and Central American region, China, Japan, the former USSR	[44,178]

TABLE 3.1 *(Continued)*

Viruses and disease associated	Country(s)	References
Tobacco ringspot nepovirus	Spreads in the North American region, China	[44,179]
Tobacco streak ilarvirus	Probably distributed worldwide, spreads in the North American region and the pacific region, Australia, Canada, Peru, and the USA	[44,14,180]
Tobacco stunt varicosavirus	Spreads in Japan	[44]
Tomato spotted wilt ilarvirus	Probably distributed worldwide	[44,142,143]
Tulare apple mosaic ilarvirus	France and the USA (California)	[44,181]
Turnip crinkle tombusvirus	Spreads in the United Kingdom and the former Yugoslavia	[44,182]
Watermelon mosaic-2 potyvirus	Probably distributed worldwide	[148,183–185]
Wild cucumber mosaic tymovirus	Spreads in the USA (California and Oregon)	[44,151,186]

TABLE 3.2 Particle Size and Shape of the Viruses Occurring on *Catharanthus roseus*.

Virus	Shape and size	References
Abelia latent tymovirus	Virions isometric, not enveloped, 25 nm in diameter, rounded in profile	[25]
Alfalfa mosaic alfamovirus	Virions bacilliform; not enveloped; mostly 30, 35, 43, and 56 nm in length; 18 nm wide	[44,156]
Apple mosaic ilarvirus	Virions isometric, not enveloped, 25 and 29 nm in diameter	[44]
Bean pod mottle comovirus	Virions isometric, not enveloped, 28 nm in diameter	[44,159]
Beet curly top Hybrigemini virus	Virions geminate, not enveloped, 18–22 nm in diameter	[39,44]
Belladona mottle tymovirus	Virions isometric, not enveloped, 27 nm in diameter	[44,160,187]
Cacao yellow mosaic tymovirus	Virions isometric, not enveloped, 28 nm in diameter	[44,161]
Carnation mottle carom virus	Virions isometric, not enveloped, 34 nm in diameter	[9,163]
Cassava green mottle nepovirus	Virions isometric, not enveloped, 27 nm in diameter	[44,48]
Cherry leaf roll nepovirus	Virions isometric, not enveloped, 28 nm in diameter	[44,164]
Citrus leaf rugose ilarvirus	Virions unusually shaped, irregularly isometric virions, 25–32 nm in diameter	[44,188]
Citrus ringspot virus	Virions filamentous; they vary in length; not enveloped; usually flexuous (extremely); of 300–500 nm, or 1500–2500 nm; circa 8–10 nm wide	[44,61]
Clover wound tumor phytoreovirus	Virions isometric, 73 nm in diameter	[44,62]
Clover yellow mosaic potexvirus	Virions filamentous, not enveloped, usually flexuous, with a clear modal length, of 540 nm, 13 nm wide	[44,67]
Cowpea severe mosaic comovirus	Virions isometric, not enveloped, 25 nm in diameter	[44,166]
Cucumber mosaic cucumovirus	Virions isometric, not enveloped, 29 nm in diameter	[44,168]
Dogwood mosaic nepovirus	Virions isometric, not enveloped, 27 nm in diameter	[44,75]
Dulcamara mottle tymovirus	Virions isometric, not enveloped, 28 nm in diameter	[44,77]

TABLE 3.2 *(Continued)*

Virus	Shape and size	References
Elm mottle ilarvirus	Virions isometric, not enveloped, 25–30 nm in diameter	[44,181]
Erysimum latent tymovirus	Virions isometric, 27 nm in diameter	[44,81]
Foxtail mosaic potexvirus	Virions filamentous, not enveloped, usually flexuous, with a clear modal length, of 500 nm	[44,170]
Humulus japonicus ilarvirus	Virions isometric, not enveloped, 24–33 nm in diameter	[44,85]
Lilac ring mottle ilarvirus	Virions isometric, not enveloped, 27 nm in diameter	[44,130]
Nandina mosaic potexvirus	Virions filamentous, usually flexuous, with a clear modal length, of 471 nm	[44,87]
Narcissus mosaic potexvirus	Virions filamentous, not enveloped, usually flexuous, with a clear modal length, of 550 nm, 13–14 nm wide	[44,172]
Okhra mosaic tymovirus	Virions isometric, not enveloped, 28 nm in diameter	[44,95]
Pea seed-borne mosaic potyvirus	Virions filamentous, not enveloped, usually flexuous, with a clear modal length, of 770 nm, 12 nm wide	[44,189]
Peach enation nepovirus	Virions isometric, not enveloped, 33 nm in diameter	[44,98]
Peanut stunt cucumovirus	Virions isometric, not enveloped, 30 nm in diameter	[44,173]
Pepper ringspot tobravirus	Virions rod-shaped, not enveloped, usually straight, with a clear modal length, of 52 and 197 nm, 22 nm wide	[44,103]
Pepper veinal mottle potyvirus	Virions filamentous, not enveloped, usually flexuous, with a clear modal length, of 770 nm, 12 nm wide	[44]
Plum American line pattern ilarvirus	Virions isometric; not enveloped; 26, 28, 31, and 33 nm in diameter; rounded in profile	[44,171]
Poplar mosaic carlavirus	Virions filamentous, not enveloped, usually flexuous (but only slightly), with a clear modal length, of 675 nm (Biddle and Tinsley, 1971)	[13,44,174]

TABLE 3.2 *(Continued)*

Virus	Shape and size	References
Potato black ringspot nepovirus	Virions isometric, not enveloped, 25 nm in diameter	[44,175]
Potato T trichovirus	Virions filamentous, not enveloped, usually flexuous, with a clear modal length, of 637 nm, 12 nm wide	[44,112]
Prune dwarf ilarvirus	Virions isometric and bacilliform; not enveloped; 19–20 nm in diameter; 20 nm in length, or 23 nm in length, or 26 nm in length, or 38 nm in length (some 20 × 73 nm)	[44,176]
Prunus necrotic ringspot ilarrvirus	Virions isometric; not enveloped; 23, 25, and 27 nm in diameter	[44,177]
Scrophularia mottle tymovirus	Virions isometric, not enveloped, 26 nm in diameter	[44,120]
Spring beauty latent bromovirus	Virions isometric, not enveloped, 28 nm in diameter	[44,118]
Tobacco mosaic satellitivirus	Virions isometric, not enveloped, less than 17 nm in diameter	[44]
Tobacco necrosis necrovirus	Virions isometric, not enveloped, 26 nm in diameter	[44,190]
Tobacco rattle tobravirus	Virions rod-shaped; not enveloped; usually straight; with a clear modal length; of 46–114 nm (T), or 180–197 B); 22 nm wide	[44,178]
Tobacco ringspot nepovirus	Virions isometric, not enveloped, 25–29 nm in diameter	[44]
Tobacco streak ilarvirus	Virions isometric; not enveloped; 27, 30, and 35 nm in diameter	[44,191]
Tobacco stunt varicosavirus	Virions rod-shaped, not enveloped, usually straight, with a clear modal length, of 300 and 340 nm, 18 nm	[44]
Tomato spotted wilt ilarvirus	Virions isometric, enveloped, 85 nm in diameter	[14,143]
Turnip crinkle tombusvirus	Virions isometric, not enveloped, 28 nm in diameter	[44]
Watermelon mosaic-2 potyvirus	Virions filamentous, not enveloped, usually flexuous, of 730–765 nm	[44,185]
Wild cucumber mosaic tymovirus	Virions isometric, not enveloped, 28 nm in diameter	[44,151,186]

Transmission: Transmitted by a vector; an insect; *Myzus persicae* and at least 13 other species; Aphididae. Transmitted in a nonpersistent manner; virus transmitted by mechanical inoculation; transmitted by grafting; not transmitted by contact between plants; transmitted by seed.

3.3.2   ALFALFA MOSAIC ALFAMOVIRUS

First reported in *Medicago sativa*; from the USA.[26]

Symptoms: Experimentally infected plants mostly show necrotic local lesions, mottles, or ringspots.[24,27]

Transmission: Transmitted by a vector; an insect; *M. persicae* and at least 13 other species; Aphididae. Transmitted in a nonpersistent manner; virus transmitted by mechanical inoculation; transmitted by grafting; not transmitted by contact between plants; transmitted by seed (50% in alfalfa seeds from individual infected plants and up to 10% in commercial seed); transmitted by pollen to the seed.[28]

3.3.3   APPLE MOSAIC ILARVIRUS

First reported in Rosa spp. and *Malus domestica*; from the USA.[29,30]

Symptoms: *C. roseus* is a diagnostically susceptible host of this virus and shows systemic chlorotic lines and rings.[31,32]

Transmission: Virus transmitted by mechanical inoculation; transmitted by grafting (of roots); possibly not transmitted by seed, but probably transmitted by pollen to the pollinated plant.[33]

3.3.4   BEAN POAD MOTTLE COMOVIRUS

First reported in *Phaseolus vulgaris* cv. Tendergreen; from Charleston, USA.[34]

Symptoms: *C. roseus* is a susceptible host of this virus. It shows severe mottling, malformed leaves, and pods.[23,35,36]

Transmission: Transmitted by a vector; an insect; *Ceratoma trifurcata, Diabrotica balteata, Diabrotica undecimpunctata howardii, Colaspis flavida, Colaspis lata, Epicauta vittata, Epilachna varivestis*; Coleoptera; virus transmitted by mechanical inoculation; transmitted by grafting; not transmitted by seed; not transmitted by pollen.

3.3.5 BEAT CURLY TOP HYBRIGEMINI VIRUS

First reported in *Beta vulgaris*; from Western USA.[37]

Symptoms: *C. roseus* is a susceptible host of this virus. The symptoms persist leaf rolling, vein clearing; leaves become dark and dull green in color.[38]

Transmission: Transmitted by a vector; an insect; *Circulifer tenellus* in North America, *C. tenellus*, *Circulifer opacipennis* in Mediterranean Basin; Cicadellidae. Transmitted in a persistent manner; virus retained when the vector molts; does not multiply in the vector; not transmitted congenitally to the progeny of the vector; not transmitted by mechanical inoculation (unless special procedures used); transmitted by grafting (by dodder from plants that are not hosts of the vector); not transmitted by seed.[39]

3.3.6 BELLADONA MOTTLE TYMOVIRUS

First reported in *Atropa belladonna*; from Germany.[40]

Symptoms: *C. roseus* is a susceptible host of this virus. Symptoms are mottling and distortion.[41,42]

Transmission: Transmitted by a vector; an insect; *Epithrix atropae*; Coleoptera; not transmitted by aphids; virus transmitted by mechanical inoculation; not transmitted by contact between plants; not transmitted by seed.[41]

3.3.7 CACAO YELLOW MOSAIC VIRUS

First reported in *Theobroma cacao*; from Giehuna, Sierra Leone.[43]

Symptoms: Diagnostically susceptible host species *C. roseus* show systemic chlorosis.[44]

Transmission: Transmitted by a vector; virus transmitted by mechanical inoculation; transmitted by grafting; not transmitted by seed.[45]

3.3.8 CARNATION MOTTLE CARMOVIRUS

First reported in *Dianthus* spp.; from the United Kingdom; by Kassanis.[46]

Symptoms: *C. roseus* is susceptible to this virus. It shows chlorotic and necrotic local lesions.

Transmission: Transmitted by means not involving a vector; virus transmitted by mechanical inoculation; transmitted by grafting; transmitted by contact between plants; not transmitted by seed.[47]

3.3.9 CASSAVA GREEN MOTTLE NEPOVIRUS

First reported in *Manihot esculenta*; in a sample from the Solomon Islands examined in Scotland.[48]

Symptoms: *C. roseus* is experimentally infected plants mostly show mottles, both in inoculated and other leaves, mosaic.

Transmission: Transmitted by means not involving a vector; virus transmitted by mechanical inoculation.[48]

3.3.10 CHERRY LEAF ROLL NEPOVIRUS

First reported in *Ulmus americana* (American elm), *Prunus avium* (cherry), and *Juglans regia* (walnut).[49–51]

Symptoms: *C. roseus* is experimentally infected plants mostly show chlorotic or necrotic local lesions, systemic necrosis or mosaic.[52,53]

Transmission: Transmitted by a vector; *Xiphinema coxi, Xiphinema diversicaudatum*, and *Xiphinema vuittenezi*.[54] Not transmitted by *Xiphinema americanum, Xiphinema bakeri, Longidorus elongatus, Longidorus leptocephalus, Longidorus macrosoma*, or *Paralongidorus maximus*[55,56]; virus transmitted by mechanical inoculation; transmitted by grafting; not transmitted by contact between plants.

3.3.11 CITRUS LEAF RUGOSE ILARVIRUS

First reported in *Citrus limon*; from the USA.[57]

Symptoms: *C. roseus* is **natural host** of citrus leaf rugose ilar virus. Symptoms persist in different plants; however, no symptoms have been described in this plant.[58]

Transmission: Transmitted by a vector; an insect; *Agalliopsis novella, Agallia constricta, Agallia quadripunctata*; Cicadellidae. Not transmitted by *Aceratagallia sanguinolenta*. Transmitted in a persistent manner; virus retained when the vector molts; multiplies in the vector; transmitted congenitally to the progeny of the vector (2% of vector eggs); not transmitted by mechanical inoculation; not transmitted by seed.

3.3.12   CITRUS RINGSPOT VIRUS

First reported in *Citrus* sp.; from Florida and California, USA.[59]
Symptoms: *C. roseus* is susceptible to this virus. Experimentally infected plants mostly show systemic mosaics, mottles, ringspots, or necrosis.[60]
Transmission: Possibly transmitted by a vector, by mechanical inoculation; transmitted by grafting.[61]

3.3.13   CLOVER WOUND TUMOR PHYTOREOVIRUS

First reported in *Melilotus officinalis* clone C10 and the vector *A.constricta*; from the USA.[62]
Symptoms: *C. roseus* is susceptible to this virus. In *C. roseus*, no symptoms have been described.[63]
Transmission: Transmitted by a vector; an insect; *A. novella, A. constricta, A. quadripunctata*; Cicadellidae. It is transmitted in a persistent manner. Virus retained when the vector molts; multiplies in the vector; transmitted congenitally to the progeny of the vector (2% of vector eggs); not transmitted by mechanical inoculation; not transmitted by seeds.

3.3.14   CLOVER YELLOW MOSAIC POTEXVIRUS

First reported in *Trifolium repens*; from the USA.[64,65]
Symptoms: *C. roseus* is susceptible to this virus. Experimentally infected plants mostly show chlorotic or necrotic local lesions, systemic mosaic.[36,66]
Transmission: Not transmitted by *Acyrthosiphon pisum, Anuraphis bakeri*.[67] Virus transmitted by mechanical inoculation.[68]

3.3.15   COWPEA SEVERE MOSAIC COMOVIRUS

First reported in *Vigna unguiculata*; from Louisiana, Arkansas, and Indiana, USA.[69]

Symptoms: *C. roseus* is a susceptible host of this virus; symptoms persist.[69,70]

Transmission: Transmitted by a vector; an insect; *Ceratoma arcuata, Ceratoma ruficornis, C. trifurcata, Ceratoma variegata, Chalcodermus bimaculatus, D. balteata, Diabrotica speciosa, Diabrotica virgifera, D. undecimpunctata, Diphaulaca* sp., *E. varivestis, Acalymma vittatum*; Coleoptera; virus transmitted by mechanical inoculation; transmitted by seed (10% in *V. unguiculata* ssp. *sesquipedalis*, 8% in *V. unguiculata*, varies with strain and cultivar); transmitted by pollen to the seed.

3.3.16   CUCUMBER MOSAIC CUCUMOVIRUS

First reported in *Cucumis sativus*; from the USA.[71]

Symptoms: *C. roseus* is susceptible to this virus. Experimentally infected plants mostly show mosaics and stunting, reduced fruit yield.[72,73]

Transmission: Transmitted by a vector; an insect; more than 60 spp. including *A. pisum, Aphis craccivora*, and *M. persicae*; Aphididae; transmitted in a nonpersistent manner; virus transmitted by mechanical inoculation; transmitted by seed.[74]

3.3.17   CUCUMBER MOSAIC VIRUS SUBGROUP IB

Symptoms: *C. roseus* is a natural host of this virus. It shows local and/ or systemic symptoms.[15]

Transmission: It is transmitted mechanically as well as by vectors. The virus isolate was efficiently sap transmitted from naturally infected periwinkle to healthy periwinkle. *M. persicae* transmitted the virus in a nonpersistent manner to healthy periwinkle. Seedlings treated with buffer only served as negative controls for inoculation.[75]

3.3.18   DOGWOOD MOSAIC NEPOVIRUS

First reported in *Cornus florida*; from Clemson, South Carolina, USA.[76]

Symptoms: Symptoms vary seasonally; symptoms mosaic.

Transmission: Not transmitted by *X. diversicaudatum*; virus transmitted by mechanical inoculation; transmitted by grafting.[77]

3.3.19 DULCAMARA MOTTLE TYMOVIRUS

First reported in *Solanum dulcamara*; from the United Kingdom in Hertfordshire.[42]

Symptoms: Symptoms persist; experimentally infected plants mostly show local lesions, mosaics, mottles, and distortion.[78]

Transmission: Transmitted by a vector; an insect; *Psylloides affinis*; Coleoptera. Transmitted in a semipersistent manner; virus transmitted by mechanical inoculation; not transmitted by contact between plants; transmitted by seed.

3.3.20 ELM MOTTLE ILARVIRUS

First reported in *Ulmus minor*, from Germany.[79]

Symptoms: *C. roseus* is susceptible to this virus. Symptoms persist (but restricted to a few branches. It shows chlorotic mosaic symptom.[80]

Transmission: Virus transmitted by mechanical inoculation; transmitted by grafting; transmitted by seed.[81]

3.3.21 ERYSIMUM LATENT TYMOVIRUS

First reported in *Erysimum helveticum*, *Erysimum perovskianum*, *Erysimum pulchellum*, *Erysimum sylvestre*, *Erysimum crepidifolium*, *Fibigia clypata*, *Arabis ludoviciana*, and *Barbarea vulgaris*; from Germany.[82]

Symptoms: *C. roseus* is susceptible to this virus experimentally infected plants mostly show local lesions, vein clearing, and mosaics.[82]

Transmission: Transmitted by a vector; an insect; *Phyllotreta* spp.; Coleoptera. Transmitted in a semipersistent manner; virus transmitted by mechanical inoculation; not transmitted by contact between plants; not transmitted by seed; not transmitted by pollen.[82]

3.3.22 FOXTAIL MOSAIC POTEXVIRUS

First reported in *Setaria italica* and *Setaria viridis*; from the USA.[83]

Symptoms: *C. roseus* is susceptible to this virus experimentally infected plants mostly show leaf mosaic.[84]

Transmission: Virus not transmitted by mechanical inoculation; transmitted by seed.[84]

3.3.23   HUMULUS JAPONICAS ILARVIRUS

First reported in *Humulus japonicus*; from the United Kingdom in seedlings grown from seed imported from the People's Republic of China.[85]
 Symptoms: Experimentally infected plants mostly show necrotic local lesions, chlorotic mottle, or mosaic in systemic leaves.[85]
 Transmission: Virus transmitted by mechanical inoculation; transmitted by seed.

3.3.24   LILAC RING MOTTLE ILARVIRUS

First reported in *Syringa vulgaris*; from the Netherlands.[86]
 Symptoms: necrotic local lesions and systemic mottling.[86]
 Transmission: Virus transmitted by mechanical inoculation; by grafting; by contact between plants and by seed.[86]

3.3.25   NANDINA MOSAIC POTEXVIRUS

First reported in *Nandina domestica*; from California, USA.[87]
 Symptoms: systemic mottling and malformation.[88]
 Transmission: Not transmitted by *M. persicae*; virus transmitted by mechanical inoculation.

3.3.26   NARCISSUS MOSAIC POTEXVIRUS

First reported in *Narcissus pseudonarcissus*; from the Netherlands and the United Kingdom.[44]
 Symptoms: *C. roseus* is susceptible to this virus; it shows symptomless systemic infection.[44]
 Transmission: Virus transmitted by mechanical inoculation; not transmitted by contact between plants and not transmitted by seed.

3.3.27   OKHRA MOSAIC TYMOVIRUS

First reported in *Abelmoschus esculentus*; from Cte d'Ivoire.[89]

Symptoms: *C. roseus* is susceptible to this virus experimentally infected plants mostly show regular vein chlorosis, spotting or dotting, chlorotic mosaic, chlorotic local lesions, and stunting.[90–92]

Transmission: Transmitted by a vector; an insect; *Podagrica decolorata*.[93] *Podagrica uniforma, P. sjostedti* in Nigeria.[94,95] Possibly *Bemisia tabaci* is a vector in Nigeria[94] but needs confirmation. Not transmitted by *Aphis gossypii, Chrysolagria cuprina, Lagria villosa, Medythia quaterna, Ootheca mutabilis,* and *Nisotra dilecta*; transmitted in a nonpersistent manner; virus transmitted by mechanical inoculation; transmitted by grafting; not transmitted by seed.

3.3.28   PEA SEED-BORNE MOSAIC POTYVIRUS

First reported in *Pisum sativum* by Musil.[96]

Symptoms: *C. roseus* is susceptible to this virus. Experimentally infected plants mostly show transitory vein clearing, resetting of stem and branches, leaves dark green and leaflets folded adaxially, flowers malformed and often sterile, pods small, few misshapen seeds.[97]

Transmission: Transmitted by a vector; an insect; *A. pisum, A. craccivora, Aphis fabae, Dactynotus escalanti, Macrosiphum crataegarius, Rhopalosiphum padi*; Aphididae; transmitted in a nonpersistent manner; virus transmitted by mechanical inoculation; transmitted by seed.

3.3.29   PEACH ENATION NEPOVIRUS

First reported in *Prunus persica*, from Japan.[97]

Symptoms: *C. roseus* is susceptible to this virus experimentally infected plants show local lesions; systemic leaf mottling.[98]

Transmission: Virus transmitted by mechanical inoculation; transmitted by grafting.

3.3.30   PEANUT STUNT CUCUMOVIRUS

First reported in *Arachis hypogaea*; from Virginia, USA.[99]

Symptoms: *C. roseus* is susceptible to this virus, experimentally infected plants show chlorotic local lesions; systemic spotting.[22,100]

Transmission: Transmitted by a vector; an insect; *A. craccivora, Aphis spiraecola,* and *M. persicae* but not *A. gossypii*; Aphididae. Transmitted in a nonpersistent manner; virus transmitted by mechanical inoculation; transmitted by seed.[101]

3.3.31   PEPPER RINGSPOT TOBRAVIRUS

First reported in *Lycopersicon esculentum*; from Brazil.[102]

Symptoms: *C. roseus* is susceptible to this virus, experimentally infected plants shown ringspots and yellow banding.

Transmission: Virus transmitted by mechanical inoculation; transmitted by grafting; not transmitted by contact between plants.

3.3.32   PEPPER VENIAL MOTTLE POTYVIRUS

First reported in *Capsicum annuum, Capsicum frutescens, Petunia* × *hybrida*; from Tafo, Ghana.[103]

Symptoms: *C. roseus* is susceptible to this virus experimentally infected plants show systemic mosaic, mottle, and leaf-shape malformation.[103,104]

Transmission: Transmitted by a vector; an insect; *A. gossypii, A. spiraecola, M. persicae, Toxoptera citricidus*; Aphididae; transmitted in a nonpersistent manner; virus transmitted by mechanical inoculation; transmitted by grafting; not transmitted by contact between plants.

3.3.33   PLUM AMERICAN LINE PATTERN ILARVIRUS

First reported in *Prunus americana* × *Prunus salicina* (Shiro plum); from Wenatchee, Washington, USA.[105]

Symptoms: *C. roseus* is susceptible to this virus experimentally infected plants mostly show chlorotic or necrotic local lesions; systemic vein banding, mottles, ringspots.[105,106]

Transmission: The means of natural spread is not exactly known, but if this is limited to transmission by infected pollen grains. The virus is not seed-borne. The international spread is most probably by means of infected planting material.

3.3.34   POPLAR MOSAIC CALARAVIRUS

Symptoms: *C. roseus* is susceptible to this virus experimentally infected plant show chlorotic or necrotic local lesions and systemic mosaic.[107,108]

Transmission: Virus transmitted by mechanical inoculations, by grafting, not transmitted by seed, transmitted by pollen to the pollinated plant.[109–111]

3.3.35   POTATO BLACK RINGSPOT NEPOVIRUS

First reported in *Solanum tuberosum* (potato) from Peru.[112]

Symptoms: *C. roseus* is diagnostically susceptible host species and shows *chlorotic* spots; systemic necrosis.[112]

Transmission: Virus transmitted by mechanical inoculation; transmitted by grafting; not transmitted by contact between plants; not transmitted by seed.

3.3.36   POTATO T TRICHOVIRUS

First reported in *Solanum tuberosum*; from Peru.[113]

Symptoms: *C. roseus* is susceptible to this virus experimentally infected plant is usually symptomless but occasionally induces mild leaf mottling.[113]

Transmission: Virus transmitted by mechanical inoculations, transmitted by grafting, not transmitted by contact between plants; transmitted by seeds.

3.3.37   POTATO YELLOW VEIN VIRUS

Potato yellow vein disease was first observed in Antioqua.

Symptoms: *C. roseus* is potential viral reservoirs.[20]

Transmission: Long known to be transmitted by the greenhouse whitefly (*Trialeurodes vaporariorium*), the precise identity of its causal agent has remained obscure. Here, we present evidence that a closterovirus with a bipartite genome PYVV is associated with PYVD.[20]

3.3.38   PRUNE DWARF ILAR VIRUS

First reported in *Prunus domestica*; from the USA.[114]

Symptoms: *C. roseus* is a susceptible host of this virus. It shows leaf yellowing and abscission.[115]

Transmission: Transmitted by means not involving a vector; virus transmitted by mechanical inoculation; transmitted by grafting; not transmitted by contact between plants; transmitted by seeds, transmitted by pollen to the seed, and transmitted by pollen to the pollinated plant.[116]

3.3.39   PRUNUS NECROTIC RINGSPOT ILARVIRUS

First reported in *Prunus persica*; from the USA.[117]

Symptoms: *C. roseus* is a susceptible host of this virus it shows chlorotic lines and rings.[115]

Transmission: Transmitted by means not involving a vector; virus transmitted by mechanical inoculation; transmitted by grafting; not transmitted by contact between plants; transmitted by seeds, transmitted by pollen to the seed, and transmitted by pollen to the pollinated plant.

3.3.40   SPRING BEAUTY LATENT BROMOVIRUS

First reported in *Claytonia virginica*; from Arkansas, USA.[115]

Symptoms: *C. roseus* is diagnostically susceptible host species it shows systemic necrosis and mottle.[118]

Transmission: Possibly on lawn mowers and tractors in lawns and public areas. Virus transmitted by mechanical inoculation; transmitted by grafting; possibly transmitted by contact between plants; not transmitted by seed.

3.3.41   SCROPHULARIA MOTTLE TYMOVIRUS

First reported in *Scrophularia nodosa*; from Germany.[119]

Symptoms: Experimentally infected *C. roseus* shows local lesions, systemic mottle, or mosaic.[42,120]

Transmission: Transmitted by a vector; an insect; *Cionus tuberculosis*, *Cionus scrophularia*, *Cionus hortulanus*, *Cionus alauda*; Coleoptera; virus transmitted by mechanical inoculation.

3.3.42 TOBACCO RATTLE TOBRAVIRUS

First reported in *Nicotiana tabacum*; from Germany.[121]

Symptoms: *C. roseus* is a susceptible host of this virus. It shows chlorotic or necrotic local lesions; systemic mottle.[122]

Transmission: Transmitted by a vector; a nematode; *Paratrichodorus allius*, *Paratrichodorus anemones*, *Paratrichodorus christiei*, *Paratrichodorus nanus*, *Paratrichodorus pachydermus*, *Paratrichodorus teres*, *Trichodorus minor*, *Trichodorus primitivus*, *Trichodorus viruliferus*; Trichodoridae; virus transmitted by mechanical inoculation; transmitted by grafting; not transmitted by contact between plants; transmitted by seeds.

3.3.43 TOBACCO MOSAIC SATELLITE VIRUS

First reported in *Nicotiana glauca*; from southern California, USA.[123]

Symptoms: *C. roseus* is a susceptible host of this virus. It is symptomless; symptoms may be caused by mixed infection with CMV or potyviruses.[123]

Transmission: Virus transmitted by mechanical inoculation; not transmitted by seed.[124]

3.3.44 TOBACCO NECROSIC NECROVIRUS

First reported in *N. tabacum*; from Cambridge, the United Kingdom.[125]

Symptoms: *C. roseus* is a susceptible host of this virus. It shows necrotic local lesions.[23,125]

Transmission: Transmitted by a vector, a fungus; *Olpidium brassicae*; Chytridiales; virus transmitted by mechanical inoculation; not transmitted by seed; not transmitted by pollen.

3.3.45   TOBACCO RINGSPOT NEPOVIRUS

First reported in *N. tabacum*.[126]

Symptoms: *C. roseus* is a susceptible host of this virus. It shows chlorotic local lesions and leaves malformation.[127]

Transmission: Transmitted by a vector; a nematode (and also nonspecifically by insects and mites—*A. gossypii, M. persicae, Melanopus* sp. *Epitrix hirtipennis, Thrips tabaci, X. americanum,* Dorylamidae; virus lost by the vector when it molts does not multiply in the vector; not transmitted congenitally to the progeny of the vector; does not require a helper virus for vector transmission; transmitted by mechanical inoculation; not transmitted by contact between plants; transmitted by seed.

3.3.46   TOBACCO STREAK ILAR VIRUS

First reported in *N. tabacum*; from Wisconsin, USA.[128]

Symptoms: *C. roseus* is maintenance and propagation hosts of this virus. It shows necrotic local lesions.[129–131]

Transmission: Transmitted by a vector; an insect; *Frankliniella occidentalis* and *T. tabaci*; Thysanoptera[132]; virus transmitted by mechanical inoculation, by grafting, not transmitted by contact between plants, transmitted by seeds, transmitted by pollen to the pollinated plant.[133]

3.3.47   TOBACCO STUNT VARICOSAVIRUS

First reported in *N. tabacum*; from Hiroshima Prefecture, Japan.[134]

Symptoms: *C. roseus* is a susceptible host of this virus. It shows necrotic local yellow leaf spotting, not systemic.[135]

Transmission: Transmitted by a vector; a fungus; *O. brassicae*; Chytridiales; virus does not require a helper virus for vector transmission; transmitted by mechanical inoculation; transmitted by grafting; not transmitted by contact between plants; not transmitted by seed.

3.3.48   TOMATO-SPOTTED WILT TOSPOVIRUS

First reported in *L. esculentum*.[136,137]

Symptoms: *C. roseus* is diagnostically susceptible host species. It shows local black spots and leaves sometimes becoming yellow and abscessing; systemic mosaic and leaf deformation.[138–140]

Transmission: Transmitted by a vector; an insect; *T. tabaci*, *Thrips setosus*, *Thrips parmi*, *Frankliniella schultzei*, *F. occidentalis*, *Frankliniella fusca*, and *Scirtothrips dorsalis*; Thysanoptera; transmitted in a persistent manner; virus retained when the vector moults; multiplies in the vector, transmitted by mechanical inoculation, transmitted by grafting, not transmitted by contact between plants, not transmitted by seed, not transmitted by pollen.[141]

3.3.49 TOMATO-SPOTTED WILT ILARVIRUS

First reported in *L. esculentum*.[142]

Symptoms: *C. roseus* is diagnostically susceptible host species. It shows local black spots, leaves sometimes becoming yellow and abscessing; systemic mosaic and leaf deformation.[14]

Transmission: Transmitted by an insect vector; transmitted in a persistent manner; transmitted by grafting; not transmitted by contact between plants; not transmitted by seed; not transmitted by pollen.[143]

3.3.50 TURNIP CRINKLE CARMOVIRUS

First reported in *Brassica campestris* ssp. *rapa*; from Scotland.[144]

Symptoms: *C. roseus* is a susceptible host of this virus. It shows local chlorotic lesions; no systemic infection.[145]

Transmission: Transmitted by a vector; an insect; *Phyllotreta* (nine species) and *Psylloides* (two species). Transmitted in a nonpersistent manner; virus transmitted by mechanical inoculation; transmitted by contact between plants; not transmitted by seed.

3.3.51 TURTLE APPLE MOSAIC ILAR VIRUS

First reported in *Malus sylvestris*; from the USA.[146]

Symptoms: *C. roseus* is maintenance and propagation hosts; it shows brown local lesions; no systemic infection.[147]

Transmission: Virus transmitted by mechanical inoculation; not transmitted by seed.

3.3.52   WATERMELON MOSAIC-2 POTYVIRUS

First reported in *Citrullus lanatus*.[148]
Symptoms: *C. roseus* is a susceptible host of this virus. It shows systemic mosaic and occasional leaf malformation.[36]
Transmission: Transmitted by a vector; an insect; *M. persicae, A. craccivora*; at least 29 species of aphids transmit watermelon mosaic-2 potyvirus,[149] Aphididae. Virus transmitted by mechanical inoculation; not transmitted by seed.[150]

3.3.53   WILD CUCUMBER MOSAIC TYMOVIRUS

First reported in *Marah macrocarpus*; from California, USA.
Symptoms: *C. roseus* is diagnostically susceptible host species. It shows symptomless systemic infection.[151,152]
Transmission: Transmitted by a vector; an insect; *Acalymma trivittata*; Coleoptera; virus transmitted by mechanical inoculation; transmitted by grafting; not transmitted by contact between plants.

3.3.54   ZANTEDESCHIA MILD MOSAIC VIRUS

Symptoms: *C. roseus* is a susceptible host of this virus. It shows mild mosaic system.
Transmission: Virus transmitted by mechanical inoculation, transmitted by grafting, not transmitted by contact between plants.[153]

3.4   RESULTS AND DISCUSSION

C. roseus is a high-utility plant, apart from its natural supply of anticancerous compounds; it is a popular ornamental plant. It is also a model plant for studies in plant pathology and in biotechnology.[154] The present chapter has compiled the existing information on virus afflictions of *C. roseus* including their names, author name, distribution, transmission, physical properties, and particle size (Tables 3.1 and 3.2). There are 54 viruses reported in this study that infects *C. roseus* naturally or artificially. These information are based on symptoms reported, host-range study, and mode of transmission. The study shows that the number and distribution of the virus are also increasing. The virus first reported in one country has been transmitted in another country, for example, groundnut bud necrosis virus has been recently reported in India.[155] The transmission and distribution of virus should be controlled so that the infection of new viruses in a particular country can be restricted. The quality and quantity of raw material is the basic necessity for efficient drugs; knowledge of virus afflictions will provide the right direction in the conservation and protection of this marvelous plant from various viruses.

3.5 CONCLUSION

The demand for natural products and plant-based medicines is increasing very fast due to advantages of natural product over synthetic drugs. *C. roseus* is an amazing herb due to its wide applications. It is necessary to know about the type of virus infections, which occur in *C. roseus* so that useful precautions must be taken before the use of particular plant for drug purpose. The present study has provided record of 54 reported virus in this anticancerous plant. The collectors should consider these virus afflictions so that the quality and quantity of raw drugs should not be affected. This study may be useful for the future aspirants to further work on different prospect on viral diseases on *C. roseus*.

ACKNOWLEDGMENTS

The authors are thankful to the Head of the Department of Botany, D.D.U. Gorakhpur University, Gorakhpur, for encouragement and providing necessary facilities. One of the authors Dr. Deepa Srivastava is thankful to UGC, New Delhi, for providing postdoctoral fellowship.

KEYWORDS

- *Catharanthus roseus*
- medicinal plant
- alkaloids
- virus
- transmission
- particle morphology

REFERENCES

1. El-Sayed, M.; Verpoorte, R. Catharanthus Terpenoid Indole Alkaloids: Biosynthesis and Regulation. *Phytochem. Rev.* **2007,** *6*, 277–305.
2. Verpoorte, R. Exploration of Nature's Chemodiversity: The Role of Secondary Metabolites as Lead for Drug Development. *Drug Dev. Today* **1998,** 3, 232–238.
3. Punia, S.; Kaur, J.; Kumar, R.; Kumar, K. *Catharanthous roseus*: A Medicinal Plant with Potential Antitumor Properties. *Int. J. Res. Aryuveda Pharm.* **2014,** *5* (6), 652–665.
4. Manganey, P.; Andriamialisoa, R. Z.; Langlois, Y.; Langlois, N.; Pottier, P. Preparation of Vinblastine, Vincristine and Leurosidine: Antitumor Alkaloids from *Catharanthus* spp (Apocyanaceae). *J. Am. Chem. Soc.* **1979,** *101*, 2243–2245.
5. Svoboda, G. H. The Role of the Alkaloids of *Catharanthus roseus* (L.) G. Don (*Vinca rosea*) and Their Derivatives in Cancer Chemotherapy. In *Workshop Proceedings Plants: The Potentials for Extracting Protein, Medicines, and Other Useful Chemicals. Congress of the United States, Office of Technology Assessment,* Washington, DC, 1983; pp 154–169.
6. Cragg, G. M.; Newman, D. J. Plants as Source of Anticancer Agents. *J. Ethnopharmal.* **2005,** *100*, 72–79.
7. Ghosh, R. K.; Gupta, I. Effect of *Vinca rosea* and *Ficus racemososus* on Hyperglycemia in Rats. *Indian J. Anim. Health* **1980,** *19*, 145–148.
8. Chattopadhyay, R. R.; Sarkar, S. K.; Ganguly, S.; Banerjee, R. N.; Basu, T. K. Hypoglycemic and Antihyperglycemic Effect of Leaves of *Vinca rosea* Linn. *Indian J. Physiol. Pharmacol.* **1991,** *35* (3), 145–151.
9. Singh, H. P.; Hallan, V.; Raikhy, G.; Kulshrestha, S.; Sharma, M. L.; Ram, R.; Gang, I. D.; Zaidi, A. A. Characterization of an Indian Isolate of Carnation Mottle Virus Infecting Carnations. *Curr. Sci.* **2005,** *88* (4), 594–601.
10. Wiart, C. *Catharanthus roseus* G. Don. In *Medicinal Plants of Southeast Asia*, 2nd ed.; Prentice Hall: Kuala Lumpur, 2002; pp 224–225.
11. Van De Heijden, R.; Jacobs, D. I.; Snoejer, W.; Hallard, D.; Verpoorte, R. The Catharanthus Alkaloids: Pharmacognosy and Biotechnology. *Curr. Med. Chem.* **2004,** *11*, 607–628.

12. Verpoorte, R.; Contin, A.; Memelink, J. Biotechnology for the Production of Plant Secondary Metabolites. *Phytochem. Rev.* **2002**, *1*, 13–25.
13. Espinha, L. M.; Gaspar, J. O. Partial Characterization of CMV Isolated from *Catharanthus roseus*. *Fitopatol. Bras.* **1997**, *22* (2), 209–212.
14. Chatzivassiliou, E. K.; Weekes, R.; Morris, J.; Wood, K. R.; Barker, I.; Katis, N. I. Tomato Spotted Wilt Virus (TSWV) in Greece: Its Incidence Following the Expansion of *Frankliniella occidentalis*, and Characterisation of Isolates Collected from Various Hosts. *Ann. Appl. Biol.* **2000**, *137*, 127–144.
15. Samad, A.; Ajayakumar, V.; Gupta, M. K.; Shukla, A. K.; Darokar, M. P.; Alam, M. Natural Infection of Periwinkle (*Catharanthus roseus*) with CMV Subgroup 1B. *Australas. Plant Dis. Notes* **2008**, *3*, 30–34.
16. Ilyas, M.; Nawaz, K.; Shafiq, M.; Haider, M. S.; Shahid, A. A. Complete Nucleotide Sequences of Two Begomoviruses Infecting Madagascar Periwinkle (*Catharanthus roseus*) from Pakistan. *Arch. Virol.* **2013**, *158* (2), 505–510.
17. Seabra, P. V.; Alexandre, M. A. V.; Rivas, E. B.; Duarte, L. M. L. Occurrence of a Potyvirus in *Catharanthus roseus*. *Arq. Inst. Biol.* **1999**, *66*, 116.
18. Maciel, S. C.; da Silva, R. F.; Reis, M. S.; Jadao, A. S.; Rosa, D. D.; Giampan, J. S.; Kitajima, E. W.; Rezende, J. A. M.; Camargo, L. E. A. Characterization of a New Potyvirus Causing Mosaic and Flower Variegation in *Catharanthus roseus* in Brazil. *Sci. Agric. (Piracicaba, Braz.)* **2011**, *68* (6), 687–690.
19. Huang, C. H.; Chang, Y. C. Identification and Molecular Characterization of Zantedeschia Mild Mosaic Virus (ZaMMV), a New Calla Lily-Infecting Potyvirus. *Arch. Virol.* **2005**, *150*, 1221–1230.
20. Salazar, L. F.; Muller, G.; Querci, M.; Zapata, J. L.; Owens, R. A. Potato Yellow Vein Virus: Its Host Range, Distribution in South America and Identification as a Crinivirus Transmitted by *Trialeurode vaporariorum*. *Ann. Appl. Biol.* **2000**, *137* (1), 7–19.
21. Chatzivassiliou, E. K.; Livieratos, I.; Jenser, G.; Katis, N. I. Ornamental Plants and Trips Population Associated with Tomato Spotted Wilt Virus in Greece. *Phytoparasitica* **2000**, *28* (3), 257–264.
22. Waterworth, H. E.; Monroe, R. L.; Kahn, R. P. *Phytopathology* **1973**, *63*, 93.
23. Thornberry, H. H. Index of Plant Virus Diseases. *U.S. Dept. Agric. Hdbk.* **1966**, *307*, 313.
24. Hull, R. Alaflafa Mosaic Virus. *Adv. Virus Res.* **1969**, *15*, 365–433.
25. Waterworth, H. E.; Kaper, J. M.; Koeing, R. Purification and Properties of Tymovirus from *Abelia*. *Phytopathology* **1975**, *65*, 891–895.
26. Weimer, J. L. *Phytopathology* **1931**, *21*, 122.
27. Schmelzer, K.; Schmidt, H. B.; Beczner, L. *Biol. Zbl.* **1973**, *92*, 211.
28. Mugal, S. M.; Zadjali, A. D.; Matrooshi, A. R. Occurrence, Distribution and Some Properties of Alfalfa Mosaic Alfamo Virus in the Sultanate of Oman. *Pak. J. Agric. Sci.* **2003**, *40* (1–2), 87–91.
29. White, R. P. *Plant Dis. Reptr.* **1928**, *12*, 33.
30. Bradford, F. C.; Joley, L. *J. Agric. Res.* **1933**, *46*, 901.
31. Fulton, R. W. *Phytopathology* **1952**, *42*, 413.
32. Posnette, A. F.; Ellenberger, C. E. *Ann. Appl. Biol.* **1963**, *45*, 74.
33. Dursunoglu, S.; Ertunc, F. Distribution of Apple Mosaic Ilarvirus (APMV) in Turkey. In *ISHS Acta Horticulture 781: XX International Symposium on Virus and Virus Like Diseases of Temperate Fruit Crops—Fruit Tree Disease*, 2006.

34. Zaumeyer, W. J.; Thomas, H. R. *J. Agric. Res.* **1948**, *77*, 81.
35. Skotland, C. B. *Plant Dis. Reptr.* **1958**, *42*, 1155.
36. Hampton, R.; Beczner, L.; Hagedorn, D.; Bos, L.; Inouye, T.; Barnett, O. W.; Musil, M. Meiners, J. *Phytopathology* **1978**, *68*, 989.
37. Ball, E. D. *Bull. Bur. Entrol. U.S. Dept. Agric.* **1909**, *66*, 33.
38. Bennett, C. W. *Monogr. Am. Phytopathol. Soc.* **1971**, *7*, 81.
39. Mumford, D. L. A New Method of Mechanically Transmitting Curly Top Virus. *Phytopathology* **1972**, *62*, 1217–1218.
40. Bode, O.; Marcus, O. *Proc. 4th Int. Cong. Crop Protect*, Hamburg, 1959, Vol. 1; pp 375.
41. Paul, H. L.; Bode, O.; Jankulova, M.; Brandes, J. *Phytopathol. Z.* **1968**, *61*, 342.
42. Guy, P. L.; Dale, J. L.; Adena, M. A.; Gibbs, A. *J. Plant Pathol.* **1984**, *33*, 337.
43. Blencowe, J. W.; Brunt, A. A.; Kenten, R. H.; Lovi, N. K. *Trop. Agric. Trin.* **1963**, *40*, 233.
44. Brunt, A. A.; Crabtree, K.; Dallwitz, M. J.; Gibbs, A. J.; Watson, L. *Viruses of Plants: Description of List from the VIDE Database*; CAB International: Wallingford, UK, 1996.
45. Baker, C. A. A Virus Related to *Clover Yellow Mosaic Virus* Found East of the Mississippi River in *Verbena canadensis* in Florida. *Am. Phytopathol. Soc.* **2004**, *88* (2), 223–232.
46. Kassanis, B. *Ann. Appl. Biol.* **1955**, *43*, 103.
47. Carrington, J. C.; Morris, T. J. *Virology* **1985**, *144*, 1.
48. Lennon, A. M.; Aiton, M. M.; Harrison, B. D. *Ann. Appl. Biol.* **1987**, *10*, 545.
49. Schuster, C. E.; Miller, P. W. *Phytopathology* **1933**, *23*, 408.
50. Swingle, R. V.; Tilford, P. E.; Irish, C. F. *Phytopathology* **1941**, *31*, 22.
51. Swingle, R. V.; Tilford, P. E.; Irish, C. F. *Phytopathology* **1943**, *33*, 1196.
52. Tomlinson, J. A.; Walkey, D. G. A. *Ann. Appl. Biol.* **1967**, *59*, 415.
53. Walkey, D. G. A.; Cooper, J. I. *Rep. Nat. Veg. Res. Stat.* **1972**, 100.
54. Flegg, J. J. M.; Rep, E. *Malling Res. Stat.* **1969**, 155.
55. Van Hoof, H. A. *Neth. J. Plant Pathol.* **1971**, *77*, 30.
56. Jones, A. T.; McElroy, F. D.; Brown, D. J. F. *Ann. Appl. Biol.* **1981**, *99*, 143.
57. Fawcett, H. S. *Citrus Diseases and Their Control*; McGraw Hill: New York, London, 1936; p 656.
58. Garnsey, S. M.; Gonsalves, D. Citrus Leaf Rugose Ilarvirus. *CMI/AAB Descr. Plant Viruses* **1976**, *164*, 4.
59. Wallace, J. M.; Drake, R. J. Citrus Stunt and Ringspot Two Previously Undescribed Virus Disease of Citrus. In *Proc. 4th Conf. IOCV Univ.*; Florida Press, Gainesville, 1968; pp 177–183.
60. Timmer, L. W.; Garnsey, S. M.; McRitchie, J. J. *Plant Dis. Reptr.* **1978**, *62*, 1054.
61. Derrick, K. S, Lee, R. F.; Hewitt, B. G.; Barthe, A.; da Graca, J. V. Characterization of Citrus Ringspot Virus. *Florida Agric. Exp. Stat. J. Ser.* **1988**, *R-00626*, 386–390.
62. Black, L. M. *Proc. Am. Phil. Soc.* **1944**, *88*, 132.
63. Hillman, B. I.; Anzola, J. V.; Halpern, B. T.; Cavileer, T. D.; Nuss, D. L. *Virology* **1991**, *185*, 896.
64. Johnson, F. *Phytopathology* **1942**, *32*, 103.
65. Pratt, M. J. *Can. J. Bot.* **1961**, *39*, 655.
66. Rao, D. V.; Hiruki, C.; Matsumoto, T. *Phytopathol. Z.* **1980**, *98*, 260.

67. Purcifull, D. E. *Plant Dis. Reptr.* **1968**, *52*, 759.
68. Bos, L. Clover Yellow Mosaic Potexvirus. *CMI/AAB Descr. Plant Viruses* **1973**, *111*, 4.
69. Perez, C.-M. *Plant Dis. Reptr.* **1970**, *54*, 212.
70. Van Swaans, K. *Neth. J. Plant Pathol.* **1973**, *79*, 257.
71. Price, W. C. *Phytopathology* **1934**, *24*, 743.
72. Douine, L.; Quiot, J. B.; Marchoux, G.; Archange, P. *Ann. Phytopathol.* **1979**, *11*, 439.
73. Mazidah, M.; Yusoff, K.; Habibuddin, H.; Tan, Y. H.; Lau, W. H. Characterization of Cucumber Mosaic Virus (CMV) Causing Mosaic Symptom on *Catharanthus roseus* (L.) G. Don in Malaysia Pertanika. *J. Trop. Agric. Sci.* **2012**, *35* (1), 41–53.
74. Choi, S. K.; Cho, I. S.; Choi G. S. First Report of *Cucumber Mosaic Virus* in *Catharanthus roseus* in Korea. **2014**, *98* (9), 1283–1283.
75. Barnett, O. W.; Baxter, L. W. *Proc. Ann. Phytopathol. Soc.* **1976**, *3*, 249.
76. Barnett, O. W.; Reddick, B. B.; Burrows, P. M.; Baxter, L. W. Characterization of Dogwood Mosaic Nepovirus from Cornus, Florida. *Phytopathology* **1989**, *79*, 951–958.
77. Gibbs, A. J.; Hecht-Poinar, E.; Woods, R. D.; McKee, R. K. *J. Gen. Microbiol.* **1966**, *44*, 177.
78. Schmelzer, K. *Phytopathol. Z.* **1969**, *64*, 39.
79. Jones, A. T. Elm Mottle Ilarvirus. *CMI/AAB Descr. Plant Viruses* **1974**, *139*, 4.
80. Jones, A. T.; Mayo, M. A. *Ann. Appl. Biol.* **1973**, *75*, 347.
81. Shukla, D. D.; Gough, K. Erysimum Latent Tymovirus. *CMI/AAB Descr. Plant Viruses* **1979**, *222*, 4.
82. Paulsen, A. Q.; Sill, W. H. *Phytopathology* **1969**, *59*, 1043.
83. Paulsen, A. Q.; Niblett, C. L. *Phytopathology* **1977**, *67*, 1346.
84. Rouleau, M.; Bancroft, J. B.; Mackie, G. A. Partial Purification and Characterization of Foxtail Mosaic Potexvirus RNA-Dependent RNA Polymerase. *Virology* **1993**, *197* (2), 695–703.
85. Adams, A. N.; Clark, M. F.; Barbara, D. J. Host Range, Purification and Some Properties of a New Ilarvirus from *Humulus japonicas*. *Ann. Appl. Biol.* **1989**, *114* (3), 497–508.
86. Van der Meer, F. A. *Acta Hortic.* **1976**, *59*, 105.
87. Moreno, P.; Attatom, S.; Weathers, L. G. *Proc. Ann. Phytopathol. Soc.* **1976**, *3*, 319.
88. Zettler, F. W.; Hiebert, E.; Christie, R. G.; Abo El-Nil, M. M. *Acta Hortic.* **1980**, *110*, 71.
89. Givord, L.; Pfeiffer, P.; Hirth, L. *C. R. Acad. Sci. Paris* **1972**, *275*, 1563.
90. Givord, L. *Ann. Phytopathol.* **1977**, *9*, 53.
91. Givord, L. *Agronom. Trop.* **1979**, *34*, 88.
92. Lana, A. O.; Bozarth, R. F. *Phytopathol. Z.* **1975**, *83*, 77.
93. Givord, L.; Den Boer, L. *Ann. Appl. Biol.* **1980**, *94*, 235.
94. Lana, A. O.; Ajibola Taylor, T. *Ann. Appl. Biol.* **1976**, *82*, 361.
95. Atiri, G. I. *Ann. Appl. Biol.* **1984**, *104*, 261.
96. Musil, M. *Biol. Bratisl.* **1966**, *21*, 133.
97. Hampton, R.; Mink, G. I. Pea Seed-Borne Mosaic Potyvirus. *CMI/AAB Descr. Plant Viruses* **1975**, *146*, 4.
98. Kishi, K.; Abibo, K.; Takanashi, K. *Ann. Phytopathol. Soc. Jpn.* **1973**, *39*, 373.
99. Miller, L. I.; Troutman, J. L. *Plant Dis. Reptr.* **1966**, *50*, 139.
100. Milbrath, G. M.; Tolin, S. A. *Plant Dis. Reptr.* **1977**, *61*, 637.

101. Bananej, K.; Hajimorad, M. R.; Roossinck, M. J.; Shahraeen. Identification and Characterization of Peanut Stunt Cucumovirus from Naturally Infected Alfalfa in Iran. *Plant Pathol.* **1998**, *47*, 355–336.
102. Silberschmidt, K. *Phytopathol. Z.* **1963**, *46*, 209.
103. Brunt, A. A.; Kenten, R. H. Pepper Ringspot Tobravirus. *CMI/AAB Descr. Plant Viruses* **1971**, *104*, 4.
104. Ladipo, J. L.; Roberts, I. M. *Ann. Appl. Biol.* **1977**, *87*, 133.
105. Kirkpatrick, H. C.; Cheney, P. W.; Linder, R. C. *Plant Dis. Reptr.* **1964**, *48*, 616.
106. Paulsen, A. Q.; Fulton, R. W. Hosts and Properties of a Plum Line Pattern Virus. *Phytopathology* **1968**, *58*, 766–772.
107. Van der Meer, F. A. *Acta Hortic.* **1980**, *110*, 211.
108. Berg, T.M. Studies on Poplar Mosaic virus and its relation to the host. Meded. Landbouwhogeschool Wageningen **1964**, *64*(11), 1–77.
109. Corte, A. *IST*; Bot. Univ. Lab. Crittogam.: Pavia, 1959, vol 17; p 1.
110. Blattny, P.; Svobodova, B.; Bojansky, L.; Prochazkova. *Acta Musei Nat. Pragae XVIII*, **1962**, *2*, 47.
111. Cooper, J. I.; Edwards, M. L. *Ann. Appl. Biol.* **1981**, *99*, 53.
112. Salazar, L. F.; Harrison, B. D. Host Range and Properties of Potato Black Ringspot Virus. *Ann. Appl. Biol.* **1978**, *90* (3), 375–386.
113. Salazer, L. F.; Harrison, B. D. Potato T Trichovirus. *CMI/AAB Descr. Plant Viruses* **1978**, *187*, 4.
114. Thomas, H. E.; Hildebrand, E. M. *Phytopathology* **1936**, *26*, 1145.
115. Fulton, R. W. *Phytopathology* **1957**, *47*, 683.
116. Al-Rwahnih, M.; Myrta, A.; Herranze, M. C.; Pallas, V. Tracking American Line Pattern Virus in Plum by ELISA and Dot-Blot Hybridization During a Whole Year. *J. Plant Pathol.* **2004**, *86*, 167–169.
117. Cochran, L. C.; Hutchins, L. M. *Phytopathology* **1941**, *31B*, 860.
118. Valverde, R. A. Spring Beauty latent virus: A New Member of the Bromovirus Group. *Phytopathology* **1985**, *75*, 395–398.
119. Hein, A. *Phytopathol. Z.* **1959**, *36*, 290.
120. Bercks, R. Scrophularia Mottle Tymovirus. *CMI/AAB Descr. Plant Viruses* **1973**, *113*.
121. Boning, K. *Z. ParasitKde* **1931**, *3*, 103.
122. Schmelzer, K. *Phytopathol. Z.* **1957**, *30*, 281.
123. Zaitlin, M.; Israel, H. W. *CMI/AAB Descr. Plant Viruses* **1975**, *151*, 5.
124. Dodds, J. A. Satellite Tobacco Mosaic Virus. *Annu. Rev. Phytopathol.* **1998**, *36*, 295–310.
125. Price, W. C. *Am. J. Bot.* **1940**, *27*, 530.
126. Fromme, F. D.; Wingard, S. A.; Priode, C. N. *Phytopathology* **1927**, *17*, 321.
127. Bowyer, J. W.; Atherton, J. G. *Phytopathology* **1972**, *61*, 1451.
128. Johnson, J. *Phytopathology* **1936**, *26*, 285.
129. Berkeley, G. H.; Phillips, J. H. H. *Can. J. Res.* **1943**, *21*, 181.
130. Fulton, R. W. *Phytopathology* **1948**, *38*, 421.
131. Kaiser, W. J.; Wyatt, S. D.; Pesho, G. R. *Phytopathology* **1982**, *72*, 1508.
132. Sdoodee, R.; Teakle, D. S. *Plant Pathol.* **1987**, *36*, 377.
133. Abtahi, F. S.; Koohi, H. Host Range and Characterization of Tobacco Streak Virus Isolate from Lettuce in Iran. *Afr. J. Biotechnol.* **2008**, *7* (23), 4260–4264.
134. Hidaka, Z. *Ann. Phytopathol. Soc. Jpn.* **1950**, *15*, 40.

135. Hiruki, C. *Can. J. Bot.* **1975**, *53*, 2425.
136. Brittlebank, C. C. *J. Agric. Victoria, Aust.* **1919**, *17*, 213.
137. Samuel, G.; Bald, J. G.; Pittman, H. A. *Bull. Council Sci. Ind. Res.*; Melbourne, 1930, 44; p 65.
138. Klinkowski, M.; Uschdraweit, H. A. *Phytopathol. Z.* **1952**, *19*, 269.
139. Smith, K. M. *Textbook of Plant Virus Diseases*, 2nd ed.; Churchill: London, 1957.
140. Best, R. *J. Adv. Virus Res.* **1968**, *13*, 65.
141. Parrella, G.; Gogssalons, P.; Gebre-Selassie, K.; Vovlas, C.; Marchoux, G. An Update of the Host Range of Tomato Spotted Wilt Virus. *J. Plant Pathol.* **2003**, *85* (4), 227–264.
142. Ie, T. S. Tomato Spotted Wilt Ilarvirus. *CMI/AAB Descr. Plant Viruses* **1970**, *39*, 4.
143. Converse, R. H.; Lister, R. M. The Occurrence and Some Properties of Black Berry Latent Virus. *Phytopathology* **1969**, *59*, 25–333.
144. Dawson, W.O., Helf, M.E. Host Range Determinant of Plant viruses. Annual review of Plant Biology. **1992**, *43*(1), 527–555.
145. Broadbent, L.; Heathcote, G. D. *Ann. Appl. Biol.* **1958**, *46*, 585.
146. Yarwood, C. E. *Hilgardia* **1955**, *23*, 613.
147. Thomas, B. J. *Ann. Appl. Biol.* **1984**, *105*, 213.
148. Webb, R. E.; Scott, H. A. Isolation and Identification of Watermelon Mosaic Virus 1 and 2. *Phytopathology* **1965**, *55*, 895–900.
149. Edwardson, J. R.; Christie, R. G. *Fla. Agric. Exp. Stat. Monogr.* **1986**, *14*, 454.
150. Srivastava, D.; Panday, N.; Tiwari, A. K.; Shukla, K. Identification of a Potyvirus Associated with Mosaic Disease of *Catharanthus roseus* in Gorakhpur and Its Histo-pathological Effects Medicinal Plants. *Int. J. Phytomed. Relat. Ind.* **2012**, *4* (1), 23–27.
151. Lindberg, G. D.; Hall, D. H.; Walker, J. C. A Study of Melon and Squash Virus. *Phytopathology* **1956**, *46*, 489–495.
152. Milne, K. S.; Grogan, R. G.; Kimble, K. A. *Phytopathology* **1969**, *59*, 819.
153. Huang, C. H.; Hu, W. C.; Yang, T. C.; Chang, Y. C. Zantedeschia Mild Mosaic Virus, a New Widespread Virus in Calla Lily, Detected by ELISA, Dot-Blot Hybridization and IC-RT-PCR. *Plant Pathol.* **2006**, *56* (1), 183–189.
154. Nejat, N.; Valdiani, A.; Cahill, D.; Tan, Y. H.; Maziah, M.; Abiri, R. Ornamental Exterior versus Therapeutic Interior of Madagascar Periwinkle (*Catharanthus roseus*): The Two Faces of a Versatile Herb. *Sci. World* **2015**, *982412*, 1–19.
155. Basavaraj Kumar A.; Holkar S. K.; Jain, R. K.; Mandal B. First Report of Groundnut Bud Necrosis Virus Infecting Periwinkle *(Catharanthus roseus)* in India. *Plant Dis.* **2017**, *101* (8), 1559.
156. Bos, L.; Jaspars, E. M. J. Alfalfa Mosaic Virus. *CMI/AAB Descr. Plant Viruses*, **1971**, *46*, 4.
157. Jaspars, E. M. J.; Bos, L. Alfalfa Mosaic Virus. *CMI/AAB Descr. Plant Viruses*; Association of Applied Biologists: Wellesbourne, UK, 1980; vol 229 (No 46 revised), pp 1–7.
158. Fulton, R. W. Apple Mosaic Ilarvirus. *CMI/AAB Descr. Plant Viruses* **1972**, *83*, 4.
159. Bancroft, J. B. Purification and Properties of Bean Pod Mottle Virus and Associated Centrifugal and Electrophoretic Components. *Virology* **1962**, *16*, 416–427.
160. Paul, H. L. Belladona Mottle Tymovirus. *CMI/AAB Descr. Plant Viruses* **1971**, *52*, 3.
161. Brunt, A. A. Cacao Yellow Mosaic Tymovirus. *CMI/AAB Descr. Plant Viruses* **1970**, *11*, 4.
162. Holling, M.; Stone, O. M. Carnation Mottle Carmovirus. *CMI/AAB Descr. Plant Viruses* No. 7, 1970, 4 pp.

163. Hollings, M.; Stone, O. M. Investigation of Carnation Viruses. *Ann. Appl. Biol.* **1964**, *53*, 103–108.

164. Christoff, A. *Phytopathol. Z.* **1958**, *31*, 381.

165. Smith, C. E. *Science* **1924**, *60*, 268.

166. De Jager, C. P. Cowpea Severe Mosaic. *CMI/AAB Descr. Plant Viruses* **1979**, *209*, 5.

167. Doolittle, S. P. A New Infectious Mosaic Disease of Cucumber. *Phytopathology* **1916**, *6*, 145–147.

168. Price, W. C. *Phytopathology* **1935**, *25*, 947.

169. Gibbs, A. J.; Harrison, B. D. Cucumber Mosaic Cucumovirus. *CMI/AAB Descr. Plant Viruses* **1970**, *1*, 4.

170. Short, M. N. Foxtail Mosaic Potexvirus. *CMI/AAB Descr. Plant Viruses* **1983**, *264*, 3.

171. Fulton, R. W. American Plum Line Pattern Virus. *CMI/AAB Descr. Plant Viruses* **1984**, *280*. Association of Applied Biologists: Wellesbourne, UK.

172. Mowat, W. P. Narcissus Mosaic Potexvirus. *CMI/AAB Descr. Plant Viruses* **1971**, *45*, 3.

173. Mink, G. I. Peanut Stunt Cucumovirus. *CMI/AAB Descr. Plant Viruses* **1972**, *92*, 4.

174. Koenig, R. Poplar Mosaic Carlavirus. *CMI/AAB Descr. Plant Viruses* **1982**, *259*, 4.

175. Salazar, L. F.; Harrison, B. D. Potato Black Ringspot Nepovirus. *CMI/AAB Descr. Plant Viruses* **1979**, *206*, 4.

176. Fulton, R. W. Prune Dwarf Ilarvirus. *CMI/AAB Descr. Plant Viruses* **1970**, *19*, 3.

177. Fulton, R. W. Prunus Necrotic Ringspot Ilarvirus. *CMI/AAB Descr. Plant Viruses* **1970**, *5*, 4.

178. Harrison, B. D. Tobaccos Rattle Tobravirus. *CMI/AAB Descr. Plant Viruses* **1970**, *12*, 4.

179. Stace-Smith, R. Tobacco Ringspot Nepovirus. *CMI/AAB Descr. Plant Viruses* **1970**, *17*, 4.

180. Fulton, R. W. Tobacco Streak Ilarvirus. *CMI/AAB Descr. Plant Viruses* **1985**, *307*, 5.

181. Fulton, R. W. Tulare Apple Mosaic Ilarvirus. *CMI/AAB Descr. Plant Viruses* **1971**, *42*, 3.

182. Holling, M.; Stone, O. M. Turnip Crinkle Tombusvirus. *CMI/AAB Descr. Plant Viruses* **1972**, *109*, 3.

183. Greber, R. S. Viruses Infecting Cucurbits in Queensland. *Qd. J. Agric. Sci.* **1969**, *26*, 145–171.

184. Schmelzer, K.; Milicic, D. Zur Kenntnis der Verbreitung des Wassermelonen mosaik Virus in Europa und seinei Fahigkit zur Biltong von Zelleinschlusskoren (on the Distribution of Watermelon Mosaic Inclusion Bodies). *Phytopathol. Z.* **1966**, *57*, 8–16.

185. Hiebert, E.; Edwardson, J. Watermelon Mosaic-2 Potyvirus. *CMI/AAB Descr. Plant Viruses* **1984**, *293*, 7.

186. Van Regenmortel, M. H. V. Wild Cucumber Mosaic Tymovirus. *CMI/AAB Descr. Plant Viruses* **1972**, *105*, 4.

187. *Hiremath, C. N.; Munshi, S. K.; Murthy, M. R. N. Structure of Belladonna Mottle Virus: Cross-Rotation Function Studies with Southern Bean Mosaic Virus. Acta Crystallogr., Sect. B: Struct. Sci.* **1990**, *46* (4), 562–567.

188. Garnsey, S. M. *Phytopathology* **1975**, *65*, 50.

189. Inouye, T. *Ann. Phytopathol. Soc. Jpn.* **1967**, *33*, 38.

190. Kassanis, B. Tobacco Necrosis Necrovirus. *CMI/AAB Descr. Plant Viruses* **1970**, *14*, 4.

191. Jones, A. T.; Mayo, M. A. *Ann. Appl. Biol.* **1975**, *79*, 297.

CHAPTER 4

Important Diseases of Papaya and Their Integrated Disease Management

DIGANGGANA TALUKDAR,[1] UTPAL DEY,[2*] and G. P. JAGTAP[3]

[1]Department of Plant Pathology and Microbiology,
College of Horticulture, Central Agricultural University, Sikkim, India

[2]ICAR Division of Crop Production, ICAR Research Complex for
NEH Region, Umiam 793103, Meghalaya, India

[3]Department of Plant Pathology, College of Horticulture, VNMKV,
Parbhani, Maharashtra, India

*Corresponding author. E-mail: utpaldey86@gmail.com

ABSTRACT

With high nutritive quality, papaya is globally an important tropical fruit that originated from Mexico and South America and later on cultivated gradually in all parts of the world. This plant holds a significant part in agricultural export. However, this plant is prone to biosecurity threat due to exposure with plenty of diseases leading to reduction in yield as well as marketable quality. This chapter covers some of the very important diseases which include fungal, bacterial, viral, and phytoplasma. This chapter elaborately discusses about some economically important disease with proper etiology, epidemiology, and symptoms which would help to identify the diseases with much ease. The fungal diseases included in this chapter are, namely, black spot, anthracnose, damping-off of seedlings, phytopthora blight, and powdery mildew. Bacterial diseases include bacterial leaf spot, internal yellowing, and purple stain rot. Viral diseases include papaya ring spot virus, papaya lethal yellowing virus, Meleira disease, and phytoplasma diseases include dieback and bunchy top. However, the most vital part of this chapter is the management practices that maintain the production and productivity of papaya. In order to get effective control along with the maintenance of sustainability and ecological balance of environment, integrated diseases management is the most effective management principle. Integrated diseases management process includes physical, cultural, biological, and finally chemical methods.

4.1   INTRODUCTION

Papaya, scientifically *Carica papaya*, is one of the important tropical fruit having lots of nutritional values covering a significant place in agricultural export in many of the countries which are in developing position as it provides livelihoods to thousands of people. With the pace of time, the tropical fruits demand in market has been mounting gradually over the past 20 years. Being deliciously tasty, it has been a rich source of antioxidants, nutrients like vitamin B, vitamin C, falate, carotenes, flavonoids, minerals, and fiber (Rivera et al., 2010). Asia was leading in papaya production during 2008–2010 producing about 52.55% of papaya crop globally. Then it was led by South America in 2nd position and Africa holding 3rd position, as shown in Figure 4.1 (FAOSTAT, 2012). But in the year 2014–15, India became the leading producer of papaya, Brazil being 2nd and Indonesia being 3rd (source: www.statistics.com/world-papaya-production). This plant is attacked by several fungal, bacterial, and viral diseases and the frequency of attack varies depending on the geographical region and its varied climatic conditions. In postharvest process also like transport, packaging, etc. This plant is attacked by several diseases for which the market quality is diminished to many fold.

4.2   FUNGAL DISEASES

4.2.1   BLACK SPOT

One of the most widespread papaya disease is the black spot disease. When it infects, it causes great damage to the whole plant in the orchards. It reduces the market quality to a great extent.

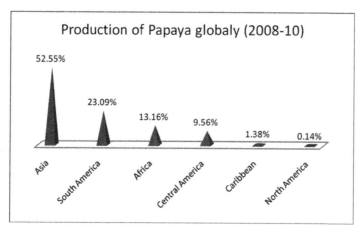

FIGURE 4.1 Production of papaya in global area during the year 2008–2010.
Source: FAOSTAT (2012).

4.2.1.1 ETIOLOGY

The causal organism of black spot is *Asperisporium caricae* (Speg.) Maubl., which is characterized by conidiophores of short length which are clumped together covering the stomata surface. Conidiophores are not branched and are hyaline to olive brown in color. The conidium is bicellular smooth but dry with ellipsoidal, pyriform, or clavate in shape with hyaline to pale brown color measuring 14–26 mm × 7–10 mm (Ventura et al., 2004).

4.2.1.2 SYMPTOMS

It is characterized by formation of round, light-brown necrotic, circular to irregular spots almost 1/4th in surrounded by yellow halo on the upper surface of the leaves (Hine et al., 1965). Gray to black colored powdery growth develops in the regions adjacent to the spots on the lower surface of the leaves. When the lesions coalesce, leaf senescence and defoliation takes place. Circular watery portions were observed on the fruits which later on become brown in color which almost attains 5 mm of diameter. The lesions remain confined in the epidermal layers and with length of time the skin in that affected area become corky.

4.2.1.3   EPIDEMIOLOGY

The disease intensity of black spot is more in temperature within 23–27°C, with winds and heavy rainfall or overhead irrigation and powerful winds. Most of the infection occurs during winter and spring being seasonal that develops the diseases lesions and spreading of fungal spores from the elder leaves to the healthy leaves. The fungi penetrate through stomata. The symptoms are noticeable after 8–10 days of inoculation (Holliday, 1980). The fruits are infected during immature stage. The disease lesion liberates new spores when the fruits are fully matured.

4.2.1.4   CONTROL MEASURES

4.2.1.4.1   Cultural Practices

The infected diseased leaves and older leaves with high severity of the disease should be detached.

4.2.1.4.2   Chemical Control

Fungicides must be applied at the initial growth stage as soon as the first symptoms appear on the plants. Lesions first appear on the older leaves at initial stage. The most critical period for control of the disease are the first 5 months after planting and this is the time where climate is the most favorable parameter for disease development. Although the fungicides effectively control black spot, but it has been noticed that during the period of prolong rain, these fungicides are not all fruitful to control the disease. The experimental evaluations have recommended that the two groups of fungicides, that is, triazole and strobilurin must be used based on their better efficacy in the experiment.

4.2.2   ANTHRACNOSE

Anthracnose is one the most prevalent diseases affecting the papaya plant and is considered as principle postharvest disease occurring in almost all parts of the world (Venturia et al., 2004). This fungus damages all parts of the plant, thus reducing the market value. The absence of control measures can cause yield loss of up to 100% in a number of orchards.

4.2.2.1 ETIOLOGY

This disease is caused by *Colletotrichum gloeosporiodes* (Penz.) Penz. Sacc. (its teleomerphic phase is *Glomerella cingulata* (Ston.) Spauld. and Schrenk as suggested by Costa et al., 2001; and Holiday, 1980. This fungi produces acervuli in subcuticular and subepidermal area of leaves producing conidial masses covering the lesion in the centers with pinkish to orange color. Having septate conidiophores, the fungi produce pale brown to hyaline conidia. The prefect stage, that is, *Glomerella cingulata* develops perithecia on various parts of the host which might be solitary or aggregated, globose to obpyriform in shape, dark brown to black colored. The asci produce eight spores and clavate to cylindrical in shape. The asci produce oval-, cylindrical-, or fusiform-shaped ascospores.

4.2.2.2 SYMPTOMS

The symptoms first appear on the ripen fruits as minute round dark portions. With time, these minute dark spots expands, forming round with slightly depressed lesions. The lesions expand with the maturity of the fruits reaching up to 2 in. size in diameter. The margin of the lesions appears dark in color and the central portion turns brown. The fungus develops large mass of spores in the central portion, turning the color to become orange or pink. Sometimes, concentric rings are formed around the spores for which the lesions appear bull's eye (Hine et al., 1963).

4.2.2.3 EPIDEMIOLOGY

The infection starts at the initial stages of the fruits but the symptoms become visible only at the maturity level. The main source of innoculum is the ascospores produced upon the infected fruits and the fallen dried leaves. The water droplets (either rain or irrigation) are the sources of innoculum. Temperature ranges from 20°C to 30°C and relative humidity of 95% is conducive for the pathogen (Quimio, 1973).

4.2.2.4   CONTROL MEASURES

4.2.2.4.1   Physical Treatment

Dipping the affected fruits for 20 min at 120°F in hot water, suggested by Akamine and Arisumi (1953) and Hine et al, (1965), gives good result.

4.2.2.4.2   Cultural Treatment

Removal of infected mature fruits and fallen leaves from the ground contributes the removal of innoculum. Harvesting should be done when the skin color changes from dark green to light green and yellow streak starts appearing from the base upwards of the fruits. Avoid injury during harvest, transport, and storage (Venturia et al., 2004).

4.2.2.4.3   Chemical Treatment

Spraying of protective fungicides like chlorothanolil and mancozeb on the entire fruit as well as flowers in every 7–14 days during rainy seasons and every 14–28 days during dry conditions (Tatagiba et al., 1999) proved to be fruitful in achieving control. Other fungicides like benomyl have been proved to be effective against this fungus (Ventura and Balbino, 1995).

4.2.3   DAMPING–OFF OF SEEDLINGS

This is a complex disease involving four fungi, namely, *Pythium aphanidermatum, P. ultimum, Phytopthora parasitica,* and *Rhizoctonia* sp., affecting damping-off on papaya seedlings.

4.2.3.1   EPIDEMIOLOGY

If the temperature reaches upto 85°F or higher, infection of *Pythium aphanidermatum* reaches up to the highest population and creates a serious problem, whereas other fungus causes problem at soil temperature lower than 85°F (Hine, 1965). The damping-off is favored by clayey, poorly drained, and inadequate aeration. The use of mulch increases the incidence of the disease (Ram et al., 1983).

4.2.3.2 SYMPTOMS

Damping-off consists of water soaking and then collapse of the stem into the surface of soil. The underdevelopment and yellowing of the leaves observed on the plants (Hine, 1965; Ram et al., 1983).

4.2.3.3 CONTROL MEASURES

4.2.3.3.1 Physical Treatment

Pretreatment of the soil prior to planting is the efficient method to reduce the fungi present in the soil.

4.2.3.3.2 Chemical Treatment

Soil treatment with certain chemicals like chloropicrin, formaldehyde, vapam, and mylone may be effective to some extend (Hine, 1965).

4.2.4 PHYTOPTHORA BLIGHT

Genus Phytopthora is one of the most virulent pathogen causing disease in plant.

4.2.4.1 ETIOLOGY

It is caused by *Phytopthora palmivora* (Butler) parasiting the above ground portion of papaya. This pathogen produces abundant sporangia which are papillate. The sporangia produces biflagellate zoospores (10–40) when matured (Ko, 1998).

4.2.4.2 SYMPTOMS

Phytopthora blight symptoms generally appear on the stems and fruits. It is characterized by formation of minute discolored spots on the fruits, leaf, or on the stems (Hine, 1965). These minute spots widen and may often girdle the young stem which develops wilting at the top of the plants and ultimately dies. The fruit gets infected on the tree itself causing shriveling, turning dark brown, and collapses to the ground. These types of food can be called mummified fruits which turn brownish black in color, weighing light, and have stone-like consistency.

4.2.4.3 EPIDEMIOLOGY

Young plant is more prone to this disease. Wet and cool environment with elevated moisture in soil prop up disease severity. High temperature between 28°C and 32°C, high humidity, and poorly drained soils favors this disease. Rain drops and wind help in dissemination of the disease (Ramirez et al., 1998).

4.2.4.4 CONTROL MEASURES

4.2.4.4.1 Cultural Control

Avoid clayey with poorly drained soils, plant on mounted soils the seedlings, and sterilize the soils before planting. Use organic material which enriches the soil which helps in prevention of this disease (Ventura et al., 2004).

4.2.4.4.2 Biocontrol

Application of *Trichoderma viride* (15 g/mL) along with decomposed FYM at the time of planting surrounding the root zone of the plants proved to be very effective (Nelson, 2008).

4.2.4.4.3 Chemical Control

From the date of planting of the seedlings, soil must be treated with copper oxychloride for amounting 3 g/L and this treatment must be carried out at every 15 days interval and continue after the formation of fruits. Systemic metalaxyl fungicides like Ridomil Gold Copper as curative agent can be used (Nelson, 2008).

4.2.5 POWDERY MILDEW

This disease is the causal organism for dozens of fruits specially occurring in orchard nurseries more prone in shades and in cool weather (Tsay et al., 2011). With its high severity, the disease affects the leaf area creating obstacle in proper functioning of the photosynthesis which reduces the fruits yields to many folds.

4.2.5.1 ETIOLOGY

Three species *Oicium caricae* (conidia of elliptical shape measuring about 24–30 μm × 17–19 μm), *O. indicum* (conidia of barrel shape measuring 31–47 μm × 12–33 μm), and *O. caricae-papayae* (36–44.4 μm × 15.6–21.6 μm) has been reported to cause powdery mildew of papaya. Conidia are hyaline, granular, and having 3–5 spores (Liberato et al., 1996).

4.2.5.2 EPIDEMIOLOGY

This disease is favored by 80–85% humidity with 24–26°C range of temperature (Annonymous, 2002). Brief period of relative humidity is required for germinations of spores. The spores present on the leaves are disseminated by winds (Ventura et al., 2004).

4.2.5.3 CONTROL MEASURES

4.2.5.3.1 Chemical Measures

Powdery mildew can be effectively controlled by fungicides like Triflumizole belonging to an imidazole at the rate: 15 g/100 L. Sulfur can be used in field conditions (Tatagiba et al., 1998).

4.3   BACTERIAL DISEASE

4.3.1   BACTERIAL LEAF SPOT

This disease has been reported from all over the world (Cook, 1975; Funanda et al., 1998).

4.3.1.1   ETIOLOGY

The causal pathogen is a Gram-negative, rod-shaped bacterium called *Pseudomonas carica-papayae* having three to six polar flagella. When grown in nutrient agar medium, the bacterium shows fluorescent but circular, flat, and grayish white colonies (Robbs, 1956).

4.3.1.2   SYMPTOMS

Pathogen produces minute spots on the lower surface of the leaves. These spots are circular to angular with dark greenish in color but always soaked in water. Later on with advance of time, the spots expand turning into light brown in color. On the affected leaves, the spots coalace turning into necrotic irregular-shaped areas progressively (Robbs, 1956; Funanda et al., 1998).

4.3.1.3   EPIDEMIOLOGY

This disease is favored by high humidity (Robbs, 1956; Funanda et al., 1998).

4.3.1.4   CONTROL MEASURES

4.3.1.4.1   Cultural Measures

Removal of effected parts is recommended (Robbs, 1956; Funanda et al., 1998).

4.3.2   INTERNAL YELLOWING AND PURPLE STAIN ROT

Erwinia cloacae cause internal yellowing. Purple stain rot is a sporadic disease caused by a bacteria *Erwinia herbicola* turning pulp rot of fruits in ripening stage.

4.3.2.1   ETIOLOGY

Both *Erwinia* sp. is a facultative anerobic, definitely Gram-negative and rod-shaped bacteria having peritrichous-type flagella distribution.

4.3.2.2   SYMPTOMS

In case of internal yellowing, pulp surrounding the seed cavity turns soft and then slowly rots releasing bad odor. The color of the pulp turns dazzling yellow to lemon green color (Nishijima et al., 1987). In case of purple stain rot, the pulp of the fruit surrounding the seed cavity turns soft and its color turns into deep red (Nishijima et al., 1987).

4.3.2.3   CONTROL MEASURES

4.3.2.3.1   Physical Treatment

Thermal treatment is effective in treating these both diseases. Sanitation has proved to eliminate the disease to some extent (Nishijima et al., 1998).

4.4   VIRAL DISEASES

4.4.1   PAPAYA RINGSPOT VIRUS

It is one of the most destructive diseases which are causing havoc in the whole world (Silva et al., 2000).

4.4.1.1 ETIOLOGY

PRSV is a viral disease belonging to the family Potyviridae and Potyvirus being genus. It is a flexous rod of 760–800 nm long and 12 nm in diameter. PRSV is having two distinctive biotypes, namely, *Papaya ringspot virus-PRSV-p* affecting papaya and *Papaya ringspot virus-PRSV-w* affecting cucurbits or watermelon which is very difficult to differentiate on the basis of coat protein sequence as informed by Souza Jr and Gonsalves (1999). Around 54–60°C is the thermal inactivation point of both the virus.

4.4.1.2 SYMPTOMS

The younger leaves are the sites where the first symptoms appear which exhibits a yellowing along with mosaic areas appearing on the leaves that decrease vigor and fruit production. Blister appears on the greenish area of the spots escaping the yellowish parts. Vein clearing, motling, puckering, and deformation of the leaves take place. In extreme infection, shoe string-like leaf distortion of the leaves appears on the plants, round concentric rings also appears on the fruits surface and the whole plant remain stunted in growth as described by Gonsalves et al., (2007) and Ventura et al. (2004).

4.4.1.3 EPIDEMIOLOGY

This disease can be disseminated mechanically or by grafting, although aphid acts as the main vector source for disease transmission. In Brazil, six aphids have been reported to transfer the disease and they are *Myzus persicae, Toxoptera citricidus, Aphis gossypii, A. fabae, A. coreopsidis,* and *Aphis* sp. In the whole worldwide, 20 species of aphids have been tested to be vector of it. These aphids very quickly spread the disease from infected plant to the healthy plant, and this way the disease is spread rapidly in the papaya orchards (Nishijima et al., 1989). The alternate host surrounding the papaya plants has been the store house for the aphids.

4.4.1.4 CONTROL MEASURES

Once the virus is established in the papaya orchard, it is very difficult to control them; hence, many cultural practices are followed to reduce the population of the viruses (Costa et al., 2000). The cultural practices carried out are as follows:

1. Care should be taken that disease-free seedlings of papaya must be planted and that to under insect-free environment. Affected plant must be immediately rough out. As a barrier, sorghum and maize must be raised around the seedlings and always avoid planting cucurbits around papaya (Wijeendra et al., 1995).
2. Establish mosaic-prone nurseries and orchards at far distance from papaya orchards (Costa et al., 2000).
3. Cleanliness environment as well as provision of properly balanced fertilizer avoids the formation of aphid colonies in the weeds (Costa et al., 2000).
4. Care should be taken to avoid planting of papaya seedlings in rows in similar direction to that of winds which may errand distribution of aphid infestation (Costa et al., 2000).
5. Cross-protection has been employed in many countries (Lima et al., 2001).
6. Transgenic plant with cp proteins has been developed in Cornell University and 55-1 line has been developed that protect against this virus (Fitch et al., 1992).

4.4.2 PAPAYA LETHAL YELLOWING VIRUS

This particular viral disease was first time reported in the year 1983 in Brazil from a place called Pernambuco. Afterwards this disease was found from other parts in the world too (Lima et al., 2001). It has been believed that this virus causes great damage in commercial level production.

4.4.2.1 ETIOLOGY

This is single-stranded RNA of ca. 1.6×10^6 Da, with a single protein capsid of 36 kDa (Amaral et al., 2006). They belong to family Sobemo-virus (Silva et al., 2000).

4.4.2.2   SYMPTOMS

Yellowing of the leaves on the top third of the stem takes which later fall off. Necrotic lesions on the leaf petioles as well as on the bottom side in the leaf veins takes place. With the progress of the disease, the stem becomes twisted and the leaves become chlorotic and ultimately the whole plant become wilts and dies. In fruits, circular spots appear, and pulp on maturation becomes stony (Ventura et al., 2004).

4.4.2.3   EPIDEMIOLOGY

Mechanical transmission takes place for this virus from one plant toward another plant. ELISA test confirms the prevailing of the virus particles on the seeds surface area as reported by Camarach et al. (1997). The virus can exist and survives in the infected rhizospheric soils, thus infecting the new disease-free seedlings at the time of planting. Virus may also be disseminated by irrigation water.

4.4.2.4   CONTROL MEASURES

Mostly cultural practices are recommended to control this disease (Ventura et al., 2004).

4.4.2.4.1   Cultural Control

1. Periodic infections and eradicate the affected plants.
2. Care must be taken to mount the nurseries and orchards at a distance from other orchards.
3. Eradicate old papaya plants that serve as inoculums for the virus.
4. Disinfect the tools used to clean the plants with chlorine bleach solution 10%.

4.4.3   MELEIRA DISEASE

Another name of this disease is sticky disease creating major problem in cultivated areas. First report of the occurrence of this disease was reported in Brazil (Rodrigues et al., 1989; Abreu et al., 2015).

4.4.3.1 ETIOLOGY

PMeV is an isometric doublestranded RNA (ds RNA) with genomic size of 12 kb (Abreu et al., 2015).

4.4.3.2 SYMPTOMS

This disease is characterized by the formation of oxidized latex on the fruit surface as well as on leaves. The latex on the diseased fruits are translucent and watery type unlike viscous in healthy fruits. The latex in contact with the oxygen present in the atmosphere reacts and forms sticky portion on then fruits surface (Abreu et al., 2015).

4.4.3.3 EPIDEMIOLOGY

The virus is transmitted by injection of the latex of the diseased plants into the stem of the healthy papaya that obtains disease symptoms approximately 45 days after inoculation. The silver whitefly (*Bemesia argentifolii* Bell. & Perring) is associated with the transmissiom of the virus (Vidal et al., 2000).

4.4.3.4 CONTROL MEASURES

4.4.3.4.1 Cultural Control

1. Carrying out weekly observations and rouging of infected plants.
2. Never collect seeds from infected plants and orchards with high disease incidence.
3. Disinfect all materials used in the process of thinning and harvesting of fruits.
4. nstall the nurseries far from other orchards.

4.5 PHYTOPLASMA DISEASE

4.5.1 DIEBACK

Dieback is one of the most serious phytoplasma diseases since 1922 and caused loss upto 100% in the plantations in Australia, thus disturbing the whole papaya industry in market (Glennie et al., 1976).

4.5.1.1   ETIOLOGY

Siddique et al. (1998) performed polymerase chain reaction with specific primers and reported that this disease is caused by DNA phytoplasma.

4.5.1.2   SYMPTOMS

The characteristic symptoms of dieback are an appearance of the inner crown leaves with leaves shriveled and dying. The larger crown leaves rapidly, develops chlorosis, and then necrosis. The entire crown dies within 1–4 weeks and then stems dies back from the top. Simmonds (1965) reported that necrotic areas develop on the phloem and there is reduction of latex flow in the affected plants. The floral parts turn green and phyllody-like structure appears, thus reducing the fruit production to a great extent (Ventura et al., 2004).

4.5.1.3   EPIDEMIOLOGY

Phloem feeding leaf hoppers and plant hoppers are responsible for spread of the disease from disease plant to healthy plant. *Orosius argentatus* Evans, a leaf hopper transmits this disease. This pathogen is also reported to be transmitted by dodder (*Cuscuta* sp.) (Cook, 1975). Compared with older plants, young plants are more vulnerable to this disease and extensive losses have been reported in first 12 months of the plantation.

4.5.1.4   CONTROL MEASURES

4.5.1.4.1   Cultural Control

Ratooning and rouging reduces the phytoplasma dissemination. Ratooning of the dieback affected plants helps in management of the disease to some extent (Guthrie et al., 1998).

4.5.2   BUNCHY TOP

Bunchy top disease of papaya is a devastating disease and it is prevalent in northern parts of South America (Nishijima et al., 1998).

4.5.2.1   ETIOLOGY

It is controversial matter of deciding the pathogen involved in the disease. According to Nishijima et al., 1998; Cook, 1975, and Ventura et al., 2004, phytoplasma cause the disease. While in accordance with the few other researchers, rickettsia-like organism is the causal organism of this disease (Davis et al., 1998).

4.5.2.2   SYMPTOMS

The affected plants are stunted in growth. Initial symptom is characterized by slender mottling of the upper leaves. With the length of time, chlorotic appearance in the interveinal areas on the infected leaves with marginal necrosis can be seen. Length on the intermodal areas is shorter. Appearance of spots with oil particles are found on the stems and petioles in their upper parts. In infected plants if the fruits set, they taste a bit bitter (Cook, 1975; Nishijima, 1998).

4.5.2.3   EPIDEMIOLOGY

Bunchy top of papaya is transmitted by *Empoasca papaya* and *Empoasca stevensi*, the two leafhoppers (Davis et al., 1998; Webb and Davis, 1987). Ventura et al. (2004) reported that grafting is also a method that can transmit the disease.

4.5.2.4   CONTROL MEASURES

4.5.2.4.1   Physical Method

1. Therapeutic measure like application of antibiotic can effectively control the pathogen.
2. Drenching the soil with tetracycline hydrochloride combined with root dip treatments has been successful to control the disease.

4.5.2.4.2   Chemical Method

Application of insectides to kill insect vectors proved to be efficient according to Nishijima et al., 1998.

4.6 CONCLUSION

The considerable warning factor for the decreasing papaya production is the various diseases attacking the plants. Therefore, in order to achieve the flourishing papaya production and its industrial and marketable utility, proper understanding of the causal agents and its epidemiology (interaction of the host plant, pathogen, and environmental factors) are of paramount value. The proper knowledge and emphasis of these biotic agents, diagnostic symptoms, and the epidemiological factors would definitely help in deciding the effective management procedures. The control of the diseases can be achieved by the use of varied synthetic chemical. However, legally, in present situations globally, in order to protect the environment as well as the integrity of human beings good heath, solo use of synthetic chemicals is almost banned. In this pretext, integrated diseases management procedures where amalgamation of physical, cultural, biological, and chemical methods is used according to the nature of pathogen causing diseases. This chapter fulfils the entire above-mentioned context that may be supportive for future prospective in knowledge of the diseases along with proper management measures.

KEYWORDS

- **bacterial**
- **biosecurity**
- **ecological balance**
- **fungal**
- **integrated disease management**
- **phytoplasma**
- **viral**

REFERENCES

Abreu, P. M.; Antunes, T. F.; Magna–Alvarez, A.; Perez-Brito, D.; Tapia-Tassell, R.; Fernandes, A. A.; Fernandes, P. M. A Current Overview of the Papaya Meleira Virus, an Unusual Plant Virus. *Viruses* **2015**, *7* (4), 1853–1870.

Amaral, P. P.; Resende, R. O.; Junior, T. S. Papaya Lethal Yellowing Virus (PLYV) Infects *Vascocella cauliflora*. *Notas Fitopatologicas* **2006**, 517–519.

Akamine, E. K.; Arisumi, T. Control of Postharvest Storage Decay of Fruits of Papaya (*Carica papaya* L.) with Special Reference to the Effect of Hot Water. *Proc. Amer. Hort. Sci.* **1953**, *61*, 270–274.

Annonymous. Diseases of Crops: Papaya. Disponivel em: http://www.krishiworld.com/ html.diseases_crops.html. (accessed Dec 30, 2002).

Camarach, R. F. A.; Lima, J. A. A.; Pio-Ribeiro, G. Presenca de "Papaya Yellowing Virus" Emsementes de Frutos Infectados de Mamoeiro, *Carica papaya*. *Fitopatologia Brasileira* **1997**, *22*, 333.

Cook, A. A. *Diseases of Tropical and Subtropical Fruits and Nuts.* New York: Hafner Press, 1975, p 317.

Costa, H.; Ventura, J. A.; Tatagiba, J. S. Mosaico do Mamoerio: Uma Ameaca a Cultura no Espirito Santo. Vitoria: INCAPER, 2002, p 4.

Costa, H.; Ventura, J. A.; Rodrigiues, C. H.; Tatagiba, J. S. Ocorrencia e Patogenicidade de Glomerella Cingulata em Mamao no Norte do Estado do Espirito Santo. *Fitopatologia Brasileira* **2002**, *26*, 328–338.

Davis, M. J.; Ying, Z.; Brunner, B. R.; Pantoja, A.; Ferwerda, F. H. Ricketsial Relative Associated with Papaya Bunchy Top Disease. *Curr. Microbiol.* **1998**, *36* (2), 80–84.

FAOSTAT. Crop Production. 2012a. http://faostat.fao.org/ site/567/default.aspx#ancor

Fitch, M.; Manshardt, R.; Golsalves, D.; Slightom, J.; Sanford, J. Virus Resistant Papaya Plants Derived from Tissues Bombarded with the Coat Protein Gene of Papaya Ringspot Virus. *Biotechnology* **1992**, *10*, 1466–1472.

Funanda, C. K.; Yorinori, M. A.; Rodrigues, A.; Auler, A. M.; Leite Jr., R. P.; Ueno, B. Ocorrencia da Mancha Foliar Bacteriana Causada por *Pseudomonas Syringae* Pv. *Carica* Papaya em Mamoeiro no Estado do Parana. *Fitopatologia Brasileira* **1998**, *23*, 209.

Glennie, J. D.; Chapman, K. R. A Review of Dieback: A Disorder of the Papaw (*Carica papaya* L.) in Queensland. *Queensl. J. Afri. Anim. Sci.* **1976**, *33*, 177–188.

Gonsalves, D.; Suzuki, J. Y.; Tripathi, S.; Ferreira, S. A. Papaya Ring Spot Virus (Potyviridae). In *Encyclopedia of Virology;* Mahy, B. W. J.; van Regenmortal, M. H. V., Eds.; Elsevier Ltd: Oxford, UK, 2007.

Guthrie, J. N.; White, D. T.; Walsh, K. B.; Scott, P. T. Epidemiology of Phytoplasma-associated Papaya Diseases in Queensland, Australia. *Plant Dis.* **1998**, *82*, 1107–1111.

Hine, R. B.; Holtzman, O. V.; Raabe, R. D. *Diseases of Papaya (Carica papaya L.) in Hawai.* Bull. 136, Hawai Agricultural Experiment Station, University of Hawai, 1963.

Holliday, P. *Fungus Diseases of Tropical Crops.* Cambridge: Cambridge University Press, 1980, p 607. https://www.statista.com/statistics/578039/world-papaya-production

Ko, W. H. Papaya Disease Caused by Fungi: Hytopthora Fruit Rot and Root Rot. In: "*Compedium of Tropical Fruits Diseases*;" Ploetz, R. C.; Zentmeyer, G. A.; Nishijima, W. T.; Rohrbach, K. G.; Ohr, H. D., Eds., 2nd ed., American Phytopathological Society: St Paul, 1998; pp 61–62.

Liberato, J. R.; Costa, H.; Ventura, J. A. *Indice de Doencas de Plantas do Estado do Espirito Santo.* Vitoria-ES: EMCAPA, 1996, p 110.

Lima, R. C. A.; Lima, J. A. A.; Souza Jr., M. T.; Pio-Ribeiro, G.; Andrade, G. P. Etiologia e Estrategias de Controle de Viroses do Mamoeiro no Brasil. *Fitopatologia Brasileira* **2001**, *26*, 689–702.

Nelson, S. Phytopthora Blight of Papaya. *Plant Dis.* **2008**, 1–7.

Nishijima, K. A.; Couey, H. M.; Alvarez, A. M. Internal Yellowing, Bacterial Disease of Papaya Fruits. *Plant Dis.* **1987**, *71*, 1029–1034.

Nishijima, W. T. Miscellaneous Papaya Diseases. In *"Compendium of Tropical Fruits Diseases;"* Ploetz, R. C.; Zentmeyer, G. A.; Nishijima, W. T.; Rohrbach, K. G.; Ohr, H. D., Eds., 2nd ed., American Phytopathological Society: St. Paul, 1998; pp 69–70.

Pernezny, K.; Litz, R. E. Common diseases of Papaya. IFAS Extension, University of Florida. 2003. http://edis.ifas.ufl.edu.

Quimio, T. H. Temperature as a Factor for Growth and Sporulation of Anthracnose Organism of Papaya. *Philippine Agriculturist* **1973**, *57*, 245–253.

Ram, A.; Oliveira, M. L.; Sacramento, C. K. Podridao das Raizes do Mamoeiro Associada com Pythium sp. Na Bahiya. *Fitopatologia Brasileira* **1983**, *10* (2), 581.

Ramirez, L.; Duran, A.; Mora, D. Combat Integrado de la Podricon Radical de la Papaya (*Phytopthora* Sp.) A Nivel de Vivero. *Agronomia Mesoamericana* **1998**, *9* (1), 72–80.

Rivera Pastrana, D. M.; Yahia, E. M.; Gonzalez-Aguilar, G. A. Phenolic and Carotenoid Profiles of Papaya Fruit (*Carica Papaya* L.) and their Contents Under Low Temperature Storage. *J. Sci. Food Agric.* **2010**, *90* (14), 2358–2365.

Robbs, C. F. Uma Nova Doenca Bacteriana do Mamoeiro. *Revista da Soceidade Brasileira de Agronomia* **1956**, *12*, 73–76.

Rodrigues, C. H.; Ventura, J. A.; Marin, S. L. D. Ocorrencia e Sintomas da Melera do mamoeira (*C. papaya*) no Estado do Espirito Santo. *Fitopatologia Bradileira* **1989**, *14*, 118.

Siddique, A. B. M.; Guthrie, J. N.; Walsh, K. B.; White, D. T.; Scott, P. T. Histopathology and Within-plant Distribution of the Phytoplasma Associated with Australia Papaya Dieback. *Plant Dis.* **1998**, *82*, 1112–1120.

Silva, A. M. R.; Kitajima, E. W.; Resnde, R. O. Nucleotide and Amino Acid Analysis of the Polymerase and the Coat Protein Genes of the Papaya Lethal Yellowing Virus. *Virus Rev. Res.* **2000**, *11*, 196.

Simmonds, J. H. Papaw Diseases. *Queensl. Agric. J.* **1965**, *91*, 666–667.

Souza Jr., M. T.; Gonsalves, D. Genetic Engineering Resistance to Plant Virus Disease, an Effort to Control Papaya Ringspot Virus in Brazil. *Fitopatologia Brasileira* **1999**, *24*, 485–502.

Tatagiba, J. S.; Liberato, J. R.; Zambolin, L.; Costa, H.; Ventura, J. A. Avaliacao de Fungicidas no Controle da Antracnose e da Podridao Penducular do Mamoeriro. *Summa Phytopathologica* **1999**, *24*, 57.

Tatagiba, J. S.; Costa, A. N.; Ventura, J. A.; Costa, H. Efeito do Boro e Calico na Incidence da Antracnose em Fructose de Mamoeiro. *Fitopatologia Brasileira* **1998**, *23*, 285–286.

Tsay, J. G.; Chen, R. S.; Wang, H. L.; Wend, B. C. First Report of Powdery Mildew Caused by *Erysiphe Diffusa, Oidium Neolycopersici* and *Podospaera Xanthii* On Papaya in Taiwan. *Plant Dis.* **2011**, *95*, 1188.

Ventura, J. A.; Costa, H.; Tatagiba, J. S. *Papaya Diseases and Integrated Control. Diseases of Fruit and Vegetables, Diagnosis and Management,* Vol III, Kluwer Academic Publishers, 2004; pp 201–268.

Ventura, J. A.; Balbino, J. M. S. Resistencia do Agente Etiologico da Antracnose Do Mamoeiro ao Benomil, no Estado do Espirito Santo. *Fitopatologia Brasileira* **1995**, *20*, 308.

Vidal, C. A.; Nascimento, A. S.; Barbosa, C. J.; Marques, O. M.; Habibe, T. C. Experimental Transmission of "Sticky Disease" of Papaya by *Bemesia Argentifolii* Bellows & Perring.

In *International Congress of Entommology*, 21, Foz do Iguacu-PR: SEB/EMBRAPA, 2000. Abstract book 2, Foz does Igauca-PR: SEB/ EMBRAPA", 2000; pp 819.

Webb, R. R.; Davis, M. J. Unreliability of Latex-flow Test for Diagnosis of Bunchy-top of Papaya Caused by a Mycoplasma Like Organism. *Plant Dis.* **1987,** *71,* 192.

Wijeendra, W. A. S.; Ranaweera, S. S.; Salim, N. The Effect of Papaya Ringspot Virus Infection on the Nitrogen Metabolism of Carica Papaya L.: Part II, Composition of Free Amino Acids in the Leaves. *Vidyodaya J. Sci.* **1995,** *5* (1), 131–138.

CHAPTER 5

Fungal Diversity under Different Agri-Systems and Their Beneficial Utilization in Plant Health Management

VIBHA*

Department of Plant Physiology, Jawaharlal Nehru Krishi Vishwa Vidyalaya, Jabalpur 482004, India

Corresponding author. E-mail: vibhapandey93@gmail.com

ABSTRACT

Rhizosphere fungi are an important group of microbes that affects plant, animal, and human life directly or indirectly. To understand their heterogeneous behavior in shaping the terrestrial ecosystem, it is mandatory to understand their distribution and probable interaction with soil and plant systems. Fungi inhabiting soil were assessed and found that the Dueteromycetous fungi dominated the total fungal population in comparison to oomycetous, zygomycetous, and ascomycetous fungi. Climatic conditions and substrate availability affected the succession of fungi in different months. There was no specific order of succession; the distribution of fungi depends on their enzymatic ability to degrade the substrate during the course of decomposition period. Out of six decomposing fungi isolated from decomposition substrate, *Trichoderma harzianum* yielded the highest percentage of crude protein (27.99%) with biomass of 375 mg, whereas the lowest protein value (17.91%) was recorded in case of *A. niger* with biomass of 422 mg. Among phosphorus solublizing fungi (PSF) isolated from calcium rich soil, maximum P-solublization was recorded from *Penicilliun citrinum* isolate (PC2) followed by *Aspergillus niger* isolates in liquid Pikovskaya's. The decomposing fungi were also used for reducing the pre-composting period of vermicompost besides making it suppressive toward soilborne pathogens of tomato. The weed plants were also decomposed by beneficial fungi and were used to reduce the tomato wilt incidence caused by *Fusarium oxysporum* f. sp. *lycopersici* (FOL). The culture filtrate of the most commonly occurring fungal species of *Aspergillus* was combined with *Trichoderma virens* to make it more effective against *Rhizoctonia solani*. Significant inhibition of mycelia growth of *R. solani* was recorded with culture filtrate mixture of *A. niger* + *T. virens*, and *A. ochraceous* + *T. virens*.

5.1   INTRODUCTION

Global fungal diversity ranges has exceeded above 1.5 million species[26] as fungi are ubiquitous and diverse. The consequences of fungal diversity are greatly marked on plant communities and ecosystems[77] through several ways especially at local scale because they participate in decomposition of crop residue, enhance soil mineral availability, disease suppression, etc. Crop debris play an important role in maintaining soil fertility and other physical properties of soil as they are principal amendment applied to the soil system. These days, agri-waste management has become an alarming issue owing to mechanization of harvesting through harvester. The management of this waste through burning has created environmental problem by way of air pollution that in turn has disturbed the rich soil biodiversity through heating. Maintenance of soil health through improving soil–water–air continuum apart from increasing water holding capacity is feasible only by crop residue decomposition. Fungi come under the important group of microbes involved in decomposition as they are more enzymatically active. They have the ability to ramify solid substrate, grow well under semi-solid fermentation condition, and colonize it quickly.[30] The factors like crop residue, fungal succession at different stage of decomposition, nutrient level of soil, and prevailing environmental conditions govern the process of decomposition.[46,60,65,69] Rice stubble takes longer time to decompose as they contain higher amount of cellulose and lignin. *Chaetomium globosum* produce the highest soluble crude protein (SCP) among eight cellulolytic fungi used for production of SCP from delignified rice straw.[16] An efficient species *Chaetomium* that is *C. thermophile*, solublize crude protein from delignified wheat straw.[67] It has been observed that the ratio of mannose, galactose, fructose, rhamnose, and ribose increase consistently with time while proportion of cellulosic glucose decrease during the process of decomposition.[49] The population of bacteria and fungi increases as the amount of applied root residue increases in soil.[64] The colonization pattern of primary fungal colonizers and successive invaders depends on the interacting factors, namely, crop residues, organic nutrients of soil, pH, aeration, depth, and climatic factors.[51,60]

Rhizosphere harbors several beneficial microbes that are receiving greater attention as they sulublize the inorganic phosphate into soluble form through the production of organic acids that simplifies the process of ion exchange reactions through acidification and chelation.[24] In addition, production of phyto-hormones such as IAA[47] by these phosphate solubilizing microorganisms (PSMs) can also increase the plants growth because auxins are important hormone for cell division. Phosphorus (P) is the second largest agricultural chemical needed by plant for its growth and development apart from being one of the most essential elements for living organisms. Moreover, there is rapid fixation of phosphorus in insoluble forms within the soil and plant utilizes only 0.1% of phosphorus present in soil. The calciorthent soil is characterized by the higher pH (of soil above 8.0) that have buffering capacity[2] to mobilize the most of the mineral P in the form of poorly soluble mineral phosphate (CaP). Fungi are more efficient than bacteria in solublizing insoluble phosphate into soluble form as has been reported by Venkateswarlu and coworkers.[78] Phosphorus solublization ability of strains of *Aspergillus* and *Penicillium* spp. are well established and are also the most commonly occurring fungi but their external incorporation in the soil is necessary for maintaining their higher number to compete with the established microbes of rhizosphere.

The plant heath and soil health problems are aggravating each day owing to excess use of chemical in agriculture sector. Hence, to manage the chemical hazard, eco-friendly approaches have become inevitable for sustainable crop production. Therefore, intense use of biological agents was proposed by the scientific community through their research as use of bioagent under sustainable agricultural could result in promising means of management of several issues of agriculture related to chemical hazards. Fungal species (like *Aspergillus* spp., *Trichoderma* spp., and nonpathogenic *Fusarium* spp.) having disease suppressing ability are predominantly found in soil. Several workers[19,20] have reported that the dominance and predominance of such fungi are responsible for making the soil suppressive toward pathogenic fungi as an inherent property.[70,76] Disease suppression achieved through mechanism of competition, antibiosis, parasitism/

predation, and induced resistance exerted by biological controlling agents contribute in the management of various diseases.[28] Biological agents, namely, *Aspergillus niger, T. harzianum,* and *Trichoderma virens* are widely used against soil-borne pathogens.[57,75] *Paecilomyces lilicans* widely recommended against root knot and cyst nematodes (as it parasitize the egg of nematodes) also inhibit the growth of *Fusarium oxysporum,*[44] whereas, *Cladosporium cladosporioides* secretes acid phosphatase and other phytohormones[19] facilitate the multiplication of rhizospheric microbes. *Trichoderma, Gliocladium, Aspergillus,* etc. are extensively explored for control of soil-borne plant pathogens[36] and are also found effective in controlling sheath blight disease. Moreover, use of biological control agents well ahead of planting along with organic amendments as food base for them can be very effective in controlling many soil-borne pathogens including *R. solani,* that has already been recognized.[28] *Trichoderma* spp predominantly colonized compost prepared from lignolytic substances[40] while low cellulosic substance and high sugar containing waste are being preferred by the isolates of *Aspergillus* and *Penicillium.*[22] *Rizactonia solani* has much narrower spectrum of biocontrol and this mycoflora does not consistently colonize composts.[23] Therefore, parasitism through suppressive compost amended substrate is critical for biological control of *R. solani* in.[22,50]

Biocidal activity botanicals against several insect pests and pathogens[27] are the characteristic properties associated with botanicals apart from having biodegradability, low mammalian toxicity, target specificity, and ability to serve as safe fungicide against many pathogenic fungi. Futhermore, triterpenoid saponin secreted by *Launaea pinnatifida* fairly showed antifungal activity against *Fusarium oxysporum.*[84] Presence of cannabidol (CBD) in different species of *Cannabis* decides the antimicrobial activity,[41] *Nicotiana plumbaginifolia* produces NpPDR1 (ATP binding cassette) that triggers the initiation of both constitutive and jasmonic acid-dependent induced defense[73] in plants against plant pathogens. Prasad and co-workers[58] documented that chloroform extracts of *Physalis minima* exhibited antifungal and antibacterial properties. Therefore, use of individual method of non-chemical control should be recommended over the combinations of more than one method to achieve effective control.[18]

Development of value-added organic amendments suppressive to soil-borne pathogens through the use of agri-friendly microbial combinations is becoming a rising need in agriculture. Combined actions of several microorganisms that reside in the gut of earthworms or on composting substrate[86] participate in formation of decomposed organic materials (i.e., vermicompost). However, microbes are found naturally in decomposing waste but their decomposing activity are comparatively slower so inoculation of such waste with fast decomposers can accelerate the process.[72] Combinations of these fungi not only reduce the predecomposition period but are also known to produce various antifungal compounds[1,53,61] that make the vermicompost suppressive toward several soilborne diseases. *Trichoderma, Penicillium, Curvularia,* nonpathogenic *Fusarium solani, Aspergillus niger,* and *Paecilomyces lilicans* produce enzymes, namely, chitinase, cellulase, hemicellulase, and xylanase,[31,37,54] which act on chitin, cellulose, and hemicellulose to break them into simpler forms. Similarly, *Cladosporium cladospoirioides* also produces the enzymes that can degrade lignin, carboxy-methyl cellulose, and xylan.[29] Metabolite produced by antagonists or by increase in populations of a specific species or selected groups of organisms that can suppress the population of harmful pathogens is known to correspond with specific suppression, imposed by beneficial pathogens against soilborne deleterious pathogens.[42,43]

5.2 EXPERIMENTAL DETAILS

The fungi were isolated from rhizospere of different crop and different soil through dilution plate technique on Rose Bengal Agar medium. Similarly, the studies of decomposing fungi were made through nylon net bag technique. The quantification of isolated fungi was made on the basis of following standard formulae:

$$\text{Frequency (\%)} = \frac{\text{Number of sample with Genus}}{\text{Total number of samples}} \tag{5.1}$$

Species Richness (S): The total number of species in the community.

Simpson's dominance index (D):

$$D = \frac{1}{\sum\limits_{i=1}^{s} P_i^2} \tag{5.2}$$

where the proportion of species (i) is relative to the total number of species (P_i), which is calculated and squared. The squared proportions for all the species are summed and the reciprocal is taken.

Shannon's diversity index (H):

$$H = -\sum_{i=1}^{S} P_i \cdot \log_2(P_i) \tag{5.3}$$

where P_i is the proportion of ith Genera that contribute to total diversity.

Pre-treated air-dried rice stubble with 1.8 L of 1% NaOH was autoclaved and loaded with isolated decomposing fungi to qualify the amount of SCP production. The phosphorus solublizing fungi (PSF) were isolated from the rhizosphere of the agricultural/horticultural crops through standard method and used in fertile calciorthent soil @ 5×10^6 cfu/mL in different combinations. The vermicompost extract, fungus loaded weed compost extract, and culture filtrate of different fungi were tested in vitro through poison food technique against major soil-borne pathogens. The fungus loaded vermicompost and weed compost were used @ 5 g/5 kg of soil to suppress the damping off and wilt incidence of tomato.

5.3 RESULTS AND DISCUSSION

5.3.1 *FUNGAL DIVERSITY UNDER DIFFERENT CROPPING SYSTEM*

5.3.1.1 *DIVERSITY OF SOIL MYCOFLORA UNDER RICE-WHEAT CROPPING SYSTEM*

The assessment of the extent of decay and colonization pattern of soil inhabiting mycoflora of paddy stubble was done by using nylon net bag technique. The highest number of fungi was recorded in dilution plate technique followed by damp chamber and under direct observation out of the three methods used for isolation of fungi. Influence of temperature, humidity, rainfall, and nutrient availability was distinct on the occurrence and colonization pattern of fungi. In the month of October, maximum (48.99 × 10^4/g dry litter) fungal population was recorded, while least (11.41 × 10^4/g dry litter) was in May. The dominance of *Mucor racemosus, Rhizopus nigricans, Chaetomium globosum,* and *Gliocladium* species were higher at the initial stage of decomposition of rice stubble that was later succeeded by the preponderance of *Aspergillus candidus, Torula graminis, Cladosporiun cladosporioides,* and *Aspergillus luchuensis* in rice wheat cropping system.[80] Higher number of fungi belonging to Deuteromycetes was isolated in comparison to ascomycetes, zygomycetes, and oomycetes from the studied (i.e., rice–wheat) cropping system.

Rainfall toward end of January supported increased microbial activity due to favorable atmospheric temperature, optimum soil moisture condition, and increase in soil pH that resulted in maximum weight loss of stubble in the month of February. The prevailing favorable atmospheric condition, especially temperature, during February has significantly contributed in weight loss of stubble owing to increased microbial activities. The close and remarkable correlation between weight loss and rate of decomposition owing to environmental factors has also been reported by earlier researchers.[7,13] Cumulative action of microbes and soil fauna along with leaching effect of rainfall were major factors for higher weight loss of decomposing substrate, which was also observed.

Distribution and colonization on substrate by soil fauna increase markedly by optimum soil moisture content has been reported by several researchers.[7,81] Zimmermann and Frey[87] reported that the increase in soil pH by the incorporation of higher biomass into the soil results in weight loss due to enhanced bioactivity microbes. The maximum fungal population was recorded in narrow ratio C:N and fresh decaying tissues do not provide enough substrate for multiplication of saprobes while senescent stubble provide enough dead tissues and the surface area for the activities of initial colonizer that further allows the succession of mycoflora, which are unable to appear initially. Hence, maximum population of fungi was recorded in October when substrate was available in simpler form. After incorporation of root residues in the soil, the bacterial and fungal population decides the rate of decomposition that remained higher in the first and second week, which gradually slowed down and finally become steady during the course of study. Decomposition of added organic starts just after its incorporation, as has been observed by the Berkenkamp and coworkers.[9] In their opinion, dead tissues of senescent stubble provide enough surface area for the activity of mycoflora. Decomposition of organic amendment are initially rapid that later become static.[66] Higher amount of root residue applied in soil increases the microbial population under favorable atmospheric condition due to availability of sufficient substrate for their multiplication.[64] Decline in fungal population during summer months was observed by several workers[15,38] due to increase in temperature and decrease in relative humidity of air. However, more recalcitrant effect of wet and warm climatic conditions on litter decomposition was recorded by Tiernan et al.[46] Nutritional levels of substrate at the last stage of its decomposition are the driving factor for fungal colonization rather than environmental conditions.[4,13]

5.3.1.2   *VARIATION IN SOIL MYCOFLORA OF PIGEONPEA CROPPING SYSTEM*

Mycofloral diversity in soil plays critical role in crop production through the process of rhizosphere biological engineering and is an indispensable part of any ecosystem. Pigeonpea cropping system is benefitted by its root resident soil microbes that participate in nitrogen fixation and phosphorus solubilizing due to favorable environment around its root system for multiplication of microbes. To undertaken the distribution of fungal diversity in calcium-rich soil that does not support the growth of several fungi as minerals gets readily fix into un-available form in such soil. Pigeonpea cropping system, under the treatment of phosphorus solubilizing bacteria (PSB), and plant growth promoting rhizobacteria (PGPR) along with Rhizobium were used for the studying the fungi. Thirty-seven species belonging to seven genera and a group of unidentified species were isolated. Dominance of *Aspergillus* and *Penicillium* genera (Fig. 5.1) were recorded from all the treatments while *Absidia* and *Cunnighmella* were recorded as the rare genera due to their lower occurrence. Although the occurrence of single species of *Periconia, Geotrichum, Pythium, Rhizopus,* and *Gliocladium* genera were recorded, their distribution was even in all the treatments. The diversity and equitability index were varied from 2.93 to 5.84 among different treatments but was almost identical in *Rhizobium* + PSB + PGPR (N) and *Rhizobium* + PGPR treatments (Fig. 5.1). The preponderance of deuteromycetes fungi was higher than mastigomycetous, zygomycetous, and *mycelia sterilia* fungi were reported by the author[56] in other cropping systems too. Several researchers suggested that the fungi belonging to class Duteromycetes are strong colonizer of the decaying substrate owing to their ability to compete better than other group of fungi besides better adaptability to diverse climatic, whereas, those of Ascomycetes and Phycomycetes were not found as strong as former group.[60,65] An order of succession of a natural substrate by fungus depends on array of different organic and inorganic nutrients released as a result of interaction between each individual and substratum vis-à-vis competition between invading fungi.[32]

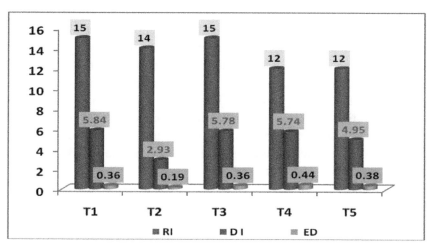

T1, Rhizobium + PSB (Local); T2, Rhizobium + PGPR (L) + PSB (L); T3, Rhizobium + PSB + PGPR (N); T4, Rhizobium + PGPR; and T5, Control; DI, diversity index, E_D, equitablility; and RI, richness index.

FIGURE 5.1   (See color insert.) Diversity and evenness of fungi under different treatments.

5.3.2   DECOMPOSITION MYCOFLORA

5.3.2.1   MYCOFLORA ASSOCIATED WITH DECOMPOSITION OF RICE STUBBLE

During the period of decomposition, total of 29 fungal species were recorded from rice stubble mixed with soil isolated by dilution plate technique. Substrate availability and climatic factors played greater role in monthly variation in fungal population and was found closely related to each other. Although the highest (47.68×10^4) fungal population of rice stubble mixed with soil was recorded in the month of October, the maximum (25.32%) moisture content of decomposed rice stubble mixed with soil was in the month of August. In addition, the lowest (16.88×10^4) population was recorded in the month of May while minimum (5.35%) moisture content of decomposed substrate was recorded in the month of April. During the studied period of decomposition, there was variation in pH from 6.8 to 7.2. Abiotic variables and nature of substrate are generally the factors that decide dynamics of fungal community.[74] Rainy and winter seasons are found most suitable for fungal growth and activity owing to favorable temperature and relative humidity. Hence, higher fungal count was recorded at the initiation of winter, which was succeeded by the higher moisture content of stubble mixed in soil at the end of rainy season. The decline in the number of fungi may be due to several factors such as low moisture content of soil, extremely low or high relative air humidity, and wide fluctuation in air temperature during these months. Several workers reported that these factors do not favor the sporulation on the substrate.[15,46] The type of plant materials and initial pH of the substrate added as organic matter to soil led to change in soil pH.[83]

Dominance of Dueteromycetous fungi constituting 75.86% of total fungal population followed by zygomycetous, oomycetous,s and asco-mycetous was recorded during the course of study. The dominating fungi were *Rhizopus stolonifer, Aspergillus flavus*, and *Trichoderma harzianum*, while *Pestalotia mangiferae, Torula graminis*, and *Alternaria solani* were obtained as rare fungal species.

The members of lower group of fungi and ascomyceteous fungi were noted as weak colonizers, whereas fungi imperfecti were strong colonizers. The initial colonizers like *Alternaria alternata, Cladosporium cladosporioides,* and *Curvularia lunata* were weak colonizers, while species of *Mucor, Rhizopus,* and various species of *Aspergillus* and *Penicillium* were frequently isolated saprophytic fungi. Shaukat and coworkers[67] isolated the species of *Alternaria, Aspergillus, Fusarium, Penicillum, Trichoderma, Mucor, Myrothecium,* and *Rhizoctonia* more frequent fungi from *Avicennia marina* amended than unamended soil. The deuteromycetous fungi exhibited their ability to utilize complex nutrients from litter as initial colonization than phycomycetous species. The late appearance of phycomycetous fungi during the fungal succession shows their ability to utilize simple compounds either as fungal products or as soluble forms. The sequential release of different organic and inorganic nutrients, interaction between individual fungi and substratum besides the competition between them, reflects the order of fungal succession upon a natural substrate.[32,70] The dominance of deuteromycetous fungi in the beginning of decomposition process was correlated with their cellulose utilizing capacity.[5] Similarly, other scientists have reported that Moniliales exhibit the higher enzymatic degradation potential.[76]

5.3.2.2   PRODUCTION OF SOLUBLE CRUDE PROTEIN USING CELLULOLYTIC FUNGI

SCP production ability of the six cellulolytic fungi tested on delignified cellulose as a carbon source varied widely. Though the utilization of cellulose has been observed by all of them and was considerable in the presence of *T. harzianum* that yielded highest percentage of crude protein (27.99%) with biomass of 375 mg, the lowest protein value (17.91%) was recorded in case of *A. niger* with biomass of 422 mg. The SCP production efficiency of cellulolytic fungi was in order of *T. harzianum* > *P. citrinum* > *C. lunata* > *A. flavus* > *A. alternata* > *A. niger*. The *T. harzianum* was the most potent among the tested fungi. Hamlyn[25] reported that the *A. niger, A. flavus,* and *Penicillium* spp. are the main sources of cellulase, amylase, hemicellulase, catalase, pectinase, and xylanase because of their ability to produce these enzymes. Increased availability of amorphous form of cellulose owing to delignification with sodium hydroxide may be ascribed to rapid and higher production of SCP by fungi from delignified cellulose apart from fungal ability to produce a variety of enzymes. Detailed studies by Punj and his associates[59] show that decomposition of delignified cellulose by sodium hydroxide resulted in increased production of cellulose to amorphous form, which was readily attacked by fungi. Similar reason for increase in SCP from delignified cellulose with sodium hydroxide was proved by other workers too.[16] The study of effects of incubation period and nitrogen sources on SCP production showed that the fifth day of incubation period and potassium nitrate as nitrogen source, among other nitrogen sources, was found most appropriate for SCP production by the *T. harzianum*. The maximum biomass production was achieved by *A. niger* and minimum by *T. harzianum* while reverse was recorded in case of weight loss of biomass. Such findings were supported by the work of earlier workers[12,32]. When wheat offal inoculated with the *A. niger*, there was the highest increase in protein percentage.[33]

5.3.2.3   POTENTIAL OF T. HARZIANUM IN DECOMPOSITION OF CELLULOSE

The *T. harzianum* was selected for the study of decomposition of cellulose on the basis of SCP production. The maximum SCP (28%) was produced on 5 days after incubation. However the further increase in incubation period did not enhance SCP production, rather a slight decrease in SCP production was observed. Fungal enzyme-controlled degradation responds to incubation time, pH, and temperature of the medium were responsible for such result, as has been proposed by Ofuya and Nwanjiuba.[52] Earlier workers also reported that decrease in SCP production could be due to autolysis of the fungal mycelium.[16,67]

5.3.3   MINERAL AVAILABILITY

Documentation of stimulatory effect of use of phosphorus solublizing bacteria on mungbean crop has been made by few workers, but with fungi, literature per se is scanty. Six phosphorus solublizing fungal (PSF) isolates belonging to *Penicillium citrinum, Penicillium digitatum,* and *Aspergillus niger* were isolated from different plant rhizosphere and their P-solublizing capacity established under in vitro conditions. The P-solublization on seventh day incubation was higher on solid medium compared to fifth day of incubation (Fig. 5.3) irrespective of amount of P released by different fungal strain.[90] Similarly, maximum P-solublization was recorded in case of *Penicilliun citrinum* isolate (PC2) followed by *Aspergillus niger* isolates (AN1, AN2, and AN3) in liquid medium. Increase in P-solublization with different incubation period was also reported by Mittal and co-workers.[48] Assimilation of NH_4^+ ions, production of organic acids, and phosphatase enzyme could be the factors involved in phosphorus dissolution. The release of protons during respiration or NH_4^+ assimilation is the most probable mechanism in the P-solublization, as was suggested by Illmer and Schinner,[34] whereas the role of neutral phosphatases enzyme in P-solublization was suggested by others. Singh and Reddy[71] proposed that the soluble P was removed from the culture at early period as rapidly as it was solublized and, at later stages, more P was released to fulfill the increased demand of nutrient by the fungi. Hence, six PSFs belonging to *Penicillium* and *Aspergillus* genera were tested for their phosphorus solublizing potential under in vivo conditions. The *Penicillium citrinum* isolate 6 exhibited the maximum and constant

P-solublization potential in the presence of tri-calcium phosphate (TCP) on Pikovskaya's medium throughout the incubation period (Fig. 5.2). Three fungi, namely, *A. niger* isolate 2, *A. niger* isolate 3, and *Penicilliun citrinum* isolate 6 were selected for pot experiments according to their performance in laboratory experiment. They enhanced the nutrient availability as well as promoted the growth of mungbean crop (Table 5.1). Beneficial effect of species belonging to genera *Penicillium* and *Aspergillus* has already been documented by several workers on various crops as they posses' ability to bring insoluble soil phosphate into soluble form.[62,88] Higher availability of N, P, and K in soil after the incorporation of *Aspergillus niger* isolate 3, *Aspergillus niger* isolate 2 + *Aspergillus niger* isolate 3, and *A. niger* isolate 2 + *P. citrinum* isolate 6, respectively (Table 5.2), was recorded in mungbean crop. Similarly, the Mn availability was significantly enhanced by applying the combination of *A. niger* isolates with PSB. However, Zn and Cu availability was markedly enhanced by the use of *A. niger* isolate 2 inoculation (Table 5.3). Inoculation of fungal combination of effective strains of *Aspergillus* and *Penicillium* enhanced the growth and yield parameters (Table 5.4) of mungbean crop in calcium-rich soil. The *Penicilliun citrinum* isolate 2 (PC2) was the best P-solublizer in both solid and liquid medium in the presence of TCP and also yielded the maximum dry mycelia mat compared to *Aspergillus niger* isolates. The *Penicillium* is one of the major genera of fungi that produce phosphatases and phytase.[6] The higher P-solublization potential of *Penicillium* sp. compared to *Aspergillus* sp was recorded by Salih and coworkers.[63] However, the higher P-solublization potential of *Aspergillus* species compared to *Penicillium* was documented by Yadav and his associates.[85] Release of N from the decaying mycelia mat, organic acids, and production of other phytohormones as result of host-pathogen reaction could be the reason for increase in N content in soil after the harvest of the crop in *A. niger* isolate 3 and phosphorus solublizing bacteria-treated plot. Kucey and co-workers[39] reported that phosphorus-solublizing microbes produces considerable amount of N and plant growth promoting substances in the rhizosphere apart from providing P. Significant enhancement of the soil P availability after the harvest of the crop by inoculation of combination of *A niger* stain 1 and *A niger* stain 2 suggests that the production of organic acids and enzymes by the *Aspergillus* strains might have promoted the P availability in soil. Phytases produced by soil microorganism[6] and release of organic acids like citric acid, oxalic acid, malic acid, and gluconic acid with ion chelation[20] property play an important role in solublizing of inorganic P. The *A. niger* isolate 2 enhanced the zinc and copper solublization while manganese and iron solublization has enhanced after the incorporation of *A. niger* 2 + PSB and *P. citrinum* 6 + *A. niger* 3, respectively. The release of organic acid could not be the only mechanism of micronutrient solublization, it could be due to secretion of other matter apart from organic acid. Moreover, several fungi produce the coprogens and dimerum acids that are potent chelators of iron.[17,45] The solublization of Zn is regulated by oxidative dissolution, while Mn and Cu is regulated by reducing mechanism.[3] The stimulatory effect on shoot length and root length was recorded in combined use of *Aspergillus* and *Penicillium* species, whereas the dry seed weight was significantly enhanced by the application of PSB alone or in combination with *A. niger* isolates. The availability of P due to secretion of several organic acids and enzymes (phytase, acid phosphatase, etc.) by fungi that release phosphate from soil P that could be attributed to increase in short and root length as phosphorus is needed for cell division and cell enlargement by the plants. Mittal et al.[48] also obtained the similar result in case of chickpea. Higher dry weight of seed compared to others was recorded in PSB alone or in combination with fungal strain. Bacteria multiply more rapidly than fungi and their number enable them to solubilize more P, which could be the reason for the result obtained.

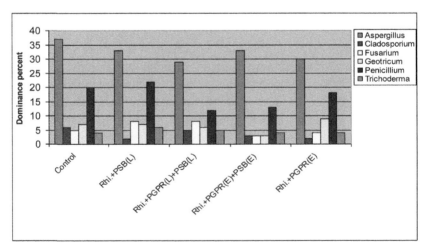

FIGURE 5.2 **(See color insert.)** Dominant fungal genera under different biological treatments.

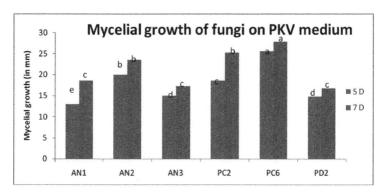

FIGURE 5.3 (See color insert.) Mycelial growth of *Aspergillus* and *Penicillium* spp. on PKV medium on fifth and seventh day of incubation.

TABLE 5.1 Effect of P-solublizing Fungi on Major Nutrients Availability Under Moong Crop.

Treatment	N	P	K
AN2	174.9[ef]	24.96[h]	126.6[ab]
PC6	162.3[g]	34.86[ef]	119.5[bc]
AN3	258.5[a]	37.56[d]	126.5[ab]
PSB	226.5[b]	44.90[b]	126.5[c]
AN2+ PC6	197.5[c]	42.03[c]	129.8[a]
AN2+ AN3	203.8[c]	47.66[a]	107.3[d]
AN2+PSB	181.2[e]	43.70[b]	128.8[a]
PC6+ AN3	174.7[ef]	34.26[f]	119.6[bc]
AN3+ PSB	194.1[cd]	27.86[g]	120.2[bc]
PC6+ AN3+PSB	165.5[fg]	63.13[e]	120.7[bc]
Control (uninoculated)	184.5[de]	21.53[i]	117.2[c]
CD at 5%	11.27	1.40	8.08
CV	3.44	2.34	3.94

AN, *Aspergillus niger*; PC, *Penicillium citrinum*; PSB, phosphorus solubilizing bacteria. Values marked with common letter were not statistically different ($p < 0.05$) in DMRT.

5.3.4 DISEASE MANAGEMENT

5.3.4.1 MANAGEMENT OF FUSARIUM WILT OF TOMATO BY WEEDS AND MYCOFLORA PROCESSED WEED COMPOST

To suppress the *Fusarium* wilt of tomato caused by *Fusarium oxysporum* f. sp. *lycopersici* (FOL), seven weeds extracts and weed composts extract predecomposed by beneficial agents was evaluated. Mycelia growth inhibition of beneficial agents has been recorded with the weed extracts. The maximum range (5.3–34.3 mm) of inhibition by weed extract was on *Trichoderma virens*, but to a lower extent (of 16.0–46.7 mm) on *Trichoderma harzianum* (Table 5.5). However, narrow range of inhibition was recorded in case of *Aspergillus niger* (21–41 mm) and *Cladosporium cladosporioides* (22.7–32.0 mm). Significant inhibitory effect of all the weed extracts and compost extract of *Cannabis sativa* loaded individually with *T. virens, T. harzianum, A. niger,* and *C. cladosporioides* was found against mycelia growth of the test pathogen (FOL; Fig. 5.4). The wilt incidence was markedly reduced to 20.02% and 18.34% by the use of predecomposed *Parthenium hysterophorus* compost prepared from individual loading of *T. harzianum* and *A. niger,* respectively (Table 5.6). Similarly, suppression of tomato wilt incidence caused by *Fusarium oxysporum* f. sp. *lycopersici* to 23.34% by *Physalis minima* compost prepared from inoculation of *P. lilacinus* was also recorded.[89] It has been reported that phenolic compounds in *P. hysterophorus*[9] and alkaloids in *P. minima*[58] have antifungal activity. Production of secondary toxic metabolite/induction of SAR in host plant by synergistic interaction of antifungal compounds in weed composts and soil microbial community structure might be the reason for such result. Elicitors like β-1, 3 glucanase, protease, polygalacturonidase, and a-amylase produced by *Trichoderma* spp. and *A. niger*[10] not only enhance decomposition but also acts as inducer for SAR in host plant. The high nutrient of weed composts prepared from decomposition promoting beneficial fungi were found suppressive toward *Fusarium* wilt of tomato as these composts support bioagent's activity and their anti-microbial compound suppress the pathogenic fungi.

TABLE 5.2   Effect of Incubation Period on Solublization of TCP by Fungi.

Treatments	Soluble phosphorus(μg mL^{-1})				Dry mycelia
	2nd	4th	6th	9th	weight (mg mL^{-1})
AN 1	45.0^b	44.67^f	365.0^c	478^b	2.9^f
AN 2	40.3^d	85.0^c	365.0^c	478^b	5.07^d
AN 3	38.0^e	560.0^a	489.0^a	457^c	6.82^b
PC2	33.0^f	63.0^e	398.0^b	335^d	5.77^c
PC6	43.0^c	136.0^b	296.33^d	521^a	7.1^a
PD2	47.0^a	65.33^d	245.0^e	121.6^e	3.75^e
CD at 5%	1.67	1.87 0.66	1.67	1.67	0.14
CV	2.29		0.26	0.23	1.86

AN, *Aspergillus niger;* PC, *Penicillium citrinum;* PD, *Penicillium digitatum* ($n = 4$); TCP, tri-calcium phosphate. Value marked with common letters was not statistically different (p < 0.05) in Duncan's multiple range test.

TABLE 5.3 Effect of P-solublizing Fungi on Minor Nutrients Under Moong Crop.

Treatment	Mn (ppm)	Zn (ppm)	Cu (ppm)	Fe (ppm)
AN2	4.5^c	0.52^a	0.69^a	5.35^{bc}
PC6	4.52^c	0.27^{fg}	0.52^{ef}	5.01^{ef}
AN3	4.95^b	0.34^c	0.54^e	5.4^{bc}
PSB	4.28^d	0.19^h	0.44^g	4.91^f
AN2+ PC6	4.69^c	0.35^c	0.65^{ab}	5.45^b
AN2+ AN3	4.77^c	0.32^{cde}	0.48^{fg}	5.00^{ef}
AN2 + PSB	5.59^a	0.33^{cd}	0.52^{ef}	5.15^{de}
PC6 + AN3	5.2^b	0.42^b	0.58^{cd}	5.89^a
AN3 + PSB	5.81^a	0.3d^{ef}	0.56^{de}	5.00^{ef}
PC6 + AN3 + PSB	4.89^c	0.26^g	0.48^{fg}	4.32^g
Control (uninoculated)	5.14^b	0.29^{efg}	0.61^{bc}	5.35^{bc}
CD at 5%	0.35	0.05	0.04	0.16
CV	4.08	5.9	4.54	1.93

AN, *Aspergillus niger*; PC, *Penicillium citrinum*; PSB, Phosphorus solublizing bacteria.
Values marked with common letter was not statistically different ($p < 0.05$) in DMRT

TABLE 5.4 Effect of P-solublizing Fungi on Biometric Parameters of Moong Crop.

Treatment	Shoot length (cm)	Root length (cm)	Dry weight/ plant (gm)	Dry seed weight/ pod (gm)
AN2	24.6	8.0	2.6	0.85
PC6	25.0	8.0	2.0	0.42
AN3	17.0	7.9	1.7	0.66
PSB	24.8	8.9	3.0	0.75
AN2 + PC6	20.0	16.3	3.0	0.53
AN2 + AN3	24.5	16.3	3.2	0.22
AN2 + PSB	21.0	7.1	1.1	0.76
PC6 + AN3	27.0	13.3	1.6	0.48
AN3 + PSB	16.8	6.1	1.2	0.74
PC6 + AN3 + PSB	21.6	5.3	1.6	0.24
Control (uninoculated)	24.6	8.0	2.6	0.55
CD at 5%	1.66	0.53	0.02	0.02
CV	4.52	3.32	0.66	2.35

AN, *Aspergillus niger*; PC, *Penicillium citrinum*; PSB, phosphorus solublizing bacteria.
Values marked with common letter was not statistically different in ($p < 0.05$) DMRT

TABLE 5.5 Effect of Weed Extracts on Mycelial Growth of Beneficial Fungi and *Fusarium Oxysporum* F. Sp. *Lycopersicae* (FOL).

Treatments	Mycelial growth of fungi (mm)					
	T. virens	T. harzianum	P. lilacinus	A. niger	C. cladosporioides	FOL
T1	34.3 ± 3.2^b	33.0 ± 3.0^{cd}	34.3 ± 2.1^a	21.3 ± 1.5^{de}	24.3 ± 1.5^{bc}	22.0 ± 1.0^f
T2	26.0 ± 1.0^c	26.7 ± 1.5^e	27.0 ± 1.0^b	27.0 ± 3.0^c	22.7 ± 1.5^c	27.7 ± 2.5^{de}
T3	26.7 ± 1.5^c	28.3 ± 7.6^{de}	27.7 ± 3.2^b	33.7 ± 2.5^b	26.0 ± 3.6^b	25.7 ± 1.5^e
T4	22.0 ± 1.0^d	36.0 ± 1.0^c	22.7 ± 2.5^c	27.0 ± 1.0^c	32.0 ± 1.0^a	31.7 ± 1.5^c
T5	5.3 ± 1.2^e	16 ± 1.0^f	16.0 ± 1.0^d	24.3 ± 4.0^{cd}	25.7 ± 1.0^{bc}	29.0 ± 1.0^d
T6	22.0 ± 3.0^d	46.7 ± 1.5^b	6.7 ± 0.6^e	41.0 ± 3.6^a	31.7 ± 0.6^a	36.7 ± 1.5^b
T7	21.0 ± 3.6^d	32.0 ± 1.0^{cd}	21.0 ± 1.0^c	26.0 ± 1.0^c	24.3 ± 2.1^{bc}	36.7 ± 0.6^b
T8	41.7 ± 1.5^a	62.7 ± 2.5^a	14.0 ± 2.0^d	19.3 ± 1.2^e	14.3 ± 2.1^d	47.0 ± 1.3^a
CD at 5%	3.9	5.5	3.3	4.3	3.3	2.6
CV	9.03	9.07	8.89	9.14	7.57	4.60

TABLE 5.6 Effect of Weeds' Compost Prepared from Beneficial Fungi on Wilt Disease Incidence.

Treatments	Disease incidence (percent) at different time intervals					
	10 days	20 days	30 days	40 days	50 days	60 days
TV(CS)	2.50[b]	5.00[c]	10.00[b]	16.67[bcd]	26.67[bc]	23.37[cb]
TH(CS)	2.50[b]	5.00[c]	7.50[b]	20.00[bc]	46.79[a]	48.58[a]
TH(PH)	5.00[b]	13.35[a]	13.35[b]	15.01[c]	18.36[cd]	20.02[c]
PL(PM)	2.50[b]	7.50[c]	10.86[b]	15.01[c]	21.69[c]	23.34[c]
AN(PH)	5.00[b]	7.50[c]	7.50[b]	11.67[d]	16.67[d]	18.34[c]
CC(NP)	2.50[b]	5.00[c]	7.50[b]	23.36[b]	33.34[b]	35.00[b]
TV(PH)	2.50[b]	5.00[c]	7.50[b]	16.68[bcd]	35.03[b]	38.39[b]
Control (compost)	10.00[a]	16.68[a]	23.36[a]	37.71[b]	53.42[a]	56.76[a]
CD at 5%	3.76	8.23	9.14	8.11	9.64	10.83

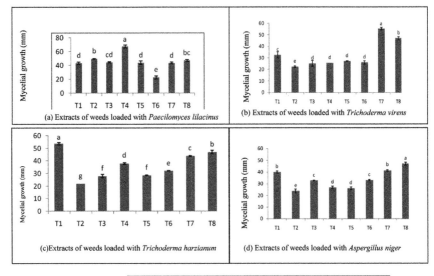

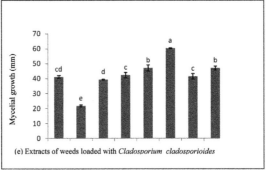

FIGURE 5.4 (See color insert.) Effect of extracts of weeds loaded with beneficial fungi on mycelial growth of *Fusarium oxysporum* f. sp. *lycopersicae* (FOL). Values marked with common letter was not statistically different (p < 0.05) in DMRT

5.3.4.2 EFFECT OF SELECTIVE MYCOFLORA AMENDED VERMICOMPOST ON SUPPRESSION OF ROOT ROT PATHOGENS OF TOMATO

Fungus-amended vermicompost supported higher earthworm population, their weight and size, and vermicast recovery in comparison to unamended control (Table 5.7). *Pythium aphanidermatum* causing damping-off disease of tomato was suppressed by aqueous extracts of various combinations of fungus-amended vermicompost with few exceptions where mycelia growth promotion was recorded, whereas inhibitory effect on mycelial growth of *Fusarium oxysporum* f. sp. *lycopersicae* by same combinations was found. Fungus-amended vermicompost inhibited the mycelial growth of *Trichoderma virens*. In case of *Paecilomyces lilacinus*, these vermicompost inhibited as well as promoted the mycelial growth of the fungus (Table 5.8). Fungal combinations of *Aspergillus niger* + *Trichoderma virens Paecilomyces lilacinus* + *Humicola grisea* used for vermicompost preparation had promoted all the biological parameters of the plant besides being found to be highly suppressive toward pre-emergence disease incidence (Table 5.9 and Fig. 5.5). Similarly, vermicompost prepared from fungal combinations of *A. niger* + *T. virens* + *P. lilacinus* + *H. grisea* + *C. cladosporioides* + *P. purpurogenum* had significantly reduced the post-emergence disease incidence.[79] Microbes present in compost and their synergistic interaction with artificially inoculated fungi led not only to enhanced nutrient removal efficiency of microbes but also the production of secondary toxic metabolites that may result in disease suppression. Antagonism of soil-borne pathogens by representative genera of *Trichoderma, Gliocladium, Penicillium,* non-pathogenic *Fusarium* spp., and *Sporidesmium* have been identified by several workers.[14,21] The propagules of *Pythium* have small amounts of reserved nutrients and depend on exogenous carbon sources for germination to infect host plant, and are described as highly sensitive to microbial nutrient competition and antibiosis.[28] However, cellulolytic and oligotrophic fungi, bacteria, and actinomycetes were involved in composts that suppress *Fusarium* wilt in tomato, as suggested in study by Borrero et al. in 2004. *Pythium* propagules need an exogenous carbon source for germination as it has small amounts of reserved nutrients, and cannot sustain microbial nutrient competition and antibiosis.

5.3.4.3 EFFECT OF FUNGAL METABOLITES AND AMENDMENTS ON MYCELIAL GROWTH OF RHIZOCTONIA SOLANI

Combining organic resources against soil-borne plant pathogens for getting the green food is an urgent need of hour. The cultural filtrates of *Aspergillus* species and *Trichoderma virens*, and organic amendment extracts were tested for their efficacy against *Rhizoctonia solani* through plate inhibition technique. Under co-culture of *A. niger* with *R. solani*, the minimum growth was attained by test fungi. However, the culture filtrate of *A. ochraceous* significantly reduced the maximum mycelial growth of

TABLE 5.7 Cast Generated During Vermicomposting Process and Biometrical Parameters of Earthworms at the End of Vermicomposting.

Treatments	Castings generated at different time intervals (g)			Biometrical parameters of earthworms at the end of 60 days		
	30 day	45 day	60 day	Weight (g)	Average size (cm)	Worm number
T1	233.1 ± 0.13	233.1 ± 0.12	203.2 ± 0.19	12.4 ± 0.7	9.1 ± 0.2	20.3 ± 1.2
T2	214.86 ± 0.19	214.43 ± 0.22	182.06 ± 0.18	12.00 ± 0.0	6.6 ± 0.1	22.7 ± 0.6
T3	217.30 ± 0.23	217.43 ± 0.20	176.26 ± 0.32	20.3 ± 0.4	9.0 ± 0.0	28.0 ± 1.0
T4	223.10 ± 0.34	223.10 ± 0.30	205.33 ± 0.32	15.9 ± 0.1	9.4 ± 0.5	28.0 ± 1.0
T5	217.43 ± 0.08	217.43 ± 0.05	214.06 ± 0.12	15.2 ± 0.3	8.0 ± 0.1	24.0 ± 1.0
T6	212.43 ± 0.03	212.43 ± 0.09	211.43 ± 0.07	19.6 ± 0.1	9.1 ± 0.1	23.7 ± 2.5
T7	212.36 ± 0.22	212.36 ± 0.28	223.63 ± 0.24	16.4 ± 0.4	9.0 ± 0.0	20.7 ± 1.5
CD at 5%	1.97	1.97	4.5	0.62	0.34	2.46

Values marked with common letter was not statistically different $p < 0.05$) in(DMRT.

TABLE 5.8 Effect of Fungal Loaded and Nonloaded Vermicompost Extracts on Mycelial Growth of Pythium Aphanidermatum, Fusarium Oxysporum, Trichoderma Virens, and Paecilomyces Lilacinus.

Treatment	Growth in mm			
	P. aphanidermatum	F. oxysporum	T. virens	P. lilacinus
T1	47.3 ± 0.5	38.2 ± 0.3	46.1 ± 0.3	90.0 ± 0.0
T2	44.9 ± 0.2	29.4 ± 0.3	36.2 ± 0.3	81.1 ± 0.1
T3	55.8 ± 0.0	33.5 ± 3.0	41.7 ± 0.2	80.6 ± 0.5
T4	21.3 ± 0.0	37.3 ± 0.2	28.3 ± 0.3	90.0 ± 0.0
T5	44.3 ± 0.0	27.8 ± 0.3	20.7 ± 0.4	58.1 ± 0.1
T6	86.7 ± 0.0	25.3 ± 0.3	38.3.3 ± 0.3	90.1 ± 0.8
T7	74.3 ± 0.0	29.3 ± 0.5	33.1 ± 0.2	90.0 ± 0.9
CD at 5%	0.33	2.03	0.52	0.36

Values marked with common letter was not statistically different (p < 0.05) in DMRT.

TABLE 5.9   Effect of Different Treatments on Growth Parameters of Tomato Plant.

Treatments	No. of leaves	Height (cm)	Shoot length (cm)	Root length (cm)	Total biomass (g)	Shoot biomass (g)	Root biomass (g)
T1	14 ± 1.5^a	25.6 ± 0.5^a	26.1 ± 0.2^b	1.9 ± 0.2^a	1.1 ± 0.0^b	0.9 ± 0.0^a	0.14 ± 0.02^b
T2	16 ± 2.0^a	25.2 ± 0.4^a	24.4 ± 0.7^a	1.9 ± 0.2^a	1.0 ± 0.0^a	0.9 ± 0.0^a	0.13 ± 0.02^b
T3	21 ± 1.5^b	25.4 ± 0.6^a	23.9 ± 0.4^a	2.3 ± 0.2^b	1.7 ± 0.1^c	1.6 ± 0.1^b	0.07 ± 0.02^a
T4	31 ± 1.7^e	42.2 ± 0.5^e	39.7 ± 0.7^e	3.5 ± 0.1^c	4.2 ± 0.0^g	3.9 ± 0.1^e	0.24 ± 0.04^c
T5	24 ± 1.0^c	35.8 ± 0.3^c	33.1 ± 0.3^c	2.4 ± 0.1^b	2.6 ± 0.0^e	2.1 ± 0.0^c	0.22 ± 0.03^c
T6	27 ± 1.5^d	39.3 ± 0.7^d	36.2 ± 0.2^d	2.8 ± 0.1^c	3.0 ± 0.0^f	2.9 ± 0.1^d	0.14 ± 0.02^b
T7	22 ± 1.5^{bc}	34.0 ± 0.2^b	32.5 ± 0.6^c	2.3 ± 0.1^b	2.5 ± 0.0^d	2.1 ± 0.0^d	0.20 ± 0.01^c
CD at 5%	2.76	0.81	0.85	0.25	0.06	0.12	0.04
CV	7.16	1.43	1.57	5.77	1.56	3.45	13.72

Values marked with common letter was not statistically different in($p < 0.05$) DMRT.

the *R. solani* followed by *A. niger*, *A. fumigatus*, *A. flavus*, and *A. terreus* (Table 5.10). Castor cake extract promoted the mycelia growth of *T. virens* among tested organic amendment, however, vermicompost extract was found highly suppressive toward *R. solani*. The effect of time on efficacy of organic amendment extracts increased markedly. Significantly higher inhibition potential of culture filtrate mixture of *A. niger* + *T. virens* and *A. ochraceous* + *T. virens* was recorded against *R. solani* in comparison to the other combinations by author.[91]

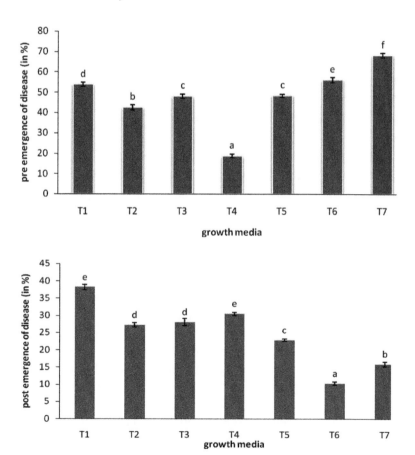

T1, no fungal inoculation, T2, AN+ TV+ PL; T3, AN+TV+PL+CC; T4, AN+TV+PL+ HG; T5, AN+TV+PL+ HG+CL; T6, AN+TV+PL+ CC+PP+FS; T7, all fungal inoculants except CC.

FIGURE 5.5 **(See color insert.)** Treatments effect on disease indices of plant pathogenic fungal consortium on tomato seedlings.

TABLE 5.10　Antagonistic Effect of *Aspergillus* spp on *R. Solani* Through Dual Culture and Poison Food Method.

Aspergillus spp.	Time (h)[a]		Growth of R. solani (mm)[b]		Mean
	48	96	Dual	Metabolite	
A. niger + T. virens	41.1	38.5	36.1	43.5	39.8
A. terreus + T. virens	40.1	62.8	49.5	53.5	51.5
A. ochraceous + T. virens	34.6	50.5	42.8	42.3	42.5
A. fumigatus + T. virens	41.1	48.7	39.5	50.4	44.9
A. flavus + T. virens	39.0	51.5	38.0	52.5	45.2
Control	49.1	58.5	53.8	53.8	53.8
Mean	40.9	51.7	43.3	49.3	
CD (P = 0.05)	0.62		0.62		0.44

[a]Interaction effect of *Aspergillus* spp. × time.

[b]Interaction effect of *Aspergillus* spp. × methods.

Hyphal interaction between the pathogenic fungi and beneficial organism is the first step for successful parasitism in biological control of pathogen. The highest growth suppression due to myco-parasitism by *A. niger* against *R. solani* was proved under dual culture. Lysis of the host cell due to secretion of metabolites by fast growing and sporulating fungi could be the reason for such inhibition. An effective biological control of *R. solani* by way of antibiosis, overgrowth, and hyperparasitism by *A. niger* was also reported. *Aspergillus* species are well known for secreting toxigenic compound around its surrounding. *A. ochraceous* secretes ochratoxin, the presence of ochratoxin in the culture filtrate might be attributed to the highest growth inhibition of *R. solani*. Ochratoxin B derivative is non-toxigenic against human being, although, there is evidence that suggests that the ochratoxin being injurious to human health. Literature per se is scanty to support the findings, but interaction between enzymes of actively growing bioagent with enzymatically active *R. solani* might got triggered that resulted in mycelia growth suppression of test pathogen. The castor and karanj cakes were the best supporting organic amendment extract for the growth of *T. virens* even at different time intervals. Presence of organic nitrogen, carbon in available form (e.g., humic acid), and absence of toxic compounds in organic substrate could be the probable reason behind this finding. *Trichoderma* spp. predominantly colonizes the substrate prepared from lignocellulosic material like tree bark.[40] Higher suppression of mycelial growth of both fungi in mixed cake extracts might be due to higher accumulation of nitrogen in amide and amino acid form. Organic amendments of high nitrogen content became toxic because of release of ammonia and nitrous acid that have a potential to suppress both the soil-borne pathogen (and beneficial organisms) through the toxic effect. Vermicompost has also inhibited the mycelia growth of *R. solani* and was next best alternate to mixed cake extracts. Low cellulose content, low nitrogen content, and presence of heavy metal ions in vermicompost may be ascribed for poor growth of *R. solani*. The *R. solani* is highly competitive saprophytic pathogenic fungi. It multiplies on fresh organic material with higher cellulose content, but does not colonize on mature compost with low cellulose content.[28] Growth supportive effect of mixture of metabolites of *A. niger* + *T. virens* and castor and karanj cakes extract was observed on *T. virens* while the same combination had opposite effect on *R. solani*. The suppressive effect of mixed metabolites of *A. niger* + *T. virens* and *A. ochraceous* + *T. virens* was recorded against *R. solani*. The inhibition of test pathogen might be due to the production of anti-biotics, humic acid like substances, maturity index of the amendment, and the hyphal interaction between *T. virens* and *R. solani* in dual culture. *A. niger* not only produces the aflatoxin but also the humic acid that are highly stable organic matter and does not break down further. Hence, test pathogen could not grow properly as the nutrients become a limiting factor for its growth. *Rhizoctonia* species can be suppressed by competition as well as parasitism in growing media.[28] Production of lytic enzyme by *T. virens* led to the mycelium disintegration of *R. solani*, which was running toward the bioagent. The *T. virens* fungi produces a peptide metabolite like gliotoxin that may contribute to antagonistic interaction with disease pathogen, as was reported by Wilhite et al.[82]

5.4 CONCLUSION

Agricultural soil contains abundant of soil microbes that account for about 5% of the biomass of total soil microbes. Rhizosphere, the most active region of the soil owing to cross talk between plant root system and its resident fungal communities through several chemical compounds or signals that are still not properly understood are the most valuable treasury for changing agriculture system. Over the past few decades, these rhizo-sphere communities are getting more attention due to their dynamic role in soil amelioration, decomposition, nutrient solublization, plant heath management, etc. The co-evolution of plant and soil fungi is a reason for their dynamic interactions in any ecosystem. Thus, the fungal community structure varies between different plant species and also between different agricultural treatments given for enhancing crop production. Hence, use of these fungal inoculants for residue management, micronutrient solublization, and plant health management is an energy-efficient, environment-friendly, and economically viable approach for sustainable crop production.

KEYWORDS

- **fungal community**
- **decomposition**
- **nutrient solublization**
- **disease suppression**
- **soil fertility**

REFERENCES

1. Abdel-Motal, F. F.; Nassar, M. S. M.; El-Zayat, S. A.; Magdi, A.; El-Sayed, M. A.; Ito, S-I. Antifungal Activity of Endophytic Fungi Isolated from Egyption Henbane (*Hyoscayamus muticus* L.). *Pak. J. Bot.* **2010**, *42*, 2883–2894.
2. Ae, N.; Arihara, J.; Okada, K.; Yoshihara, T.; Johansen, C. Phosphorus Uptake by Pigeon Pea and Its Role in Cropping Systems of the Indian Subcontinent. *Science* **1990**, *248*, 477–480.
3. Altomare, C.; Norvell, W. A.; Bjorkman, T.; Harman, G. E. Solublization of Phosphate and Micronutrients by the Plant Growth Promoting and Biocontrol Fungus *Trichoderma Harzianum* Rifai 1295-22. *Appl. Environ. Microbiol.* **1999**, *65*, 2926–2933.
4. Ambus, P.; Jensen, E. S. Nitrogen Mineralization and Denitrification as Influenced by Crop Residue Particle Size. *Plant Soil* **1997**, *197*, 261–270.
5. Aneja, K. R. Biology of Litter Decomposition Fungi as Decomposers of Plant Litter. In *Prospective in Mycology and Plant Pathology;* Agnihotri, V. P., Sarbhoy, A. K., Kumar, D., Eds.; Malhotra Publishing House: New Delhi, 1988; pp 389–394.
6. Aseri, G. K.; Jain, N.; Tarafdar, J. C. Hydrolysis of Organic Phosphate Forms by Phosphatases and Phytase Producing Fungi of Arid and Semi Arid Soils of India. *Am.-Eur. J. Agric. Environ. Sci.* **2009**, *5*, 564–570.

7. Beare, M. H.; Wilson, P. E.; Fraser, P. M.; Butler, R. C. Management Effects on Barley Straw Decomposition, Nitrogen Release and Crop Production. *Soil Sci. Soc. Am. J.* **2002**, *66*, 848–856.

8. Berkenkamp, A.; Priesack, E.; Munch, J. C. Modelling the Mineralization of Plant Residue on the Soil Surface. *Agronomie* **2002**, *22*, 711–722.

9. Belz, R. G.; Van der, L. M.; Reinhadt, C. F.; Hurle, K. In *Soil Degradation of Parthenin—Does It Contradict a Role in Allelopathy of the Invasive Weed Parthenium Hysterophorus L,* Proceedings 14th EWRS Symposium, Hamar, Norway, 1988; p 166.

10. Benítez, T.; Rincón, A.; Carmen, L. M.; Codón, A. C. Biocontrol Mechanisms of *Trichoderma* Strains. *Int. Microbiol.* **2004**, *7*, 249–260.

11. Chahal, D. S.; Gray, W. D. Growth of Cellulolytic Fungi on Wood Pulp. *Ind. Phytopathol.* **1969**, *12*, 79–91.

12. Chhonkar, P. K.; Tarafdar, J. C. Accumulation of Phosphatase in Soil. *J. Ind. Soc. Soil Sci.* **1984**, *32*, 266–272.

13. Coockson, W. R.; Beare, M. H.; Wilson, P. E. Effect of Prior Crop Residue Management Decomposition. *Appl. Soil Ecol.* **1998**, *7*, 179–188.

14. Cotxarrera, L.; Trillas, G. M. I.; Steinberg, C.; Alabouvette, C. Use of Sewage Sluge Compost and *Trichoderma asperellum* Isolates to Suppress Fusarium Wilt of Tomato. *Soil Biol. Biochem.* **2002**, *34*, 467–476.

15. Cruz, A. G.; Garcia, S. S.; Rojas, F. J. C.; Ceballos, A. I. O. Foliage Decomposition of Velvet Bean During Seasonal Drought. *Interciencia* **2002**, *27* (11), 625–630.

16. Dhillon, G. S.; Kalra, S. S.; Singh, A.; Kahlon, S. S.; Kalra, M. S. In *Bioconversion of the Delignified Rice Straw by Cellulolytic Fungi,* Proceedings of RRAI Symposium, PAU, Ludhiana, India, 1980; pp 77–80.

17. Dori, S.; Solel, Z.; Kashman, Y.; Barash, I. Characterization of Hydroxamate Siderophores and Siderophore-mediated Iron Uptake in *Gaeumannomyces Graminisvar. Tritici. Physiol. Mol. Plant Pathol.* **1990**, *37*, 97–106.

18. Ehteshamul-Haque, S.; Abid, M.; Ghaffar, A. Efficacy of *Bradyrhizobium* Sp., and *Paecilomyces Lilacinus* with Oil Cakes in the Control of Root Rot of Mungbean. *Trop. Sci.* **1995**, *35*, 294–299.

19. El-Shora, H. M.; Metwally, M. Effect of Phytohormones and Group Selective Reagents on Acid Phosphatise from *Cladosporiun Cladosporioides. Asian J. Biotechnol.* **2009**, *1*, 1–11.

20. Gadagi, R. S.; Shin, W. S.; Sa, T. M. Malic Acid Mediated Aluminum Phosphate Solubilization by *Penicillium Oxalicum* CBPS-3F-Tsa Isolated from Korean Paddy Rhizosphere Soil. *Dev. Plant Soil Sci.* **2007**, *102*, 285–290.

21. Garbeva, P.; van Veen, G. A.; van Elas, J. D. Assessment of Diversity and Antagonism toward *Rhizactonia Solani* AG23, of Pseudomonas Species in Soil from Different Agricultural Regimes. *FEMS Microbiol. Let.* **2004**, *47*, 51–64.

22. Gorodecki, B.; Hadar, Y. Suppression of *Rhizoctonia Solani* and *Sclerotium Rolfsi* in Container Media Containing Composted Separated Cattle Manure and Composted Grape Marc. *Crop Prot.* **1990**, *9*, 271–274.

23. Grebus, N. E.; Watson, M. E.; Hoitink, H. A. J. Biological, Chemical and Physical Properties of Composted Yard Trimmings as Indicators of Maturity and Plant Disease Suppression. *Compend. Sci. Utility* **1994**, *2*, 57–71.

24. Han, H. S.; Supanjani; Lee, K. D. Effect of Co-inoculation with Phosphate and Potassium Solubilizing Bacteria on Mineral Uptake and Growth of Pepper and Cucumber. *Plant Soil Environ.* **2006**, *52*, 130–136.

25. Hamlyn, P. F. Fungal Biotechnology. *Brit. Mycol. Soc. Newslett.* **1998**, pp. 130–134.

26. Hawksworth, D. L. Fungal Diversity and Its Implications for Genetic Resource Collections. *Stud. Mycol.* **2004**, *50*, 9–17.

27. Harish, S.; Saravanakumar, D.; Radjacommare, R.; Ebenezar, E. G.; Seetharaman, K. Use of Plant Extracts and Biocontrol Agents for the Management of Brown Spot Disease in Rice. *Biocontrol.* **2008**, *53*, 555–567.

28. Hoitink, H. A. J.; Boehm, M. J. Biocontrol Within the Context of Soil Microbial Communities: A Substrate-dependent Phenomenon. *Ann. Rev. Phytopathol.* **1999**, *37*, 427–446.

29. Hong, J. Y.; Kin, Y. H.; Jung, M. H.; Jo, W. C.; Cho, J. E. Characterization of Xylanase of *Cladosporiumn Cladosporioides* H1 Isolated from Janggyeong Panjeon in Haeinsa Temple. *Mycobiology* **2011**, *39*, 306–309.

30. Hudson, H. J. Fungal Saprophytism. In *Studies in Biology;* Edward Arnold: London, 1971; Vol. 32.

31. Jørgensen, H.; Kutter, J. P.; Olsson, L. Separation and Quantification of Cellulases and Hemicellulases by Capillary Electrophorosis. *Ann. Biochem.* **2003**, *317*, 85–93.

32. Hobbies, E. A.; Watrud, L. S.; Maggard, S.; Shiroyama, T.; Rygiewicz, P. T. Carbohydrate Use and Assimilation by Litter and Soil Fungi Assessed by Carbon Isotopes and BIOLOG (R) Assays. *Soil Biol. Biochem.* **2003**, *35* (2), 303–311.

33. Iyayi, E. A. Changes in the Cellulose, Sugar and Crude Protein Contents of Agro-industrial By-products Fermented with *Aspergillus Niger, Aspergillus Flavus* and *Penicillium* Sp. *Afr. J. Biotechnol.* **2004**, *3*, 186–188.

34. Illme r, P.; Schinner, E. Solubilization of Inorganic Calcium Phosphate. Solubilization Mechanisms. *Soil Biol. Biochemis.* **1995**, *27*, 257–263.

35. Khan, A.; William, K. L.; Nevalainen, H. K. M. Effect of *Paecilomyces lilacinus* Protease and Chitinase on the Eggshell Structures and Hatching of *Meloidogyne javanica* Juvenile. *Biol. Cont.* **2004**, *31*, 346–352.

36. Khan, A. A.; Sinha, A. P. Influence of Different Factors on the Activity of Fungal Bioagents to Manage Sheath Blight of Rice in Nursery. *Ind. Phytopathol.* **2005**, *58*, 289–293.

37. Khan, A.; Williams, K. L.; Nevalainen, H. K. M. Effects of *Paecilomyces lilacinus* Protease and Chitinase on the Eggshell Structures and Hatching of *Meloidogyne javanica* Juveniles. *Biol. Cont.* **2004**, *31*, 346–352.

38. Khanna, P. K. The Succession of Fungi on Some Decaying Grasses. PhD Thesis, Banaras Hindu University, Varanasi, India, 1964.

39. Kucey, R. M. N.; Janzen, H. H.; Legget, M. E. Microbial Mediated Increase in Plant Available Phosphorus. *Adv. Agro.* **1989**, *42*, 199–228.

40. Kuter, G. A.; Nelson, E. B.; Hoitink, H. A. J.; Madden, L. V. Fungal Populations in Container Media Amended with Composted Hardwood Bark Suppressive and Conducive to *Rhizoctonia* Damping-off. *Phytopathology* **1983**, *73*, 1450–1456.

41. Leizer, C.; Ribnicky, D.; Poulev, A.; Dushenkov, S.; Raskin, I. The Composition of Hemp Seed Oil and Its Potential as an Important Source of Nutrition. *J. Nutraceut. Funct. Med. Foods* **2000**, *2*, 35–53.

42. Liebman, J. A.; Epstein, L. Activity of Fungistatic Compounds from Soil. *Phytopathology* **1992**, *82*, 147–153.

43. Lockwood, J. L. Evolution of Concepts Associated with Soilborne Plant Pathogens. *Ann. Rev. Phytopathol.* **1998**, *26*, 93–121.

44. Mansoor, F.; Sultana, V.; Ehteshamul-Haque, S. Enhancement of Biocontrol Potential of *Pseudomonas aeruginosa* and *Paecilomyces lilacinus* Against Root Rot of Mungbean by a Medicinal Plant *Launaea nudicaulis* l. *Pak. J. Bot.* **2007**, *39*, 2113–2119.

45. Manulis, S.; Kashman, Y.; Barash, I. Identification of Siderophores and Siderophore-mediated Uptake of Iron in *Stemphylium botryosum*. *Phytochemistry* **1987**, *26*, 1317–1320.

46. Mc Tiernan, K. B.; Couteaure, M. M.; Berg, B.; Berg, M. P.; de Anta, R. C.; Gallardo, A.; Kratz, W.; Piussi, P.; Remacle, J.; De Santa, A. V. Changes in Chemical Composition of *Pinus sylvestris* Needle Decomposition Along a European Coniferous Forest Climate Transect. *Soil Biol. Biochem.* **2003**, *35* (6), 801–812.

47. Mehnaz, S.; Lazarovits, G. Inoculation Effects of *Pseudomonas Putida, Gluconacetobacter azotocaptans*, and *Azospirillum lipoferum* on Corn Plant Growth Under Green House Conditions. *Microb. Ecol.* **2006**, *51*, 326–335.

48. Mittal, V.; Singh, O.; Nayyar, H.; Kaur, J.; Tewari, R. Stimulatory Effect of Phosphate-solubilizing Fungal Strains (*Aspergillus awamori* and *Penicillium citrinum*) on the Yield of Chickpea (*Cicer arietinum* L. cv. GPF2). *Soil Biol. Biochem.* **2008**, *40*, 718–727.

49. Murayama, S. Changes in Monosaccharide Composition during the Decomposition of Straws Under Field Conditions. *Soil Sci. Plant Nutr.* **1984**, *30*, 367–381.

50. Nelson, E. B.; Kuter, G. A.; Hoitink, H. A. J. Effect of Fungal Antagonist and Compost Age on Suppression of *Rhizoctonia* Damping-off in Container Media Amended with Composted with Hardwood Bark. *Phytopathology* **1983**, *73*, 1457–1462.

51. Nikhra, K. M. Studies on Fungi from Jabalpur Soils with Special Reference to Litter Decomposition. PhD Thesis, Jabalpur University, India, 1981.

52. Ofuya, C. O.; Nwajiuba, C. U. Microbial Degradation and Utilization of Cassava Peel. *World J. Microbiol. Biotechnol.* **1990**, *6*, 144–148.

53. Ogihara, J.; Kato, Ju.; Oishi, K.; Fujimoto, Y.; Eguchi, T. Production and Structural Analysis of PP-V, a Homologue of Monascorubramine, Produced by a New Isolate of *Penicillium* Sp. *J. Biosci. Bioeng.* **2000**, *90*, 549–554.

54. Okunowo, W. O.; Gbenle, O. G.; Osuntoki, A. A.; Adekunle, A. A.; Ojokuku, S. A. Production of Cellulolytic and Xylanolytic Enzymes by a Phytopathogenic *Myrothecium roridum* and Some Avirulent Fungal Isolates from Water Hyacinth. *Afr. J. Biotechnol.* **2010**, *9*, 1074–1078.

55. Pandey, V.; Sinha, A. Mycoflora Associated with Decomposition of Rice Stubble Mixed with Soil. *J. Plant Protec. Res.* **2008**, *48*, 247–253.

56. Vibha; Jha, P. K.; Nidhi. Effect of Introduced Beneficial Inocula of Native and Exotic Bioagents on Microbial and Dominant Fungal Population in Pigeonpea Rhizosphere of Calciorthent Soil. *Int. J. Agri. Sci.* **2012**, *8* (2), 329–334.

57. Papavizas, G. C. *Trichoderma* and *Gliocladiu*: Biology and Ecology for Biocontrol. *Ann. Rev. Phytopathol.* **1985**, *23*, 23–54.

58. Prasad, S. H. K. R.; Swapna, N. L.; Rajasekhar, D.; Anthonamma, K.; Prasad, M. Preliminary Phytochemical and Antimicrobial Spectrum of Cultured Tissues of *Physalis minima* (L.). *Int. J. Chem. Sci.* **2009**, *7*, 2719–2725.

59. Punj, M. L.; Kochar, A. S.; Bhatia, I. S. Studies on the Carbohydrate Polymers of Roughages and Their Metabolism in the Rumen: II. Lignin and Structural Carbohydrates. *Ind. J. Anim. Sci.* **1971,** *41,* 531–536.

60. Rai, J. P.; Sinha, A.; Govil, S. R. Litter Decomposition Mycoflora of Rice Straw. *Crop Sci.* **2001,** *21* (3), 335–340.

61. Rubio, M. B.; Hermosa, R.; Reino, J. L.; Callado, I. G.; Monte, E. Thctfl Transcription Factor of *Trichoderma harzianum* is Involved in 6-Pentyl-2-Hpyran-2-One Production and Antifungal Activity. *Fungal Gen. Biol.* **2009,** *46,* 17–27.

62. Rudresh, D. L.; Shivprakash, M. K.; Prasad, R. D. Effect of Combined Application of *Rhizobium*, Phosphate Solubilizing Bacterium and *Trichoderma* Spp. on Growth, Nutrient Uptake and Yield of Chickpea (*Cicer aritenium* L.). *Appl. Soil Ecol.* **2005,** *28,* 139–146.

63. Salih, H. M.; Yahya, A. I.; Abdul-Rahman, A. M.; Munam, B. H. Availability of Phosphorus in a Calcareous Soil Treated with Rock Phosphate or Super Phosphate or Effected by Phosphate Dissolving Fungi. *Plant and Soil* **1989,** *20,* 181–185.

64. Sameni, M.; Pour, A. T. Effect of Root Residue of Liquorice on Different Characteristics of Soil from Southern Region of Iran. I. Microbial Characteristics. *Commun. Soil Sci. Plant Anal.* **2001,** *32,* 3259–3275.

65. Santro, A. Vde.; Rutigliano, F. A.; Berg, B.; Fioretto, A.; Puppi, G.; Alfuni, A. Fungal Mycelium and Decomposition of Needle Litter in Three Contrasting Coniferous Forests. *Acta Oecologia* **2002,** *23,* 247–259.

66. Sariyildiz, T.; Anderson, J. M. Decomposition of Sun and Shade Leaves from Three Deciduous Tree Species, as Affected by Chemical Composition. *Biol. Fert. Soil* **2003,** *37,* 137–146.

67. Sekhon, A. K. Studies on Utilization of Wheat Straw for Production of Cellulose and Protein by Thermophillic Fungi. MSc Thesis, PAU, Ludhiana, India, 1975.

68. Shaukat, S. S.; Siddiqui, I. A.; Mehdi, F. S. *Avicennia marina* (Mangrove) Soil Amended Changes the Fungal Community in the Rhizosphere and Root Tissue of Mungbean and Contributes to Control of Root Knot Nematodes. *Phytopathol. Mediter.* **2003,** *42* (2), 135–140.

69. Simoes, M. P.; Madeira, M.; Gazariani, L. Decomposition Dynamics and Nutrient Release of *Cistus salvifolius* L. and *Cistus ladanifer* L. Leaf Litter. *Revista de Ciencias Agrarias* **2002,** *25,* 508–520.

70. Singh, A. Studies on Fungal Decomposition of Sunhemp (*Crotolaria juncea* L.) in Soil. PhD Thesis, BHU, Varanasi, India, 2001.

71. Singh, H.; Reddy, M. S. Effect of Inoculation with Phosphate Solubilizing Fungus on Growth and Nutrient Uptake of Wheat and Maize Plants Fertilized with Rock Phosphate in Alkaline Soils. *Eur. J. Soil Biol.* **2011,** *47,* 30–34.

72. Singh, A.; Sharma, S. Composting of Crop Residue through Treatment with Micro-organism and Subsequent Vermicompost. *Biores. Technol.* **2002,** *85,* 107–111.

73. Stukkens, Y.; Bultreys, A.; Grace, S.; Trombik, T.; Vanham, D.; Boutry, M. NP RDPR1 a Pleotropic Drug Resistance-type ATP Binding Cassette Transporter from *Nikotiana Plumbaginifolia*, Plays a Major Role in Plant Pathogen Defense. *Plant Physiol.* **2005,** *139,* 341–352.

74. Thorman, M. N.; Currah, R. S.; Bayley, S. E. Succession of Microfungal Assemblages in the Decomposing Peat Land Plants. *Plant Soil* **2003,** *250* (3), 323–333.

75. Ullah, M. H.; Khan, M.; Aslam, S. S. T.; Habib, A. Evaluation of Antagonistic Fungi Against Charcoal Rot of Sunflower Caused by *Macrophomina phaseolina* (Tassi) Goid. *Afr. J. Environ. Sci. Technol.* **2011**, *5*, 616–621.

76. Valenzuela, V. P.; Leiva, S.; Godoy, R. Seasonal Variation and Enzymatic Potential of Microfungi Associated with the Decomposition of *Northophagous pumilio* Leaf Litter. *Rev. Chil. Historia Nat.* **2001**, *74* (4), 737–749.

77. van der Heijden, M. G.; Bardgett, R. D.; van Straalen, N. M. Review The Unseen Majority: Soil Microbes as Drivers of Plant Diversity and Productivity in Terrestrial Ecosystems. *Ecol. Lett.* **2008**, *11*, 296–310.

78. Venkateswarlu, A.; Rao, V.; Raina, P. Evaluation of Phosphorus Solubilization by Microorganisms Isolated from Aridsol. *J. Ind. Soil Sci. Soc.* **1984**, *32*, 273–277.

79. Vibha; Jha, P. K.; Nidhi. Effect of Selective Mycoflora Amended Vermicompost on Suppression of Root Rot Pathogens of Tomato. *Mycol. Plant Pathol.* **2013**, *43* (3), 306–313.

80. Vibha; Sinha, A. Variation of Soil Mycoflora in Decomposition of Rice Stubble from Rice-wheat Cropping System. *Mycobiology* **2007**, *35* (4), 191–195.

81. Vijay, T.; Naidu, C. V. Studies on Decomposition and Associated Mycoflora in *Albizzia amabovine* Leaf Litter. *Ind. J. For.* **1995**, *18*, 153–157.

82. Wilhite, S. E.; Lumsden, R. D.; Straney, D. C. Peptide Synthatase Gene in *T. Virens.* *Appl. Environ. Microbiol.* **2001**, *67* (11), 5055–5062.

83. Xu, R. K.; Coventry, D. R. Soil pH Changes Associated with Lupin and Wheat Plant Materials Incorporated in a Red-brown Soil. *Plant Soil* **2003**, *250* (1), 113–119.

84. Yadav, R. A.; Chakravarti, N. New Antifungal Triterpenoid Saponin from *Launeae pinnatifida* Cass. *Ind. J. Chem.* **2009**, *48B*, 83–87.

85. Yadav, J.; Verma, J. P.; Tiwari, K. N. Solubilization of Tricalcium Phosphate by Fungus *Aspergillus niger* at Different Carbon Source and Salinity. *Trends Appl. Sci. Res.* **2011**, *6*, 606–613.

86. Yasir, M.; Aslam, Z.; Kim, W. S.; Lee, S. W.; Jeon, O. C.; Chung, Y. R. Bacterial Community Composition and Chitinase Gene Diversity of Vermicompost with Antifungal Activity. *Biores. Technol.* *100*, 4396–4403.

87. Zimmermann, S.; Frey, B. Soil Respiration and Microbial Properties Acid Forest Soil: Effect of Wood Ash. *Soil Biol. Biochem.* **2002** *34*, 1727–1737.

88. Zayed, G.; Motaal, H. A. Bioactive Composts from Rice Straw Enriched with Rock Phosphate and Their Effect on the Phosphorus Nutrition and Microbial Community in Rhizosphere of Cowpea. *Biores. Technol.* **2005**, *96*, 929–935.

89. Vibha; Nidhi. Management of Fusarium Wilt of Tomato by Weeds and Mycoflora Processed Weeds Compost. *Bioscan* **2014**, *9*, 197–202.

90. Vibha; Kumari, G. R.; Nidhi. Impact of Phosphate Solubilizing Fungi on the Soil Nutrient Status and Yield of Mungbean (*Vigna radiate* L.) Crop. *Ann. Agri. Res.* **2014**, *35*, 136–143.

91. Vibha. Effect of Fungal Metabolites and Amendments on Mycelial Growth of *Rhizoctonia Solani. J. Plant Protec. Res.* **2010**, *50*, 93–97.

CHAPTER 6

Mycoflora Associated with Paddy Varieties

SHVETA MALHOTRA*

Department of Botany, Arya Mahila PG College, Shahjahanpur-242001, UP, India

*Corresponding author. E-mail: shvetamudit@gmail.com

ABSTRACT

India is an agricultural based country in which rice is the most important cereal crop. Most of the people are depend upon it, so to increase its production it is necessary to control the diseases related to its seed and other part of the plant. Most of the seed-born diseases are caused by fungi. This chapter included three variety of paddy (dhanpant, NDR, kranti) and their percent seed germination, isolation and frequency of fungi from unsterilized and surface sterilized seeds via blotter technique was done. Effect of fungicides (Bavistin and Thiram) on percent seed germination is also studied.

6.1 INTRODUCTION

Food is the most important need of human life, which is derived from various plant parts, such as roots, stems, leaves, fruits, and seeds. Cereals are the most important seeds utilized by men. Paddy (*Oryza sativa*) is one of the most important cereal crops cultivated in all warm continents (Asia, Africa, America, Australia, and India) for the grains, which constitute the staple food of millions. The origin of cultivated rice is not known. India, particularly South India, could be its original home. It is an annual grass, also known by various names such as dhan, chawal, chal, tandul, bhat, etc. Its cultivation is mainly concentrated in Asian countries which include about 90% of the world's cultivation. In India, its cultivated area is 43 million ha and its total production is 80 million tons. Its contribution is about 42% of the total food grains producing in the country (Chandra, 1996; Hegde, 2000). The average temperature during the rice season ranges between 21°C and 35°C which should be high at planting and falling gradually at the time of harvesting. The chief soil types on which rice is grown in India are alluvial soils, laterite soils, alkaline soils, and black soils. The pH of rice soils is on the acid side. There are three distinct seasons when rice is harvested, aus or autumn rice; aman or winter rice; and boro, spring or summer rice. The most important crop for the whole of India is the aman which coincides with southwest monsoon and may be planted in June–July and harvested in September–October. The maturation period ranges between 150 and 180 days. The aus crop is planted in May–June and harvested in October. The maturation period varies from 90 to 120 days. This crop is grown under upland condition. The upland rice accounts 17% (7.1 million/ha) of our total rice area and contributes 10% in total production. More than 85% of the upland rice area is concentrated in Assam, Bihar, Madhya Pradesh, Orissa, West Bengal, and Uttar Pradesh (Mishra, 1999; Gourangkar, 2002, Alam et al. 2014). The boro is planted in December–January and harvested in March–April.

It is estimated that once the population in India crosses 1.38 billion in 2025 AD, the country will have to import about 60 million tons of food grain annually. During this stage, the annual demand for food will be increased by 325 million tons per year. This causes ecocrisis (Hegde, 2000). So there is a great need to increase the crop production but there are many factors responsible for the decreased production of paddy and maize during the growth of crop.

Cereals are attacked by various types of pathogens such as fungi, bacteria, viruses, and nematodes, etc. Out of several causes, fungi play the most significant role in determining the yield and quality of grains of paddy and maize (Christensen and Mirocha, 1976; Malavolta et al., 1979; Lambat et al., 1985; Jayaweera et al., 1988; Waghary et al., 1988; Brekalo et al., 1992; Kalyansundaram, 1998; Harnandez et al., 1999; Franco et al., 2001; Nagugi et al., 2002; Maksimov et al., 2002; Teplyakov and Teplyakova, 2003; Surekha et al., 2011). The losses due to fungi are marked by reduction in yields. Such diseases cause reduction in germination and result in poor stand of crop. The seedlings emerging from diseased seeds lack vigour and are easily attacked by pre and postemergence damping-off of pathogens (Dharamvir, 1974; Haque et al., 2007). Such losses are marked by the reduction in quality of produce. Discoloration results in poor quality of grain or seed and is an important degrading factor. In seeds for sowing, such disorders may indicate presence of seed-borne infection. In grains for human consumption or for industrial purposes, discoloration means poor quality and adversely affects its market value. Many-seed borne fungi infect seed coat causing conspicuous necrotic grey to black discoloration rendering the grain poor in appearance and low in milling quality. In paddy, seed-borne fungi are able to penetrate the tissue rendering it weak and prone to breaking during polishing (Kim, 1992; Castano et al., 1992; Pandey et. al., 2000: Prakash and Rao, 2002; Mohamood et al., 2012). Grain discoloration in cereals is caused by various fungal species of *Drechslera, Curvularia, Fusarium, Pyricularia, Cladosporium,* and *Trichoconiella.*

Fungi present on seeds or grains bring about physical and biochemical changes during storage. Discoloration of paddy grains cause reduction in starch content from 2.26% to 20.37% in different categories of discoloration. Besides this, the amount of glycerol, carbohydrates, fatty acids, and amino acids also changes which results in unpleasant odor, change in color, and other biochemical changes in diseased seeds (Misra, 1987; Utobo et al., 2011).

Mycotoxins are a group of highly toxic metabolites produced by some fungi on stored agricultural commodities including food grains. Various diseases are caused by mycotoxins, viz., alimentary toxic aleukia and yellow rice toxin. Some mycotoxins have carcinogenic properties. Mycotoxins have been reported from cereals and a number of fungi present on seeds and grains such as species of *Aspergillus, Penicillium, Alternaria, Fusarium, Curvularia, Trichoderma, Rhizopus,* and *Mucor* are capable of producing mycotoxins, sometimes their accumulation during storage may reach concentration above the level of mammalian tolerance (Anonymous, 1979; Mall et al., 1988; Sinha and Sinha, 1988; Waghray et al., 1988; Bilgrami and Chaudhary, 1992; Yoshizawa, 1992; Tanaka et al., 2007).

The spoilage of grains by fungi is almost exclusively the result of complex chemical processes. Therefore, control of grain spoilage can be ensured by proper storage which is aimed at preventing or retarding the invasion of such fungi or creating unfavorable conditions for their multiplication. Storage fungi find it difficult to grow and multiply if grains are stored at a moisture content of 12% or less. The optimum moisture content for storage of many types of seeds is 6–8%. But Indian climatic conditions, harvesting practices, high temperature and high moisture levels during the monsoon season, unseasonal rains, and sudden floods favor the fungal infection. In India, about 70% of the total production is still retained by the farmers in a traditional underground, aboveground, and roof storage structures wherein stored grains are not firmly protected against exposure to moisture and temperature (Prasad and Pathak, 1987; Lacicowa and Pieta, 1998; Wicklow et al., 1998; Bock et al., 1999; Sheroze et al., 2003; Abedin et al., 2012). Neergaard (1977) has classified the fungi of food grains as "field fungi" and "stored fungi" and shown that their presence on seed may cause reduction in the field of food grains discoloration, chemical and physical changes, loss of germination, nutritive value, and sometimes, poisoning of food grains. Harvesting threshing, processing include cleaning and transportation causes mechanical damage or cracks in the seed coats and seeds which make them susceptible to the attack by stored fungi (Singh and Tyagi, 1984; Prasad and Pathak, 1987; Singh et al., 1988; Nandi et al., 1989; Misra and Misra, 1991, Aly, 1992; Krishna Rao, 1992; Mishra et al., 1995; Puroshotham et al., 1996; Bhattacharya and Rhah, 2002; Gracia et al., 2002; Singh et al., 2002; Marineck et al., 2003; Jayas and White, 2003; Butt et al., 2011). As pathogen, they cause number of destructive diseases.

In rice, fungi are known to cause 55 diseases, 43 of which are seed born or seed transmittable (Neergaard, 1979; Richardson, 1979, 1981; Malvolta et al., 1979; Ou, 1985; Jayaweera et al., 1988; Waghary et al., 1988; Pramanic and Chakrabarti, 1992; Sundar and Satayavir, 1997; Shahjahan et al., 1998; Khan et al., 1999; Sharma and Saxena, 2000; Franco et al.; 2001; Khalid et al., 2001; Anwar et al., 2002; Biswas, 2002; Prakash and Rao, 2002; Biswas, 2003; Chahal et al., 2003; Sundar et al., 2003; Alam et al., 2014).

Blast is the most serious disease in rice. The pathogen of blast disease is *Pyricularia oryzae,* which causes severe panicle infection and reduction in yield to the half. In India, 75% loss of grains has been reported in susceptible cultivars (Padmanabhan, 1965; Arase et al., 1999). The rate of transmission of *P. oryzae* from seeds to seedlings is much greater in situations where seed is not covered by soil of 7–18% (Aulakh et al., 1974b; Okhovat et al., 1992; Upadhyay et al., 1996; Manandhar et al., 1998; Sawada and Itoh,

1999; TeBeest and Guerber, 1999; Fukaya et al., 2001; Anwar et al., 2002; Vijaya, 2002).

Brown spot disease of rice caused by *Bipolaris oryzae (syn. Drechslera oryzae)* has been reported in all the rice growing countries. Brown spot was one of the principal causes of Bengal famine of 1942 in India. Loss in grain weight ranged from 12% to 30% as well as loss in filed grain from 18% to 22%, depending upon the degree of cultivars. Susceptibility was recorded by Padmanabhan, 1965; Prabhu, and Vieira, 1989; Upadhyay et al., 1996; Rathaiah, 1997; Mia and Safeeulla, 1998). Heavy losses generally occurred in highly susceptible cultivars of rice infected with narrow brown leaf spot caused by *Cercospora oryzae* (Das and Sahu, 1998; Nguefack et al., 2007).

Stack-burn disease of rice caused by *Alternaria padwickii* does not cause much damage in India. Symptoms of stack-burn disease appear on seedlings, leaves, and the grains. However, 40–60% damage has been recorded by Sharma and Siddiqqi (1978) and Reddy and Khare (1978) and up to 80% damage by Cheran and Raj (1966). Singh and Maheshwari (2001), Nguefack et al. (2007) also reported the influence of stack-burn disease of paddy on seed health status.

Leaf scald caused by *Microdochium oryzae (syn. Geralachia oryzae)* has been reported from Africa, Asia, Australia, Oceania, Eruope, Central America, and West Indies (Mia et al., 1986; Manandher, 1998). Leaf scald is usually seen on mature leaves of older plants. These are large areas encircled by dark brown bands with light brown holes. Yu and Mathur (unpublished as quoted by Agarwal et al., 1989) found up to 31–41% infected seedlings.

Bakanae disease is caused by *Fusarium moniliforme* Tele (*Gibberella fujikuroi*). This fungus has been harbored mainly in the embryo but also in the empty glumes, pedicel, palea, and lemma. (Hino and Faruta, 1968; Vidhyasekaran et al., 1970; Hajra et al., 1994; Sunder and Satyavir, 1997; Bohra et al., 2001). In general, this disease renders severe damage in specific localities or seasons and 15% yield losses by this disease have been reported in India (Ou, 1995).

Sheath blight disease is mainly caused by *Rhizoctonia solani* and is considered to be of major economic importance only second to blast in Japan, China, Taiwan, Srilanka, and the United States (Gangopadhyay and Chakraborty, 1982; Lee and Rush, 1983; Krishna Roy, 1992; Cedeno et al., 1999; Chahal et al., 2003; Sundar et al., 2003). Kozaka (1970) reported losses up to 30–40% in case of severe infection of the sheath and leaf blades.

Sheath-rot disease is caused by *Sarocladium oryzae* (Syn. *Acrocylindrium oryzae*). This disease is present in Taiwan, Japan, very common in Southeast Asia and the Indian subcontinent, and the United States (Mohan and Subramanian, 1981; Ou, 1985; Laxshman, 1991; Hajarika and Phookan, 1999; Reddy et al., 2000). In India, Chakrabarty and Biswas (1978) recorded 9.6–2.6% yield reduction among the seven cultivars they examined, with an average of 14.5%. Severe infection was observed on semi-dwarf varieties by Raina and Singh (1980). Murlidharan and Venkata (1980) observed sheath rot at panicle initiation stage, panicle either remained within the leaf sheath or emerged only partially causing 85% loss in yield. This disease is associated with reduced spikelets per panicle as well as grain weight. Extensive rotting of sheaths enclosing the panicles caused significant losses and glume discoloration in India. Due to this disease, percentage of germination was reduced from 94% to 58% (Upadhayay and Diwakar, 1984; Vidhyasekaran et al, 1984; Reddy et al., 2000).

False smut disease is caused by *Ustilaginoidea virens* (syn. *Ustilago virens* Cooke; *Tilletia oryzae* Palouillard). Heavy losses have been recorded from many countries, upto 44% in India (Singh and Dube 1978; Singh et al., 1992; Wang et al., 1998; Hegde and Anahosur, 2000; Ahonsi and Adeoti, 2002). Spores of the fungus were found on the surface of paddy seeds.

Karnal smut, also known as bunt, is known to occur in Africa, Burma, China, India, Indonesia, Japan, Korea, Malaysia, and Pakistan, etc. The disease is caused by *Tilletia barclayana* (syn. *Neovossia barclayana* Bret; *Neovossia horrida* (Tak.) Padw.) and can be seen at crop maturity (Cartwright et al., 1997; Bialooki and Piotrowska, 1999). Usually only few grains in a year are attacked. Spores shed from the grains settle on the other grains or leaves and form a black covering. The surface born inoculum may be as high as 40.55 spores/g × 105 spores/g of seeds (Shetty and Shetty; 1986).

Udbatta disease is caused by *Ephelis oryzae*. This disease is seen only at the time of panicle emergence. Mycelium of *Ephelis oryzae* is present in the embryo, as well as conidia on the seed surface. This disease has been reported from India, China, Hongkong, New Caledonia, and West Africa. In India, it occurs particularly in the hilly regions of Orissa state. Usually, 2.3% of ear head infection is observed, but in years of severe incidence, losses up to 10–11% are common in susceptible cultivars (Mohanty,

1978; Ranganthaiah and Safeeulla, 1994, 2000). In Karnatka state, Yu and Mathur (1989) reported losses up to 30% in rice cultivars (as quoted by Agarwal, 1989). Govindu (1969) considered Udbatta as an important disease in some areas of Bangalore, causing direct and indirect losses up to 1.75–3.69% in different rice cultivars.

Scab is caused by *Fusarium graminearum* (Tele. *Gibberella zeae* Schw). This is widely distributed disease of rice in tropics. This disease normally does not cause heavy damage but it may be severe under environmental conditions such as high humidity (Ou, 1985). Infected grains are lighter in weight, shrunken and brittle, and often do not germinate. If germination occurs, disease seedlings are produced. The fungus may also attack on nodes so the infected stem wilts and break. Chung et al. (1964) found that an isolate from wheat and infected rice caused postemergence blight. Dark blue perithecia of *G. zeae* can be observed on the seeds.

According to Zeigler et al. (1987), seed discoloration is another great problem of rice grains caused by certain fungi on the glumes or kernels (as quoted by Agarwal et al., 1989; Mahmood et al., 2012). High moisture content in grains (more than 15%) and high relative humidity (>65%) also favors discoloration. Fungi that are associated with discolored grains can be divided into two major groups:

Field fungi: (*Bipolaris oryzae, Alternaria padwickii, Pyricularia oryzae, Fusarium moniliforme, F. graminearum, Nigrospora oryzae, Epicoccum nigrum, Curvularia sp, Phoma sp, Sarocladium oryzae*, etc.)

Storage fungi: (*Aspergillus, Penicillium sp, Absidia sp, Mucor sp, Rhizopus sp, Chaetomium sp, Monilia sp, and Streptomyces* sp, etc.)

Discoloration of grains cause deterioration in grain quality, reduction in seed viability, and such grains on planting usually exhibit preemergence and postemergence death of seedlings. Toxin production by storage molds associated with discolored grains has also been exhibited by Christensen (1957), Vidhyasekaran and Govind Swami (1968), Duraiswamy and Mariappan (1983), Vidhyasekaran et al. (1984), Waghray et al. (1988), Jayaraman and Kalyansundaram (1989), Castano et al. (1992), Kim, (1992), Williams, (1992), Sachan and Agarwal (1993), Sachan and Agarwal (1994), Divakar and Prasad (1997), Kalyansundaram (1998), Pandey et al. (2000), Prakash and Rao, (2002). During the last few years, the agricultural strategy has acquired new dimensions. The use of high yielding varieties and hybrid has become a significant aspect of new technology which is being accompanied by a series of related changes in agricultural practices, such as massive application of fertilizer and plentiful irrigation. This has given rise to a drastic change in complexion of plant diseases. Major diseases have developed from minor ones which were previously of little or no importance. A number of seed-borne diseases have cropped up as a result of intensive agricultural programs and monoculture of crop cultivars. In our country, pre and postharvest spoilage of grain is a major problem. It is therefore essential to intensify researches on study of seed mycoflora present on fresh and stored seeds and to study effect of seed fungicides in controlling the postharvest seed mycoflora with special reference to protection of agriculture produce against fungal deterioration. As the total cultivated area of India is limited and there are no prospects of its increase, it is necessary to enhance yields of crops by reducing losses due to seed-borne diseases and postharvest spoilage of produce in storage.

6.2 MATERIALS AND METHODS

Fungi form the largest group among such microorganisms cause seed damage, seed rot, seedling rot, diseases at later stages of crop growth till maturity. Different types of fungi are associated with seeds as intraembryal, extraembryal, contaminant, or with inert matter mixed with the seed sample. Several methods have been evolved for testing seeds for associated microorganisms which have been reviewed from time to time (de Tempe, 1961, 1963, 1964; Limonard, T.,1966, 1968; Neergaard, 1977; Agarwal, V. K., 1976; Agarwal and Shrivastava, 1981, 1985; Agarwal and Sinclair, 1987; Fedavora, R. N., 1987; Gaur and Dev, 1996) Various methods are used according to ISTA rules (ISTA, 1993) for the detection of fungi based on their location in or on the seeds.

6.2.1 COLLECTION OF SEED SAMPLES

Three varieties of each maize and paddy were collected from Chandra Shekhar Azad University of Agriculture and Technology, Kanpur, in order to study the germination and fungi found associated with these varieties and their control.

Paddy: Kranti, Narendra (NDR), Dhan Pant

One variety of maize (R-9902) and one variety of paddy (NDR) were selected randomly and stored at room temperature in the laboratory for further studies. Remaining varieties of maize and paddy were studied immediately while they were fresh.

6.2.2 CHECKING OF MECHANICAL INJURIES

For checking the extent of damage in the maize seeds, they were spread on the glass plate supported on a stand in a dark room. The cracks were observed by converging light on glass plate through a microscopic lamp. By this way, cracked and intact seeds were separated from each variety of maize and percentage frequency of cracked and intact seeds was calculated.

6.2.3 PERCENT SEED GERMINATION

For checking the percent seed germination, 400 seeds of each variety of maize and paddy were taken. In maize, first intact and cracked seeds were isolated. All the petridishes lined with three layered moistened blotting papers were sterilized in autoclave at 15 lbs pressure for 10–15 min. After sterilization, these petridishes were transferred into inoculation chamber.

6.2.3.1 PERCENT SEED GERMINATION OF UNTREATED SEEDS

After transfer of petridishes in the inoculation chamber, untreated seeds were placed into petridishes (25 seeds/petridish in case of paddy, 10 seeds/petridish in case of maize) and their germination was observed at room temperature till 7 days under sterilized conditions. Their coleoptile and shoot length were noted everyday. Mean value of coleoptile length and shoot length was calculated along with percentage seed germination.

6.2.4 ISOLATION OF FUNGI

Several workers such as Christensen (1957), Neergaard and Saad (1962), and Christensen and Lopez (1962) have clearly pointed out that no single method by itself was adequate enough for the isolation of all the myco-flora that might be present in or on the seeds. In the present investigation, various techniques were employed for the isolation of the mycoflora of maize and paddy seeds.

6.2.4.1   STANDARD BLOTTER METHOD

The technique described by Neergaard and Saad (1962) is the standard blotter technique. In the present investigation, modified blotter method (Limonard, 1966) was employed. Under this technique, 400 seeds of each variety were tested in replicates of 25 seeds/petridish (paddy), 10 seeds/petridish (maize) of 9 cm diameter. Before transplanting the seeds in the petridishes, petridishes lined with three layers of blotting paper and well soaked in distilled water were sterilized in autoclave at 15 lbs pressure for 10–15 min.

6.2.4.2   ISOLATION OF MYCOFLORA PRESENT ON THE SEED SURFACE

Sterilized petridishes were then transferred into inoculation chamber for plating 10 seeds in each petridish. By this way, 40 petridishes were prepared for the study of the fungal flora present on the seed surface. All such petridishes were incubated at 22–25°C temperature for 7 days in alternating cycle of 12 h of darkness and 12 h of light (white fluorescent tube). Light was supplied by two tubes hanging horizontally, 20 cm apart from each other and 40 cm apart from the petridishes.

6.2.4.3   ISOLATION OF INTERNAL SEED MYCOFLORA

In another set of experiment, all the seeds were surface sterilized by 0.1% aqueous solution of mercuric chloride for 1 min. Then the seeds were thoroughly washed with sterilized distilled water and finally plated in the petridishes. All the procedures and conditions were kept same for both sets of experiments. After incubation of 7 days, colonies of fungi were counted and identified.

6.2.5 PERCENT FREQUENCY OF FUNGI

For determining the frequency of each fungus, colonies of fungi developed on the seeds were counted and the percentage of each fungus was calculated according to following formula:
No. of colonies of a fungus in a petridish

$$\text{Fungal frequency} = \frac{\text{No. of colonies of a fungus in a petridish}}{\text{Total no. of colonies in a petridish}} \times 100$$

The mean value of percent frequency will be calculated and presented in the tables.

6.2.6 MAINTENANCE OF PURE CULTURE

In cases where colony failed to sporulate, it was carefully transferred to P.D.A. slants under aseptic conditions and incubated at different temperatures in continuous light, continuous darkness or alternate darkness or light. Many fungi when subjected to this treatment did sporulate.

6.2.7 EFFECT OF FUNGICIDES ON SEED GERMINATION AND SEED MYCOFLORA

In this study, fungicides belonging to different groups were used, which are as follows:

1. Organic sulphur compound Thiram (tetramethyl thiuram disulphide)
2. Systemic fungicide-Bavistin [2 (Methoxy-carbamoyl)-benzimidazole]

Seeds of maize (intact and cracked) and paddy were treated with Thiram (3 g/kg) and Bavistin (2 g/kg) for checking their effect on the percent seed germination. Seeds to be treated with fungicides were transferred into 150 mL flasks containing Thiram and Bavistin powder with abovementioned rates. The flasks were shaken for 20 min on the mechanical shaker. The seeds after fungicidal treatment were removed from the flasks and 10 seeds were plated in each sterilized petridish fitted with three layered moistened blotting paper. The germination of these seeds was observed till 7 days at room temperature and coleoptile and shoot length were noted, seed mycoflora also checked on P. D. A.

6.3   EXPERIMENTAL RESULTS

6.3.2   PERCENT SEED GERMINATION IN DIFFERENT VARIETIES OF PADDY

It is clear from Table 6.1 that no germination occurred in the seeds of maize varieties, Dhan Pant, NDR, and Kranti up to 48 h. The germination started after 72 h. It was 46.60%, 76.60%, and 66.60% in Dhan Pant, NDR, and Kranti, respectively, that increased with time. It was 50.00% in Dhan Pant after 96 h which remains constant till 168 h. In NDR and Kranti, it was 80.00% and 90.00% after 96 h that remains constant in NDR till 168 h, but increased in Kranti and reached upto 93.30%.

TABLE 6.1   Showing Percent Seed Germination in Paddy Variety Dhan Pant, NDR, and Kranti.

Duration in hours	% Seed germination Dhan Pant	% NDR	% Kranti
24	–	–	–
48	–	–	–
72	46.60	76.60	66.60
96	50.00	80.00	90.00
120	50.00	80.00	93.00
144	50.00	80.00	93.30
168	50.00	80.00	93.30

6.3.2.1   FUNGI ISOLATED FROM UNSTERILIZED AND SURFACE STERILIZED SEEDS OF DIFFERENT VARIETIES OF PADDY BY BLOTTER METHOD

It is clear from Table 6.2 that large number of fungi were isolated from different varieties of paddy (Dhan Pant, NDR, and Kranti) by using blotter method (Table 6.4). Total 26 genera were isolated from paddy seeds. These

26 genera were *Absidia* sp., *Acremonium* sp., *Alternaria* sp., *Aspergillus* sp., *Bipolaris* sp., *Cercospora* sp., *Chaetomium* sp. *Cladosporium* sp., *Curvularia* sp., *Diplodia* sp., *Drechslera* sp., *Epicoccum* sp., *Fusarium* sp., *Geotrichum* sp., *Microdochium* sp., *Mycogone* sp., *Nigrospora* sp., *Penicillium* sp., *Phoma* sp. *Pyrenochaeta* sp., *Pyricularia* sp., *Rhizoctonia* sp., *Rhizopus* sp., *Trichothecium* sp., *Trichoderma* sp., and *Verticillium* sp.

TABLE 6.2 Fungi Isloated from Unsterilized and Surface Sterilized Seeds of Different Varieties of Paddy by Using Blotter Technique.

S. no.	Fungi isolated	Unsterilized			Surface Sterilized		
		Dhan Pant	Narendra	Kranti	Dhan Pant	Narendra	Kranti
1	*Absidia corymbifera*	+	-	+	+	-	+
2	*Acremonium* sp.	-	-	+	-	-	+
3	*Alternaria alternata*	+	+	+	+	+	+
4	*A. longipes*	+	-	+	-	-
5	*A. padwickii*	+	+	+	+	+	+
6	*A. tenuis*	+	+	+	+	+	-
7	*Aspergillus candidus*	+	+	-	+	-	-
8	*A. flavus*	+	+	-	+	-	-
9	*A. fumigatus*	+	-	-	-	-	-
10	*A. glaucus*	+	+	-	+	-	-
11	*A. humicola*	+	-	-	-	-	-
12	*A. luchuensis*	-	+	-	-	+	-
13	*A. minutes*	+	+	-	+	+	-
14	*A. nidulans*	+	+	-	+	+	-
15	*A. niger*	+	+	+	+	+	+
16	*A. parasiticus*	+	-	-	+	-	-
17	*A. sydowi*	+	-	-	+	-	-
18	*A. tamarii*	+	+	-	+	+	-
19	*A. versicolor*	+	+	+	+	+	+
20	*A. wentii*	+	-	-	+	-	-
21	*Bipolaris halodes*	-	+	-	-	+	-
22	*B. oryzae*	+	+	+	+	+	+
23	*Cercospora oryzae*	-	+	+	-	+	+
24	*Chaetomium* sp.	+	+	+	+	+	+
25	*Cladosporium* sp.	-	-	+	-	-	-
26	*Curvularia affinis*	+	+	-	+	+	-
27	*C. lunata*	+	+	+	+	+	+
28	*C. pallescence*	+	+	-	+	+	-

TABLE 6.2   *(Continued)*

S. no.	Fungi isolated	Unsterilized			Surface Sterilized		
		Dhan Pant	Narendra	Kranti	Dhan Pant	Narendra	Kranti
29	*C. verruculosa*	-	+	-	-	+	-
30	*Diplodia* sp.	-	-	+	-	-	+
31	*Drechslera oryzae*	-	+	+	-	+	+
32	*Epicoccum purpurescence*	+	+	+	+	+	+
33	*Fusarium equisetii*	+	+	-	+	+	-
34	*F. graminearum*	+	+	+	+	+	+
35	*F. longipes*	+	+	-	+	-	-
36	*F. moniliforme*	+	+	+	+	+	+
37	*F. scirpi*	+	-	-	-	-	-
38	*F.semitectum*	+	-	-	+	-	-
39	*Geotrichum candidum*	-	-	+	-	-	-
40	*Microdochium oryzae*	+	+	+	-	+	+
41	*Mycogone nigra*	-	-	+	-	-	+
42	*Nigrospora oryzae*	-	-	+	-	-	+
43	*N. sphaerica*	-	-	+	-	-	+
44	*N. padwickii*	+	-	+	+	-	+
45	*Penicillium citrioviride*	+	+	-	+	+	-
46	*P. crysogenum*	+	-	-	-	-	-
47	*P.nigricans*	+	-	+	-	-	+
48	*Phoma* sp.	-	+	+	-	+	+
49	*Pyrenochaeta oryzae*	-	-	+	-	-	+
50	*P. terrestris*	+	-	+	+	-	-
51	*Pyricularia oryzae*	+	+	+	+	+	+
52	*Rhizoctonia bataticola*	+	-	-	+	-	-
53	*R. oryzae*	+	+	-	+	-	-
54	*R. solani*	+	+	+	+	+	-
55	*Rhizopus arrhizus*	+	-	-	+	-	-
56	*R. oryzae*	+	-	-	+	-	-
57	*R. stolonifer*	+	-	-	+	-	-
58	*Trichothecium* sp.	-	+	+	-	+	+
59	*Trichoderma viride*	-	+	+	-	+	+
60	*Verticillium cinnabarium*	-	+	+	-	-	+

Out of 26 genera following 60 species of fungi were isolated in which *Absidia corymbifera, Acremonium* sp., *Alternaria alternata, A. longipes, A. padwickii, A. tenuis, Aspergillus candidus. A. flavus, A. glaucus. A. luchuensis, A. minutus, A. nidulans, A. niger, A. parasiticus, A. sydowi, A. tamarii, A. versicolor, A. wentii, Bipolaris halodes, B. oryzae, Cercospora oryzae, Chaetomium* sp., *Cladosporium* sp., *Curvularia affinis, C. lunata, C. pallescence, C. verruculosa, Diplodia* sp., *Drechslera oryzae, Epicoccum prupurescence, Fusarium equiseti, F. graminearum, F. longipes, F. moniliforme, F. semitectum, Geotrichum candidum, Microdochium oryzae, Mycogone nigra, Nigrospora oryzae, N. sphaerica, N. padwickii, Penicillium citrioviride, P. nigricans, Phoma* sp. *Pyrenochaeta oryzae, P. terrestris, Pyricularia oryzae, Rhizoctonia bataticola, R. oryzae, R. solani, Rhizopus arrhizus, R. oryzae, R. stolnifer, Trichothecium* sp., *Trichoderma viride and Verticillium cinnabarium* were isolated from both unsterilized and surface sterilized seeds while *Aspergillus fumigatus. A. humicola, Fusarium scirpi, Penicillium chrysogenum* were isolated only from unsterilized paddy seeds.

Out of 60 species, total 43 species were obtained from usnterilized seeds of veriety Dhan Pant (DP) in which *Absidia corymbifera, Alternaria alternata, A. longipes, A. padwickii, A. tenuis, Aspergillus candidus, A. flavus, A. glaucus A. minutus, A. nidulans A. niger A. parasiticus, A. sydowi, A. tamarii, A. versicolor. A. wentii, Bipolaris oryzae, Chaetomium* sp., *Curvularia affinis, C. lunata, C. pallescence, Epicoccum purpurescence, Fusarium equiseti, F. graminearum, F. longipes,.F. moniliforme, F. semitectum, Microdochium oryzae, Nigrospora padwickii, Pencillium citrioviride, Pyrenochaeta terrestris, Pyricularia oryzae, Rhizoctonia bataticola, R. oryzae, R. solani, Rhizopus arrhizus, R. oryzae, R. stolonifer* were isolated from both unsterilized and surface sterilized seeds while *Aspergillus fumigatus, A. humicola, Fusarium scirpi, Microdochium oryzae, Penicillium chrysogenum, P. nigricans* were obtained from unsterilized condition only.

In paddy variety, NDR out of 60 species, 35 species of fungi were obtained in which *Alternaria alternata, A. padwickii, A. tenuis, A. luchuensis, A. minutus, A. nidulans, A. niger, A. tamarii, A. versicolor, Bipolaris halodes, B. oryzae, Cercospora oryzae, Chaetomium* sp., *Curvularia affinis, C. lunata, C. pallescence, C. verruculosa, Drechslera oryzae, Epicoccum purpurescence, Fusarium equiseti, F. graminearum, F. longipes, F. moniliforme, Microdochium oryzae, Penicillium citrioviride, Phoma* sp., *Pyricularia oryzae, Rhizoctonia solani, Trichothecium* sp., *Trichoderma viride* were obtained from unsterilized and surface sterilized seed while *Aspergilllus condidus, A. flavus, A. glaucus, Fusarium longipes, Rhizoctonia oryzae* and *Verticillum cinnabarium* were obtained from only unsterilized seeds.

In paddy variety, Kranti out of 60 species, 32 species of fungi were isolated from unsterlized and surface sterilized seeds. These were *Absidia corymbifera, Acremonium* sp., *Alternaria alternata, A. padwickii, Aspergillus niger, A. versicolor, Bipolaris oryzae, Cercospora oryzae, Chaetomium* sp., *Curvularia lunata, Diplodia* sp., *Drechslera oryzae, Epicoccum purpurescence, Fusarium graminearum, F. moniliforme, Microdochium oryzae, Mycogone nigra, Nigrospora oryzae, N. sphaerica, N. padwickii, Phoma* sp., *Penicillium nigricans, P. yrenochaeta oryzae, Pyricularia oryzae, Trichothecium* sp., *Trichoderma viride, Verticillium cinnabarrium* while *Alternaria tenuis, Cladosporium* sp., *Geotrichum candidum, Pyrenochaeta terrestris, Rhizoctonia solani* were isolated from unsterilized seeds.

Alternaria alternata, A. padwickii, Aspergillus niger, A. versicolor, Bipolaris oryzae, Chaetomium sp., *Curvularia lunata, Epicoccum prupurescence, Fusarium graminearum, F. moniliforme, and Pyricularia oryzae* were obtained from all the three varieties of paddy.

6.3.2.2   FREQUENCY OF FUNGI ISOLATED FROM UNSTERLIZED AND SURFACE STERILIZED SEEDS OF DIFFERENT VARIETIES OF PADDY BY USING BLOTTER METHOD

Table 6.3 shows the percent frequency of different fungus species isolated from unsterilized and surface sterilized seeds of paddy varieties on blotter. The frequency of *Absidia corymbifera* in unsterilized seeds of varieties DP and Kranti was 20.80% and 10.00%, respectively; *Acremonium sp.,* in Kranti was 13.30; *Alternaria alternata* in DP, NDR, and Kranti was 10.00%, 15.00%, 9.60%, repectively; *Alternaria longipes* in DP was 12.00%; *Alternaria padwickii* in DP, NDR, and Kranti was 17.50%, 11.65%, and 10.90%, respectively; *Alternaria tenuis* in DP, NDR, Kranti was 15.00%, 14.44%, and 6.80%, respectively; *Aspergillus candidus* in DP and NDR was 9.305 and 8.50%, respectively; *Aspergillus flavus* in DP and NDR was 20.00% and 12.50%, respectively; *A. fumigatus* in DP was 6.60%; *A. glaucus* in DP and NDR was 8.00% and 4.12%, respectively; *A. humicola* in DP was 6.75%; *A. luchuensis* in NDR was 10.00%; *A. minutus* in DP and NDR was 14.00% and 8.95%, respectively; *A. nidulans* in DP and NDR was 8.00% and 8.00%, respectively; *A. niger* in DP, NDR, and Kranti was

15.00%, 6.10%, and 6.00%, respectively; *A. parasiticus* and *A. sydowi* in DP was 15.00% and 10.00%, respectively; *A. tamarii* in DP and NDR was 12.00% and 10.00%, respectively; *A. versicolor* in DP, NDR, and Kranti was 11.30%, 8.15%, and 6.22%, respectively; *A. wentii* in DP was 12.00%; *Bipolaris halodes* in NDR was 10.00%, *B. oryzae* in DP, NDR, and Kranti was 12.00%, 7.30%, and 7.10%, respectively; *Cercospora oryzae* in NDR and Kranti was 6.60% and 5.00, respectively; *Chaetomium* sp. in DP, NDR, and Kranti was 15.00%, 10.00%, and 9.00%, respectively; *Cladosporium sp.* in Kranti was 9.00%; Curvularia affinis in DP and NDR was 20.10% and 18.72%, respectively; *C. lunata* in DP, NDR, and Kranti was 16.00%, 12.35%, and 9.00%, respectively; *Curvularia pallescence* in DP and NDR was 15.00% and 13.30%, respectively; *C. verruculosa* in NDR was 13.30%; Diplodia sp in Kranti was 6.00%; *Drechslera oryzae* in NDR and Kranti was 10.00% and 8.00%, respectively; *Epicoccum purpurescence* in DP, NDR, and Krani was 21.00%, 15.00%, and 10.00%, respectively; *Fusarium equiseti* in DP and NDR was 11.05% and 10.00%, respectively; *F. graminearum* in DP, NDR, and Kranti was 20.00%, 11.40%, and 20.00%, respectively; *F. longipes* in DP and Kranti was 18.70% and 6.15%, respectively; *F. moniliforme* in DP, NDR, and Kranti was 15.98%, 15.00%, and 14.20%, respectively; *F. scirpi* in DP was 4.00%; *Fusarium semitectum* in DP was 10.00%; *Geotrichum candidum* in Kranti was 5.00%; *Microdochium oryzae* in DP, NDR, and Kranti was 8.00%, 15.00%, and 16.00%, respectively; *Mycogone nigra, Nigrospora oryzae, N. sphaerica* in Kranti was 8.00, 6.60, and 8.10%, respectively; *Nigrospora padwickii* in DP and Kranti was 10.00 and 10.00%, respectively; *Penicillium Citrioviride* in DP and NDR was 24.00 and 18.66%, respectively; *P. chrysogenum* in DP was 6.00%; *P. nigricans* in DP and Kranti was 8.55 and 8.00%, respectively; *Phoma* sp. in NDR and Kranti was 12.50% and 6.60%, respctively; *Pyrenochaeta oryzae* in Kranti was 10.00%; *P. terrestris* in DP and Kranti was 7.80% and 5.00%, respectively; *Pyricularia oryzae* in DP,NDR, and Kranti was 13.50%, 21.65%, and 18.00%, respectively; *Rhizoctonia bataticola* in DP was 20.00%; *R. oryzae* in DP and NDR was 15.55% and 15.00%, respectively; *R. solani* in DP, NDR, and Kranti was 20.00%, 20.00%, and 4.80%, respectively; *Rhizopus arrhizus, R. oryzae, R. stolonifer* in DP was 25.00%, 7.68%, and 22.50%, respectively; *Trichotehcium* sp., in NDR and Kranti was 5.00% and 5.15%, respectively; *Trichoderma viride* in NDR and Kranti was 6.60% and 8.00%, respectively; *Verticillum cinnabarium* in NDR and Kranti was 10.00% and 10.00%, respectively.

TABLE 6.3   Frequency of Fungi Isolated from Unsterilized and Surface Sterilized Seeds of Deferent Varieties of Paddy by Using Blotter Technique

S. no.	Fungi isolated	Unsterilized			Surface sterilized		
		Dhan Pant	Narendra	Kranti	Dhan Pant	Narendra	Kranti
1	*Absidia corymbifera*	20.80	-	10.00	18.10	-	6.70
2	*Acremonium sp.*	-	-	13.30	-	-	8.50
3	*Alternaria alternata*	10.00	15.00	9.60	7.50	9.00	9.50
4	*A. longipes*	12.00	-	-	4.78	-	-
5	*A. padwickii*	17.50	11.65	10.90	13.60	8.70	5.00
6	*A. tenuis*	15.00	14.44	6.80	10.00	8.00	-
7	*Aspergillus candidus*	9.30	8.50	-	5.03	-	-
8	*A. flavus*	20.00	12.50	-	10.00	-	-
9	*A. fumigatus*	6.60	-	-	-	-	-
10	*A. glaucus*	8.00	4.12	-	4.00	-	-
11	*A. humicola*	6.75	-	-	-	-	-
12	*A. luchuensis*	-	10.00	-	-	6.85	-
13	*A. minutes*	14.00	8.95	-	7.10	5.00	-
14	*A. nidulans*	8.00	8.00	-	4.60	6.60	-
15	*A. niger*	15.00	6.10	6.00	10.00	4.00	4.00
16	*A. parasiticus*	15.00	-	-	12.00	-	-
17	*A. sydowii*	10.00	-	-	8.00	-	-
18	*A. tamarii*	12.00	10.00	-	7.18	7.00	-
19	*A. versicolor*	11.30	8.15	6.22	10.00	7.00	4.00
20	*A. wentii*	12.00	-	-	10.00	-	-
21	*Bipolaris halodes*	-	10.00	-	-	6.60	-
22	*B. oryzae*	12.00	7.30	7.10	10.00	5.15	5.00
23	*Cercospora oryzae*	-	6.60	5.00	-	5.00	3.30
24	*Chaetomium sp.*	15.00	10.00	9.00	11.45	7.15	6.18
25	*Cladosporium sp.*	-	-	9.00	-	-	-
26	*Curvularia affinis*	20.10	18.72	-	14.16	10.76	-
27	*C. lunata*	16.00	12.35	9.00	15.00	10.00	7.18
28	*C. pallescence*	15.00	13.30	-	11.10	8.15	-
29	*C. verruculosa*	-	13.30	-	-	12.30	-
30	*Diplodia sp.*	-	-	6.00	-	-	3.50
31	*Drechslera oryzae*	-	10.00	8.00	-	5.18	4.00
32	*Epicoccum purpurescence*	21.00	15.00	10.00	16.50	8.00	6.60
33	*Fusarium equisetii*	11.05	10.00	-	7.68	7.16	-

TABLE 6.3 *(Continued)*

S. no.	Fungi isolated	Unsterilized			Surface sterilized		
		Dhan Pant	Narendra	Kranti	Dhan Pant	Narendra	Kranti
34	*F. graminearum*	20.00	11.40	20.00	8.00	5.00	7.85
35	*F. longipes*	18.70	6.15	-	12.50	-	-
36	*F. moniliforme*	15.98	15.00	14.20	10.00	11.14	10.00
37	*F. scirpi*	4.00	-	-	-	-	-
38	*F. semitectum*	10.00	-	-	7.80	-	-
39	*Geotrichum candidum*	-	-	5.00	-	-	-
40	*Microdochium oryzae*	8.00	15.00	16.00	-	10.00	8.05
41	*Mycogone nigra*	-	-	8.00	-	-	4.11
42	*Nigrospora oryzae*	-	-	6.60	-	-	7.50
43	*N. sphaerica*	-	-	8.10	-	-	7.00
44	*N. padwickii*	10.00	-	10.00	7.80	-	6.60
45	*Penicillium citrioviride*	24.00	18.66	-	15.00	11.00	-
46	*P. crysogenum*	6.00	-	-	-	-	-
47	*P. nigricans*	8.55	-	8.00	-	-	5.40
48	*Phoma sp.*	-	12.50	6.60	-	8.00	5.00
49	*Prenochaeta oryzae*	-	-	10.00	-	-	4.10
50	*P. terrestris*	7.80	-	5.00	5.00	-	-
51	*Pyricularia oryzae*	13.50	21.65	18.00	10.00	12.35	11.00
52	*Rhizoctonia bataticola*	20.00	-	-	15.00	-	-
53	*R. oryzae*	15.55	15.00	-	7.00	-	-
54	*R. solani*	20.00	20.00	4.80	15.10	10.00	-
55	*Rhizopus arrhizus*	25.00	-	-	5.00	-	-
56	*R. oryzae*	7.68	-	-	4.80	-	-
57	*R. stolonifer*	22.50	-	-	13.00	-	-
58	*Trichothecium sp.*	-	5.00	5.15	-	3.00	4.00
59	*Trichoderma viride*	-	6.60	8.00	-	5.70	6.00
60	*Verticillium cinnabarium*	-	10.00	10.10	-	-	6.60

The frequency of *Absidia corymbifera* in surface sterilized seeds of varieties DP and Kranti was 18.10% and 6.70%, respectively; *Acremonium sp* in Kranti was 8.50%; *Alternaria alternata* in DP, NDR, and Kranti was 7.50%, 9.00%, and 9.50%, respectively; *A. longipes* in DP was 4.78%; *A. padwickii* in DP, NDR, and Kranti was 13.60%, 8.70%, and 5.00%, respectively; *A. tenuis* in DP and NDR was 10.00 and 8.00%, respectively;

Aspergillus candidus, A. flavus, A. glaucus in DP was 5.03%, 10.00%, and 4.00%, respectively; *Aspergillus luchuensis* in NDR was 6.85%; *A. minutus* in DP and NDR was 7.10% and 5.00%, respectively; *A. nidulans* in DP and NDR was 4.60% and 6.60%, respectively; *A. niger* in DP, NDR and Kranti was 10.00%, 4.00%, and 4.00%, respectively; *A. parasiticus* and *A. sydowi* in DP was 12.00% and 8.00%, respectively; *A. tamarii* in DP and NDR was 7.18% and 7.00%, respectively; *A. versicolor* in DP, NDR and Kranti was 10.00%, 7.00%, and 4.00%, respectively; *A. wentii* in DP was 10.00%: *Bipolaris haldoes* in NDR was 6.60%; *B. oryzae* in DP, NDR and Kranti was 10.00%, 5.15%, and 5.00%, respectively; *Cercospora oryzae* in NDR and Kranti was 5.00% and 3.30%, respectively; *Chaetomium* sp., in DP, NDR and Kranti was 11.45%, 7.15%, and 6.18%, respectively; Curvularia affinis in DP and NDR was 14.16% and 10.76%, respectively; *C. lunata* in DP, NDR and Kranti was 15.00%, 10.00%, and 7.18%, respectively; *C. verruculosa* in NDR was 12.30%; *Diplodia sp* in Kranti was 3.50% *Drechslera oryzae* in NDR and Kranti was 5.18% and 4.00%, respectively; *Epicoccum purpurescens* in DP, NDR and Kranti was 16.50%, 8.00%, and 6.60%, respectively; *Fusarium equiseti* in DP and NDR was 7.68% and 7.16%, respectively; *F. graminearum* in DP, NDR and Kranti was 8.00%, 5.00%, and 7.85%, respectively; *F. longipes* in DP was 12.50%; *F. moniliforme* in DP, NDR and Kranti was 10.00%, 11.14%, and 10.00%, respectively; *F. semitectum* in DP was 7.80%, *Microdochium oryzae* in NDR and Kranti was 10.00% and 8.05%, respectively; *Mycogone nigra, Nigrospora oryzae, N. sphacrica* in Kranti was 4.11%, 7.50%, and 7.00%, respectively; *N. padwickii* in DP and Kranti was 7.80% and 6.60%, respectively; *Penicillium citrioviride* in DP and N.DR was 15.00% and 11.00%, respectively; *P. nigricans* in Kranti was 5.40%; *Phoma sp* in NDR and Kranti was 8.00% and 5.00%, respectively; *Pyrenochaeta oryzae* in Kranti was 4.10%; *P. terrestris* in DP was 5.00%; *Pyricularia oryzae* in DP, NDR and Kranti was 10.00%, 12.35%, and 11.00%, respectively; *Rhizoctonia bataticola* and *R. oryzae* in DP was 15.00% and 7.00%, respectively; *R. solani* in DP and NDR was 15.10% and 10.00%, respectively; *Rhizopus arrhizus, R. oryzae and R. stolonifer* in DP was 5.00%, 4.80%, and 13.00%, respectively; *Trichothecium sp* in NDR and Kranti was 3.00% and 4.00%, respectively; *Trichoderma viride* in NDR and Kranti was 5.70% and 6.00%, respectively; *Trichothecium* sp. In NDR and Kranti was 3.00% and 4.00%, respectively; *Verticillium cinnabarium* in Kranti was 6.60%.

6.4 EFFECT OF BAVISTIN ON PERCENT SEED GERMINATION OF DIFFERENT VARIETIES OF PADDY

It is clear from Table 6.4 that when the seeds were treated with Bevistin, there was no germination occuring till 48 h in all the three varieties. It started after 72 h, it was 60.00%, 79.00%, and 70.80% in DP, NDR, and Kranti, respectively, which gradually increased with the time. In DP, it was 80.60% at 120 h and 90.30% after 144 h and after that became constant. Same way in NDR, it was 80.00% after 96 h and 94.00% after 120 h and 98.00% after 144 h after that it remains constant while in kranti it was 70.00% after 72 h that gradually increased with the time it was 83.00%, 90.00%, 93.30%, and 95.00% after 96, 120, 144, and168 h, respectively.

TABLE 6.4   Effect of Bavistin on Percent Seed Germination of Different Varieties of Paddy.

Duration in hours	% Seed germination		
	Dhan Pant	**NDR**	**Kranti**
24	–	–	–
48	–	–	–
72	60.00	79.00	70.80
96	80.60	80.00	83.00
120	90.30	94.00	90.00
144	90.30	98.00	93.30
168	90.30	98.00	95.00

6.5 EFFECT OF THIRAM ON PERCENT SEED GERMINATION OF DIFFERENT VARIETIES OF PADDY

It is clear from Table 6.5 that when the seeds were treated with thiram, there occurred no germination till 48 h after that it increased gradually in DP. It was 72.00%, 83.00%, 90.50%, and 94.00% after 72, 96, 120, and 144 h, respectively, that remains same after 168 h. While in NDR it was 88.80% after 72 h and 93.00% after 96 hours after that it became constant (100.00%). Same way in Kranti, germination started after 72.00% h and it was 90.00%. After that it became constant (100.00%).

TABLE 6.5   Effect of Thiram on Percent Seed Germination of Different Varieties of Paddy.

Duration in hours	% Seed germination		
	Dhan Pant	NDR	Kranti
24	–	–	–
48	–	–	–
72	72.00	88.80	90.00
96	83.00	93.00	100.00
120	90.50	100.00	100.00
144	94.00	100.00	100.00
168	94.00	100.00	100.00

6.6 DISCUSSION

India is a vast country with agriculture-based economy, where the importance of seed pathology to agriculture has been realized but still not fully integrated with the country's efforts of the production and supply of high quality seeds. The seed mycoflora affects adversely germination and the health of the emerging seedlings. Seeds have also been recognized as an efficient vehicle for transport of the pathogens from one area to the other. Therefore, the importance of healthy seeds or seeds free from serious pathogens could not be overemphasized. Besides this, release of high yielding varieties in recent years has greatly altered the disease situation in these crops. Little is known about the seed pathology of newly developed varieties, it was, therefore, considered necessary to study certain aspects of seed mycoflora.

It is evident from Table 6.1 that stored paddy variety DP showed poor germination in comparison to remaining two fresh varieties because during storage large number of storage fungi viz. *Alternaria, Aspergillus, Curvularia, Fusarium, Mucor, Penicillium, Rhizopus,* and *Trichoderma,* etc. attack them. These fungi are capable in producing mycotoxins which affect the metabolites of grains, so gradual loss in viability of seeds occur during storage (Singh and Tyagi, 1984; Misra and Misra, 1991; Kim and Lee, 1992; Krishan Rao, 1992; Moreno et al., 2000; Orsi et al., 2000; Garcia et al., 2002; Rath and Padhi, 2004; Prasad, 2004, Surekha et. al., 2011).

It is evident from Tables 6.2 and 6.3 that when the mycoflora of three varieties of paddy was tested by blotter method, it showed variation quantitavely and qualitatively. *Alternaria alternata, A. padwickii, A. tenuis,*

Aspergillus niger, A. versicolor, Bipolaris oryzae, Chaetomium sp, Curvularia lunata, Epicoccum purpurescens, Fusarium graminearum, F. moniliforme, Microdochium oryzae, Rhizoctonia solani and *Pyricularia oryzae* were isolated from all the three varieties of paddy, that is, DP, Narendra (NDR), and Kranti in both unsterilized conditions. *Alternaria longipes, Aspergillus fumigatus, A. humicola, A. parasiticus, A. sydowi, A. wentii, Fusarium scirpi, F. semitectum, Penicillium chrysogenum, Rhizoctonia bataticola, Rhizopus arrhizus, R. oryzae, and Rhizopus stolonifer* were isolated from both unsterilized and surface sterilized seeds of paddy variety DP; *Aspergillus candidus, A. flavus, A. glaucus, A. minutus, A. nidulans, A. tamarii, Curvularia affinis, C. pallescens, Fusarium equiseti, F. longipes,* and *Penicillium citrioviride* were isolated from both unsterilized and surface sterilized seeds of paddy varieties DP and NDR. *Acremonium sp, Cladosporium* sp., *Diplodia sp, Geotrichum candidum, Mycogone nigra, Nigrospora oryzae, N. spaerica,* and *Pyrenochaeta oryzae* were isolated from both unsterilized and surface sterilized seeds of paddy varieties Kranti; *Cercospora oryzae Drechslera oryzae, Phoma sp, Trichoderma viride, Trichothecium roseum* and *Verticillium cinnabarium* were isolated from both unsterilzed and surface sterilized seeds of paddy varieties NDR and Kranti; (*Aspergillus candidus A. flavus, A. glaucus, Fusarium longipes* and *Rhizoctonia oryzae* were isolated from usnterilized seeds of paddy varieties DP and NDR and surface sterilized seeds of variety DP;) *Absidia corymbifera, Nigrospora padwickii, Penicillium nigricans* and *P. terrestris* were isolated from unsterlized seeds of paddy variety DP and Kranti and *A. luchuensis, Bipolaris halodes* and *Curvularia verculosa* were isolated from unsterilized seeds of paddy varieties NDR.

From Tables 6.4 and 6.5, it is clear that when the seeds were surface sterilized with 0.1% mercuric chloride, it reduced the number and frequency of fungi. By blotter method, 60 fungi belonging to 26 genera were isolated in unsterilized condition in which *Aspergillus fumigatus, A. humicola, Cladosporium sp, Fusarium scirpi, Geotrichum candidum* and *Pencillium chrysogenum* were not isolated from surface sterilized seeds of paddy.

Alternaria alternata, A. padwickii, Aspergillus candidus, A. flavus, A. fumigatus, A. niger, A. versicolor, Bipolaris oryzae, Cercospora oryzae, Chaetomium globosum, Cladosporium cladospoiriodes, Curvularia sp, C. affinis, C. lunata, Drechslera oryzae, Epicoccum sp, Fusarium moniliforme, (G. fujikuroi), F. semitectum, Mucor sp., *Nigrospora oryzae, Phoma* sp., *Pyricularia oryzae, Rhizoctonia oryzae, R. solani Rhizopus stolonifer* were isolated in this study have been reported by Supriaman and

Palmer (1979); Saponaro et al. (1986), Parate and Lanjewar (1987); Zakeri and Zad (1987); Grewal and Kang (1988); Vallejos and Mattos (1990); Jayanandarjah and Seneviratne (1991); Bokhary (1991); Ahmad and Raza (1992); Castano et al. (1992); Kim (1992); Kim and Lee (1992); Kim and Yu (1992); Wu et al. (1992); Ilyas and Javaid (1995); Khan et al. (1999); Zad and Khosravi (2000); Franco et al. (2001); Khalid et al. (2001) on paddy seeds from all over the world.

Absida corymbifera, Alternaria sp., *A. alternata, A. padwickii, Aspergillus sp, A. candidus, A. glaucus, A. flavus Bipolaris oryzae, Chetomium sp, Chaetomium globosum, Cladosporium sp, Curvularia affinis, C. lunata, C. pallesecence, C. verruculosa, Drechslera oryzae, Epicoccum sp, E. purpurescence, Fusarium equseti, Fusarium moniliforme, F. oxysporum, F. semitectum, Mycogone nigra, Nigrospora oryzae, Penicillum sp, P. citrioviride, P. purpurescence, Phoma sp, Pyrenochaeta sp, Pyricularia oryzae, Rhizoctonia solani, Rhizopus sp,* and *Sarocladium oryzae* were isolated in this study have been reported by Mittal and Sharma (1978); Singh and Shukla (1980); Mohan and Subramanian (1981); Singh and Raju (1981); Verma and Aulakh (1983); Jha and Prasad (1984); Jayaweera et al. (1988); Sundar and Satyavir (1988); Waghray et al. (1988); Jayaraman and Kalyansundaram (1989); Chakrabarti and Rao (1990); Maiti et al. (1991); Mia and Safeculla (1992)0; Roy (1992); Rangnathaiah and Safeeulla (1994); Vyas (1994); Tiwari (1995); Babu and Lokesh (1996); Rathaiah (1997); Kalyansundaram (1998); Sunder and Satyavir (1998); Pandey et al. (2000); Lalitha and Raveesha (2003) on paddy seeds from India.

It is clear from the results that fresh varieties showed higher germination and less fungal frequency while stored varieties showed low germination and high fungal frequency because during storage stored fungi attacked on the seeds and replace the field fungi. These stored fungi are mycotoxin in nature which affects the metabolic activities, discoloration of grains, loss in grain weight.

It is also evident that when the seeds of all the three varieties of paddy and maize were treated with Bavistin and Thiram, there was an enhancement in percent germination of seeds of all the varieties. Thiram proved to be more effective than Bavistin, Variation in the growth of the seedlings raised from untreated and treated seeds with Bavistin and Thiram is evident from Tables 6.4 and 6.5.

In paddy, same results were obtained, untreated seeds showed poor germination in comparision to treated seeds. Among the three varieties of paddy, variety DP (stored) showed poorest germination and variety Kranti showed best germination in both conditions (untreated and treated with fungicides). These results are in accordance with the finding of Wal and Vander, 1978; Prakash and Kauraw, 1983; Pathak, 1989; Kapkoti and Pandey, 1990; Vishwanathan and Narayananswamy, 1993; Milivojevic et al., 1996; Sharma and Chahal, 1996; Puzari et al., 1998; Sundar and Satyavir, 1998; Scholz et al., 1999; Akter et al., 2001; Carmona et al., 2001; Goulart and Fialho, 2001; Lu et al., 2001; Parisi et al., 2001; Bohra et al., 2001; Anwar et al., 2002; Singh et al., 2002; Thapak and Thrimurty, 2002.

KEYWORDS

- mycoflora
- paddy
- fungi
- production
- mycotoxins

REFERENCES

Agarwal, P. C.; Mortensen, C. N.; Mathur, S. B. Seed Born Diseases Seed Health Testing of Rice. *Tech. Bull. No.-3, Phytopathol. Papers* **1989,** (33), 28–29.

Agarwal, V. K. Techniques for the Detection of Seed Borne Fungi. *Seed Res.* **1976,** *4,* 24–31.

Agarwal, V. K.; Sinclair, J. B. *Principles of Seed Pathology* CRC Press, Inc.: Boca Raton, Florida, USA, 1987, pp 29–76.

Agarwal, V. K.; Srivastava, A. K. A Simple Technique for Routine Examination of Rice Seeds Lots for Rice Bunt. *Seed Technol.* **1981,** *11,* 1.

Agarwal, V. K.; Srivastava, A. K. 'Naoh Seed Soak' Method for Routine Examination of Rice Seeds for Rice Bunt. *Seed Res.* **1985,** 13, 159–161.

Ahonsi, M. O.; Adeoti; A. A. Flase Smut on Upland Rice in 8 Rice Producing Locations of Edo State, Nigeria. *J. Sustainable Agri* **2002,** *20* (3), 81–94.

Ahmad, M. I.; Raza, T. Association of *Fusarium moniliforme* Sheld with Rice Seeds and Subsequent Infection in Pakistan. *Rev. Pl. Pathol.* **1992**, *71* (8), 568.

Akter, S.; Mian, M. S.; Mia, M. A. T. Chemical Control of Sheath Blight Disease (Rhizoctonia-Solani) of Rice. *Bangladesh J. Pl. Pathol.* **2001**, *17* (1–2), 35–38.

Alam, S.; Seth, R. K.; Shukla, D. N. Screening of Some Fungi Isolation of Rice Cultivars in Different Site of Allahabad, Varanasi, Mirzapur, Jaunpur and Chandauli District in U.P. *J. Agri. Vet. Sci.* **2014**, *7* (8), 67–71.

Aly, H. Y. Control of Storage Molds of Moist Grains During Storage. *Rev. Pl. Pathol.* **1992**, *71* (9), 654.

Anonymous. Recommended Practices for the Prevention of Mycotoxins in Food, Feed and Their Products Prepared by F.A.O. and U.N.E.P. No 10, 1979, 1–71.

Anwar, A.; Bhat, G. N.; Singhara, G. S. Management of Sheath Blight and Blast in Rice Through Seed Treatment. *Ann. Pl. Protect. Sci.* **2002**, *10* (2), 285–287.

Arase, S.; Honda, Y.; Nozu, M.; Nishimura, S. Susceptibility Including Toxin Produced by *Pyricularia Oryzae. Rev. Pl. Pathol.* **1999**, *78* (2), 162.

Aulakh, K. S.; Mathur, S. B.; Neergaard, P. Seed Health Testing of Rice and Comparision of Field Incidence with Laboratory Counts of Drechslera Oryzae and *Pyricularia oryzae. Seed Sci. Technol.* **1974b**, *2*, 393–398.

Babu, H. N. R.; Lokesh, S. Seed Mycoflora of Some Paddy (*Oryza sativa* L.) Varieties in Karnataka (India). *Pl. Dis. Res.* **1996**, *11* (1), 49–51.

Bhattacharya, K.; Rhah, S. Deteriorative Changes of Maize Groundnut and Soyabean Seeds by Fungi in Storage. *Mycopathologia 155* (3), 135–141.

Bialooki, P.; Piotrowska, J. Rice Smut (*Tilletia barclayana*) in Poland. *Ochrona Roslin* **1999**, *43* (2), 12–13.

Bilgrami, K. S.; Chaudhary, A. K. Competing Mycoflora with *Aspergillus flavus* in Kernels of Rabi and Kharif Maize Crops of Bhagalpur. *Rev. Pl Pathol.* **1992**, *71* (3), 188.

Biswas, A. Evaluation of New Fungicidal Formulation for Sheath Blight Control. *J. Mycopathol. Res.* **2002**, *41* (1), 9–11.

Biswas, A. Kernel Smut Disease of Rice Current Status and Future Challenges. *Environment Ecol.* **2003**, *21* (2), 336–351.

Bock, C. H.; Jeger, M. J.; Mughogho, L. K.; Cardwell, K. F.; Mtisi, E. Effect of Dew Point Temperature and Conidium Age on Germination, Germ Tube Growth and Infection of Maize and Sorghum by *Peronosclerospora sorghii. Mycological Res.* **1999**, *103* (7), 859–864.

Bockhary, H. A. Seed Borne Fungi of Rice (*Oryza sativa* L.) from Saudi Arabia. *Zeitschrif fur pflanzentrankheiten undflanzenschultz* **1991**, *98* (3), 287–292.

Bohra, B.; Rathore, R. S.; Jain, M. L. Management of Fusarium Stalk Rot of Maize Caused by *Fusarium moniliforme*, Shelden. *J. Myco and Pl. Pathol.* **2001**, *31* (2), 245–247.

Brekalo, J.; Palaversic, B.; Rojc, M. Monitoring the Occurrence and Severity of Maize Diseases in Croatia from 1985-1989. *Rev. Pl. Pathol.* **1992**, *71* (5), 330.

Butt, A. R.; Yaseen, S. I.; Javaid, A. Seed Borne Mycoflora of Stored Rice Grains and Its Chemicals. Institute of Plant Pathology, University of the Punjab, Lahore, Pakistan. *J. Anim. Plant Sci.* **2011**, *21* (2), 193.

Carmona, M. A.; Barreto, D. E.; Reis, E. M. Effect of Iprodione and its Mixture with Thiram and Triticonazole to Control *Drechslera teres* in Barley Seeds. *Fitopatologia Brasileira* **2001**, *26* (2), 176–179.

Cartwright, R. D.; Parsons, C. E.; Ross, W. J.; Lee, F. N.; Templeton, G. E. Rice Diseases Monitoring and on Farm Rice Cultivar Evaluation in Arkansas. *Rev. Pl. Pathol.* **1997,** *78* (5), 433.

Castano, J.; Klap, K.; Zaini, Z. Etiology of Grain Discoloration in Upland Rice in West Sumatra. *Rev. Pl. Pathol.* **1992,** *71* (10), 745.

Cedeno, L.; Nass, H.; Carrero, C.; Cardona, R.; Rodriguez, H.; Aleman, L. *Rhizoctonia oryzae sativae,* Causal Agent of Aggregate Sheath Spot on Rice in Venezuela. *Rev. Pl. Pathol.* **1999,** *78* (4), 355.

Chahal, K. S.; Sokhi, S. S.; Rattan, G. S. Investigations on Sheath Blight of Rice in Punjab. *Indian Phytopathol.* **2003,** *56* (1), 22–26.

Chakrabarti, D. K.; Biswas, S. Estimation of Yield Loss in Rice Affected by Sheath Rot. Pl. Dis. Reptr. **1978,** *62,* 226–227.

Chakrabarti, S. K.; Prasad Rao, R. D. V. J. Elimination of Seed Borne Pathogens from Rice Seeds. *Indian Phytopathol.* **1990,** *43,* 250.

Chandra, D. On Farm Evaluation of Rice Varieties under Rainfed Lowland Situation. *Indian Farming* **1996,** *3,*12.

Cheeran, A.; Raj, J. S. Effect of Seed Treatment on the Germination of Rice Seeds Infected by *Trichoconis padwickii* Ganguly. *Agri. Res. J. Kerela* **1966,** *4,* 57–59.

Christensen, C. M. Deterioration of Stored Grain by Fungi. *Bot. Rev.* **1957,** *23,* 108–134.

Christensen, C. M.; Lopez, L. C. Damage by Fungi to Stored Grains in Mexico. *Agricolas, S.A.G. Mexico* **1962,** *44,* 29.

Christensen, C. M.; Mirocha, C. J. Relation of Relative Humidity to the Invasion of Rough Rice by *Aspergillus parasiticus. Phytopathology* **1976,** *66,* 204–205.

Chung, H. W.; Chung, H. S.; Chung, B. J. Studies on Pathogenecity of Wheat Scab Fungus (*Gibberella zeae*) to Various Crop Seedlings. *J. Pl. Protection Korea* **1964,** *3,* 21–25.

Das, S. R.; Sahu, S. K. Host Range of *Sphaerulina oryzina,* The Incitant of Narrow Brown Leaf Spot of Rice. *Indian Phytopathol.* **1998,** *51* (2), 196.

De Tempe, J. Routine Methods for Determining the Health Condition of Seeds in Seed Testing Station. *Proc. Int. Seed Testing Assoc.* **1961,** *26,* 27–60.

De Tempe, J. Inspection of Seeds For Adhering Pathogenic Elements. *Proc. Int. Seed Testing Assoc.* **1963,** *28,* 153–165.

De Tempe, J. Percent Development in Seed Health Testing. *Proc. Int. Seed. Testing Assoc.* **1964,** *29,* 479–486.

Dharmavir. Study of Some Problems Associated with Post Harvest Fungal Spoilage of Seeds and Grains. *Current Trends Pl. Pathol.* **1974,** 221–226.

Divakar, A. P.; Prasad, B. K. Effect of Storage Fungi of Seeds on the Seedlings Disease of Paddy. *J. Myco Pl. Pathol.* **1997,** *27* (2), 184–187.

Duraiswamy, V. S.; Mariappan, V. Biochemical Properties of Discolored Rice Grains. *Int. Rice Res.* **1983b,** 3.

Fedovora, R. N. A Rapid Method of Detecting Spores on the Surface of Seeds and Plants. *Mikologiya i Fitopatologia* **1987,** *21,* 563–565.

Franco, D. F.; Ribeiro, A. S.; Nunes, C. D.; Ferreira, E. Fungi Associated with Seeds of Irrigated Rice in Rio Grande Do Sul. *Revista Brasileira de Agrociencia* **2001,** *7* (3), 235–236.

Fukaya, T.; Hosaka, M.; Iitomi, A.; Wakahata, M.; Odashima, S.; Sthibata, S.; Kutsauzawa, T.; Maisawa, S.; Shonai, R. Occurrence and Cause of Rice Blast Diseases of Seedlings in Akita Prefecture. 52 Annual Reports of the Society of Plant Protection of North Japan, 2001, 11–13.

Gangopadhyay, S.; Chkrabarti, N. K. Sheath Blight of Rice. *Rev. Pl. Pathol.* **1982**, *61* (10), 451–460.

Garcia, M. J. DEM.; Biaggioni, M. A. M.; Ferreira, W. A.; Kohara, E. Y.; Almedia, A. M. D. E. Succession of Fungal Species in Maize Stored in an Aerated System. *Revista Brasileiara* **2002**, *27* (2), 14–22.

Gaur, A.; Dev, U. Detection Techniques for Seed Borne Fungi, Bacteria and Viruses. *Indian Phytopathol.* **1996**, *49* (4), 319–328.

Goulart, A. C. P.; Filho, W. F. B. Corn Seed Treatment with Fungicides for Pathogen Control. *Summa Phytopathologica* **2001**, *27* (4), 414–420.

Grewal, S. K.; Kang, M. S. Seasonal Carryover of Fusarium Maniliforme in Field. Causal Organism of Sheath Rot of Rice. *Phytopathologia Mediterrarla.* **1988**, *27* (1), 36–37.

Gourangkar. Upland Rice Ecosystem in India. *Indian Farming* **2002**, *10*, 13–17.

Govindu, C. H. Occurrence of *Ephelis* on Rice Variety Ir-8 and Cotton Grass in India. *Pl. Dis. Reptr.* **1969**, *53*, 360.

Hajra, K. K.; Ganguly, L. K.; Khatua, D. C. Backanae Disease of Rice in West Bengal. *J Mycopathol. Res.* **1994**, *32* (2), 95–99.

Haque, A. H. M. M.; Akon, M. A. H.; Islam, M. A.; Khalequzzaman, K. M.; Ali, M. A. Study of Seed Health, Germination and Seedling Vigour of Farmers. Seedling Vigour of Farmers Produced Rice Seeds. *Int. J. Sustain. Crop Prod.* **2007**, *2* (4), 34–39.

Hararika, D. K.; Phookan, A. K. Impact of *Sarocladium oryzae* and their Metabolities on Incidence of Sheath Rot of Rice. *J. Myco. Pl. Pathol.* **1999**, *29* (2), 268.

Hegde, N. G. Sustainable Agriculture for Food Security. *Indian Farming* **2000**, *3*, 4–11.

Hegde, Y.; Anahosur, K. H. Survival, Perpetuation and Life Cycle of *Claviceps Oryzae. Sativae* Causal Agent of False Smut of Rice in Karnataka. *Indian Phytopathol.* **2000**, *53* (1), 61–64.

Hernandez Livera, A.; Villasenor, Mir, H. E.; Barrera Gutierrez, E.; Rosas Romero, M. Effect of Foliare Diseases on Mycoflora and Quality of Seeds of Wheat. *Rev. Pl. Pathol.* **1999**, *78* (10), 904.

Hino, T.; Furuta, T. Studies on the Control of Bakanae Disease of Rice Plants, Caused by *Gibberella Fujikuroi.* II. Influence of Flowering Season on Rice Plants and Seed Transmitability through Flower Infection. *Bull. Chugoku Agri. Experimental Station* **1968**, *E2*, 97–110.

Ilyas, M. B.; Javaid, M. S. Mycoflora of Bas-370 Rice Seeds Collected From Gujranwala, Hafizabad, Sheikhpura and Sialkot Districts. *Pakistan J. Phytopathol.* **1995**, *7* (1), 50–52.

ISTA. International Rules for Seed Testing. *Seed Sci. Tech.* **1993**, *21*, 296.

Jayas, D. S.; White, N. D. G. Storage and Drying of Grains on Canada: Low Cost Approaches. *Food Control* **2003**, *14* (4), 255–261.

Jayaraman, P.; Kalyansundaram, I. Moisture Relations of Storage Fungi in Rice. *Indian Phytopathol.* **1989**, *42* (1), 67–72.

Jayaweera, K. P.; Wijesundera, R. L. C.; Medis, S. A. Seed Borne Fungi of *Oryza Sativa. Indian Phytopathol.* **1988**, *41* (3), 355–358.

Jeyanandarajah, P.; Seneviratne, S. N.; De, S. Fungi Seed Borne in Rice (*Oryza Sativa*) in Srilanka. *Seed Sci. Technol.* **1991**, *19* (3), 561–569.

Kalyansundarm, I. Storage Fungi in Rice and Their Effect. *Rev. Pl. Pathol.* **1998,** *77* (5), 523.

Kapkoti, N.; Pandey, K. N. Effect of Fungicides on the Seed Mycoflora of Rice, Maize and Wheat in Kumaun Hills. *J. Myco. Pl. Pathol.* **1990,** *20* (2), 149–150.

Khalid, N.; Anwar, S. A.; Haque, M. I.; Riaz, A.; Khan, M. S. A. Seed Borne Fungi and Bacteria of Rice and Their Impact on Seed Germination. *Pakistan J Pl. Pathol.* **2001,** *13* (1), 75–81.

Khan, T. Z.; Gill, M. A.; Nasir, M. A.; Bokhari, S. A. A. Fungi Associated with Seeds of Different Rice Varieties of Lines. *Pakistan J Phyto. Pathol.* **1999,** *11* (1), 22–24.

Kim, C. H. Fungi Associated with Discolored Rice Grains. Research Reports of the Rural Development Administration. *Crop Protect.* **1992,** *34* (1), 6–11.

Kim, J. S.; Lee,Y. W. Indentification of *Aspergillus sp and Penicillium sp* Isolated from Deteriorated Rice. *Rev. Pl. Pathol.* **1992,** *71* (8), 567.

Kim, W. G.; Yu, S. H. Disease Development and Yield Loss in Rice Caused by Sclerotial Fungi on the Paddy Field. *Rev. Pl. Pathol.* **1992,** *71* (9), 657.

Kozaka, T. Penicillium Sheath Blight of Rice Plants and its Control. *Japanese Agri. Res. Quarterly* **1970,** *5,* 12–16.

Krishana Rao, V. Impact of Various Storage System on Grain Mycoflora Quality and Viability of Paddy During Storage. *Indian Phytopathol.* **1992,** *45* (1), 44–48.

Lacicowa, B; Pieta, D. The Effect of Temperature and Rainfall on Participation of the Pathogen Causing Root and Stem Rot in Spring Barley. *Acta Agrobotanica* **1998,** *51* (1–2), 51–61.

Lakshmanan, P. A New Sheath Rot (Shr) Disease of Rice Identified in Tamilnadu. *Int. Rice Rea. Newslett.* **1991,** *16* (5), 20.

Lambat, A. K.; Nath, R.; Agarwal, R. C.; Khetarpal, P. K.; Dew, U.; Kaur, P.; Majumdar, A.; Varshney, J. L.; Mukeval, P. M.; Rani. I. Pathogenic Fungi Intercepted in Important Seeds and Planting Material During 1982. *Indian Phytopathol.* **1985,** *38* (1–2), 109.

Lee, F. N.; Rush, M. C. Rice Sheath Blight: A Major Rice Disease. *Pl. Dis,* **1983,** *67,* 829–832.

Limonard, T. A Modified Blotter Test For Seed Health. *Netherland J. Pl. Pathol.* **1966,** *72,* 319–321.

Limonard, T. Ecological Aspects of Seed Health Testing. *Proc. Int. Seed Test Assoc.***1968,** *33,* 1.

Lalitha, V.; Raveesha, K. A. Seed Mycoflora of Some Paddy (*Oryza Sativa*) Varieties Grown in Karnataka (India). Proceedings of 90th Indian Science Congress, Part III, Plant Science Section, 2003, pp 74.

Liu, X. L.; LiJain, Q.; Zhang, L.; Luo, J.; Li, X. Effect of Fungicides and Its Mixture on the Toxicity and Morphology of the Pathogens *Rhizoctonia Solani* of Rice. *Acta Phytophylacica Sinica* **2001,** *28* (4), 352–356.

Mahmood, Y. M.; Aslam Khan; Munzur Ahmad; Smi Ul- Allah. "Resilience of Agricultural System Against Crises" Management of Fungi Associated with Grain Discolouration in Rice, An Emerging Threat to Rice Crop in Pakistan. *Tropentag* **2012,** 19–21.

Maksimov, I. V.; Troshina, N. B.; Khairullin, R. M.; Surina, O. B.; Ganiev, R. M. The Effect of Common Bunt on the Growth of Wheat Seedlings and Calluses. *Russian J. Pl. Physiol.* **2002,** *49* (5), 685–689.

Maiti, D.; Vaniar, M.; Shukla, V. D. Off Season Perpetuation of *Sarocladium oryzae* Under Ononcropped Rainfed Ecosystem. *Indian Phytopathol.* **1991,** *44* (4), 454–457.

Malavolta, V. M. A.; Amaral, R. E.; De, M.; Alexander, J. Fungi Recorded on Rice Crops in Different Regions of Brazil. *Biologica* **1979,** *45* (9–10), 159–164.

Mall, O. P.; Pateria, H. M.; Chauhan, S. K. Mycoflora and Aflatoxin in Wet Harvest Sorghum. *Rev. Pl. Plathol.* **1988,** *67* (12), 547.

Manandhar, H. K.; Jorgensen, H. J. L.; Smedegaard, P. V.; Mathur, S. B. Seedling Infection of Rice by *Pyricularia oryzae* and its Transmission to Seedings. *Pl. Dis.* **1998,** *82* (10), 1093–1099.

Marineck, A.; Pinho; E. V. Der Von; Pinho; R. C. Von; Machaclo, D. A. C. Seed Quality Of Corn During Storage Effect of Harvest Time And Fungicide Treatment. *Rev. Pl. Pathol.* **2003,** *82* (5), 551.

Mia, M. A. T.; Safeeulla, K. M. Survival of Seed Borne Inoculum of *Bipolaris oryzae* the Causal Agent of Brown Spot Disease of Rice. *Seed Res.* **1992,** *26* (1), 78–82.

Mia, M. A. T.; Safeeulla, K. M.; Shetty, H. S. Seed Borne Nature of *Gerlachia oryzae*, the Incitant of Leaf Scald of Rice in Karnatka. *Ind. Phytopathol.* **1986,** *39*, 92–93.

Milivojevic, M.; Matijevic, D; Drinic, G.; Selakovic, D.; Todorovic, G. Effect of Fungicides for Seed Treatment on Expressing of Biological Traits and Yield of Maize Inbred Lines. *Pesticides* **1996,** *11* (4), 261–266.

Misra, A. K. Studies on Discolouration of Paddy Seeds and Its Control. Ph. D. Thesis, Indian Agri. Res. Ins., New Delhi, 1987.

Misra, J. K.; Misra, H. S. Fungi and Storage Loss of Paddy Varieties in Eastern U. P. A Preliminary Survey. *Narendra Deva J. Agri. Res.* **1991,** *6* (1), 204–205.

Mishra, G. N. Strategic Approaches for Bosting Upland Rice Field. *Indian Farming* **1999,** *3,* 9.

Misra, J. K.; Gergon, E.; Mew, T. W. Storage Fungi and Seed Health of Rice: A Study in Philippines. *Mycopathologia 131,* 1–24.

Mittal, R. K.; Sharma, M. R. Addition to the Mycoflora Associated with the Seeds of *Oryza sativa. Proc. Nat. Acad. Sci. India* **1978,** *48* (2), 77–78.

Mohanty, N. N. Mode of Infection of Udbatta Disease of Rice. *Rev. Pl. Pathol.* **1978,** *57* (2), 115.

Mohan, R.; Subramanian, C. L. Sheat Rot of Rice a Seed Borne Disease: MACCO. *Agri. Digest.* **1981,** *5* (9), 7.

Moreno Martinez, E.; Vazquez –Badillo, M. E.; Facio Parra, F. Temperature in Relation to Longevity of Maize Seed Stored at Low Moisture Content. *Agrociencia 34* (2), 175–180.

Muralidharan, K.; Venkata, G. Out Break of Sheath Rot of Rice. *Int. Rice Res. Newslett.* **1980,** *5,* 7.

Nandi, S. K.; Mukherjee, P. S.; Nand, B. Deterioration of Maize Grains in Storage by Fungi and Associated Mycotoxins Formation. *Ind. Phytopathol.* **1989,** *42*, 312.

Neergaard, P. *Seed Pathogogy,* vol. 1.; The MacMillan Press. Ltd.: London and Basingstoke, 1977, pp 1–839.

Neergaard, P.; Saad, A. Seed Health Testing of Rice. A Contribution of Development of Laboratory Routine Testing Methods. *Indian Phytopathol.* **1962,** *15,* 85–111.

Nguefack, J.; Nguikwie; Fotio. Fungicidal Potential of Essential Oils and Fractions from *Cymbopogon citratus, Ocimum gratissimmum* and *Thymus vulgaris* to Control

Alternaria padwickii and *Bipolaris oryzae*, Two Seed Borne Fungi of Rice (*Oryza sativa* L.). *J. Essential Oil Res.* **2007**, *17*, 581–587.

Ngugi, H. K.; King, S. B.; Abayo, G. O.; Reddy, Y. V. R. Prevalence, Incidence and Severity of Sorghum Diseases in Western Kenya. *Plant Dis.* **2002**, *86* (1), 65–70.

Okhovat, M.; Techrani, A. S.; Eshtiaghi, H. Estimating Losses from Rice Blast Disease. *Rev. Pl. Pathol.* **1992**, *71* (3), 191.

Orsi, R. B.; Correa, B.; Possi, C. R.; Schammass, E. A.; Noqueira, J. R.; Dias, S. M. C.; Malozzi, M. A. B. Mycoflora and Occurrence of Fumonisins in Freshly Harvested and Stored Hybrid Maize. *J. Stored Products Res.* **2000**, *36* (1), 75–87.

Ou, S. H. *Rice Diseases*. CAB. International Mycological Institute Kew: Surrey, UK, 1985.

Orsi, R. B.; Correa, B.; Possi, C. R.; Schammass, E. A.; Noqueira, J. R.; Dias, S. M. C.; Malozzi, M. A. B. Mycoflora and Occurrence of Fumonisins in Freshly Harvested and Stored Hybrid Maize. *J. Stored Products Res.* **2000**, *36* (1), 75–87.

Padmanabhan, S. Y. Estimating Losses from Rice Blast in India. *Rice Blast Dis* **1965**, 203–221.

Parate, D. K.; Lanjewar, R. D. Studies on Seed Mycoflora of Two Rice Cultivars Grown in Rice Tract of Vidurbha. *PKV Res. J.* **1987**, *11* (1), 47–50.

Parisi, J. J. D.; Malavolta, V. M. A.; Leonel Junior, F. L. Chemical Control of Seed Borne Fungi in Rice Seeds (Oryza Sativa L.). *Summa Phylopathologica 27* (4), 403–409.

Pandey, V.; Agarwal, V. K.; Pandey, M. P. Location and Seed Transmission of Fungi in Discolored Seeds Of Hybrid Rice. *Indian Phytopathol.* **2000**, *53* (1), 45–49.

Pathak, K. D. Chemical and Physical Treatment of Seeds for Control of Plant Diseases. *Plant Pathol.* **1989**, 42, 304.

Prabhu, A. S.; Vieira, N. R. de A. Sementes de Arroz Infect as Dreshlera Oryzae. G. Erminacao Transmissao e Control. *Buletin de Pesquisa Gopinia Embrapachnapaf* **1989**, *7*, 1–36.

Prakash; Kauraw, L. P. Compatibilities Between Certain Pesticides Used for Insect Pests and Seed Borne Fungi in Stored Paddy. *Pesticides* **1983**, *17* (11), 21–22.

Prakash, A.; Rao, J. Heteropterans and Fusarium Moniliforme Synder & Hansen Interactions to Deteriorate Grain Quality in Rice. *Seed Res. 30* (2), 339–341.

Pramanik, M.; Chakraborti, N. K. Severe Incidence of Rice Blast in West Bengal During 1990. *Rev. Pl. Pathol. 71* (10), 745.

Prasad, B. K. Storage Fungi of Crop Seeds and Their Significance. Proceedings of. 91st Indian Sci. Congress Part III, Section of Plant Sciences, 2004, pp 9.

Prasad T, Pathak, S. S. Impact of Various Storage Systems on Biodeterioration of Cereals, **1987**, *Indian Phytopathol.* 40, 39–46.

Puzari, K. C.; Saikia, U. N.; Bhattacharyya, A. Management of Banded Leaf and Sheath Blight of Maize with Chemicals. *Ind. Phytopathol.* **1998**, *51* (1), 78–80.

Raina, G. L.; Singh, G. Sheath Rot Out Break in the Punjab. *Int. Rice Res.* **1980**, *5* (2), 63–71.

Ranganathaiah, K. G.; Safeeulla, K. M. Seed Borne Nature and Mode of Infection of *Ephelis Oryzae* Syd. Incitant of Udbatta Disease of Rice. *Mysore J. Agri. Sci.* **1994**, *28* (1), 60–63.

Rath, M.; Padhi, B. Effect on Storage of Sheath Rot Fungus. Proceedings of 91st Indian Sci. Cogress, Part III. Agricultural and Forestry Sci., 2004, pp 22.

Rathaiah, Y. Grain Discoloration of Glutinous Rice by *Drechslera Oryzae. Ind. Phytopathol.* **1997**, *50* (4), 580–581.

Reddy, A. B.; Khare, M. N. Seed Borne Fungi of Rice in M.P. and Their Significance. *Indian Phytopathol.* **1978**, *31*,: 300–303.

Reddy, M. M.; Reddy, C. S.; Gopal Singh, B. Effect of Sheath Rot Disease on Qualitative Characters of Rice Grains. *J. Myco. Pl. Pathol.* **2000**, *30* (1), 68–72.

Richardson, M. J. *An Annotated List of Seed Borne Diseases,* 3rd ed, CAB International Mycological Insti: Kew, Surry, England and ISTA, Zurich Switzerland, 1979.

Roy, A. K. Sources of Seed Borne Infection of Sheath Blight in Rice. *Rev. Pl. Pathol.* **1992**, *71* (10), 746.

Sachan, I. P.; Agarwal, V. K. Effect of Seed Discoloration of Rice on Germination and Seedling Vigour. *Seed Res.* **1994**, *22* (1), 39–44.

Saponaro, A.; Portapuglia, A.; Montori, F. Some Important Seed Borne Pathogenic Fungi on Rice. *Informatore Fitopathologic* **1986**, *36* (1), 40–43.

Sawada, H.; Itoh, H. WIN R (Carpropamid KTU – 3616) A New Systemic Compound for Rice Blast Control. *Rev. Pl. Pathol.* **1999**, *78* (1), 51.

Scholz, K.; Vogt, M.; Kunz, B. Application of Plant Extracts for Controlling Fungal Infestation of Grains and Seeds During Storage. *Rev. Pl. Pathol.* **1999**, *78* (10), 904.

Shahjahan, A. K. M.; Rush, M. C.; Jones, J. P.; Groth, D. E. First Report of Occurrence of White Leaf Streak in Louissiana Rice. *Plant Dis.* **1998**, *82* (11), 1282.

Sharma, R. C.; Chahal, H. S. Seed Health Studies on Parental Lines of Hybrid Rice. *Seed Res* **1996**, *24* (2), 145–150.

Sharma, G.; Saxena, S. C. Severity of Banded Leaf and Sheath Blight of Maize Caused by *Rhizoctonia solani* in Presence of *Trichoderma* species. *J. Mycol. Pl. Pathol.* **2000**, *30* (2), 261.

Sharma, I. P.; Siddiqui, M. R. Study on Seed Mycoflora of Paddy from Assam and West Bengal. *Seed Res.* **1978**, *6* (1), 43–47.

Sheroze, A.; Rashid, A.; Shakir, A. S.; Khan, S. M. Effect of Biocontrol Agents on Leaf Rust of Wheat and Influence of Different Temperature and Humidity Levels on Their Colony Growth. *Int. J. Agri. Biol.* **2003**, *5* (1), 83–85.

Shetty, S. A.; Shetty, H. S. Seed Health Testing of Paddy Against Kernel Smut. Proceedings of. 21$_{st}$ ISTA congress Brisbane, July 1986, pp 10–19.

Singh, D.; Maheswari, V. K. The Influence of Stack Burn Disease of Paddy on Seed Health Status. *Seed Res.* **2001**, *29* (2), 205–209.

Singh, D.; Maheswari, V. K.; Gupta, A. Rice Seed Variability Influenced by Fungicide Dressing in Store. *Bhartiya Krishi Anusandhan Patrika* **2002**, *17* (2–3), 117–121.

Singh, J.; Shukla, T. N. Post Harvest Pathology of Rice Study of Microflora Associated with Seeds. *Indian Phyto. Pathol.* **1980**, *43*, 268.

Singh, R. A.; Raju, C. A. Studies on Sheath Rot of Rice. *Int. Rice Res. Newslett.* **1981**, *6* (2), 11–12.

Singh, T.; Tyagi, R. P. S. Fungal Diseases and Spoilage of Cereals and Millets in Storage. *Rev. Trop. Pl. Pathol.* **1984**,*1*, 485–499.

Singh, K.; Singh, T.; Singh, D. Seed Mycoflora of Cron from Tribal Areas of Rajasthan. *Indian J. Myco. Pl. Pathol.* **1988**, *17* (1), 34–42.

Singh, R. A.; Dube, K. S. Assessment of Loss in Seven Rice Cultivars Due to False Smut. *Ind, Phytopathol.* **1978**, *31*, 186–188.

Singh, S.; Pal, V.; Panwar, R. M. S. False Smut of Rice Its Impact on Yield Components. *Crop Res. (Hisar)* **1992**, *5*, 246–248.

Singh, N.; Kaur, N. Chahal, S. S. Detection of *Fusarium moniliforme* Causing Sheath Rot of Rice and Its Impact on Seed Health. *J. Myco Pl. Pathol. 32* (2), 155–157.

Sinha, K. K.; Sinha, A. K. (1988). Presence of Aflatoxin and *Aspergillus flavus* in Marketed Wheat. *Indian Phytopathol.* **1988**, *41*, 44–45.

Sundar, S.; Satyavir. Sensitivity of *Fusarium moniliforme* Isolates from Bakanae Disease of Rice of Memc and Carbendazim. J. Myco. Pl. Pathol. **1998a**, *28* (3), 246–250.

Surekha, M.; Kiran Saini, V.; Krishna Reddy; A. Rajender Reddy. Fungal Succession in Stored Rice (Oryza sativa Linn.) Fodder and Mycotoxin Production. *African Journal of Biotechnology* **2011**, *10* (4), 550–555.

Tanaka, K.; Sago, Y.; Zheng, Y.; Nakagawa, H.; Kushiro, M. Mycotoxins in Rice. *Int. J. Food Microbial.* **2007**, *199*, 59–66.

TeBeest, D. O.; Guerber, C. A. Role of Infested Seeds in the Epidemiology and Control of Rice Blast Disease. *Rev. Pl. Pathol.* **1999**, *78* (5), 434.

Teplyakov, B. I.; Teplyakova, O. I. Diseases of Spring Winter in West Siberia. *Zashchita iKarantin Rastenii* **2003**, *1*, 17–18.

Thapak, S. K.; Thrimurty, V. S. Effect of Fungicides on the Growth, Sporulation and Disease Severity of Rice Caused By *Sarocladium oryzae*. *J. Myco. Pl. Pathol.* **2002**, *32* (2), 266.

Tiwari, R. Seasonal Mycoflora Analysis of 4 Varieties of Rice at Gorakhpur. *J. Living World* **1995**, *2* (1), 63–67.

Upadhyay, A. L.; Singh, V. K.; Gupta, P. K. Varietal Screening for Resistance to Brown Spot (*Cochliobolus miyabeanus*) and Blast (*Magnaporthe grisea*) Diseases of Rice in Rainfed Lowlands. *Indian J. Agri. Sci.* **1996**, *66* (10), 594–596.

Upadhyay, R. K.; Diwakar, M. C. Sheath Rot in Chatisgarh, M.P., India. *Int. Rice Res. Newslett.* **1984**, *9* (5), 5.

Utobo, E. B.; Ogbodo, E. N.; Nwogbaga, A. C. Seed Mycoflora Associated with Rice and their Influence on Growth at Abakaliki, South East Agroecology, Nigeria. *Libyan Agri. Res. Center J. Int.* **2011**, *2* (2), 79–84.

Thapak, S. K.; Thrimurty, V. S. Effect of Fungicides on the Growth, Sporulation and Disease Severity of Rice Caused By *Sarocladium oryzae*. *J. Myco. Pl. Pathol.* **2002**, *32* (2), 266.

Vallejos, O. V.; Mattos, C. L.. Mycoflora of *Oryza sativa*. *Fitopatologia* **1990**, *25* (2), 54–59.

Verma, M. L.; Aulakh, K. S. Effect of Temperature and Period of Storage on Discolorate Seeds of Rice in Reaction of Mycoflora Germination Pre and Post Emergence Infection. *Ind. Phytopathol.* **1983**, *36*, 201.

Vidhyasekarn, P.; Rangnathan, K; Rajamanickam, B.; Radhakrishnan, J. Quality of Rice Grains from Sheath Rot Affected Plants. *Int. Rice Res. Newslett.* **1984**, *9*, 19.

Vidhyasekaran, P.; Subramanian, C. L.; Govindswamy, C. V. Production of Seed Borne Fungi and Its Role in Paddy Seed Spoilage. *Ind. Phytopathol.* **1970**, *23*, 518–525.

Vijaya, M. Field Evaluation of Fungicides Against Blast Disease of Rice. *Ind. J. Pl. Pathol.* **2002**, *30* (2), 205–206.

Viswanathan, R.; Narayanaswamy, P. Effect of Fungicides on the Seeds Germination, Seedling Length, Dry Matter Production and Vigour Index in Rice. *Madras Agri. J.* **1993**, *80* (1), 19–21.

Vyas, N. L. Seed Mycoflora of Paddy and Maize in Mizoram. *Ind. J. Hill Farming* **1994**, *7* (2), 220–221.

Waghray, S.; Reddy, C. S.; Reddy, A. P. K. Seed Mycoflora and Aflatoxin Production in Rice. *Ind. Phytopathol.* **1988,** *41* (3), 492.

Wang, J.; Levy, M.; Dumkle, L. D. Sibling Species of Cercospora Associated With Gray Leaf Spot of Maize. *Phytopathology 88* (12), 1209–1275.

Wal, D.; Van, Der. Seed Treatment of Maize. *Rev. Pl. Pathol.* **1978,** *57* (8), 320.

Wicklow, D. T.; Weaver, D. K.; Throne, J. E. Fungal Colonists of Maize Grain Conditioned at Constant Temperature and Humidities. *J. Stored Product Res.* **1998,** *34* (4), 355–361.

Williams, P. C.. Storage of Grains and Seeds. *Rev. Pl. Pathol.* **1992,** *71* (6), 398.

Wu, Q. N.; Liang, K. G.; Zhu, X. Y.; Wang, X. M.; Jin, J. T.; Wang, G. Y. Isolation and Identification of the Pathogen of Maize Stalk Rot in Beijing And Zhejiang. *Rev. Pl. Pathol.* **1992,** *71* (9), 654.

Yoshizawa, T. Natural Occurrence of Mycotoxins in Small Grain Cereals (Wheat, Barley, Rye, Oat, Sorghum, Millets and Rice). *Rev. Pl. Pathol.* **1992,** *71* (3), 178.

Zad, S. T.; Khosravi, V. Investigation on Important Seed Borne Fungal Diseases of Dominant Rice Cultivars in Mazandaran Iran. *Proc Int. Symposium Crop Protect. Gent. Belgium, Part II* **2000,** 587–592.

Zakeri, Z.; Zad, J. Seed Borne Fungi Associated with Some Abnormalities of Rice Seedligns. *Iranian J. Pl. Pathol.* **1987,** *23* (1–4), 7–8.

CHAPTER 7

Plant Disease Detection and Management: An Overview

S. M. YAHAYA*

Department of Biology, Kano University of Science and Technology, Wudil P.M.B. 3244, Nigeria

Corresponding author. E-mail: sanimyahya@gmail.com

ABSTRACT

Plant diseases cause substantial losses in the quality and yield of plant produced worldwide. These losses cause by various factor in which biotic and abiotic stress are the one of the major reason. Many insects, such as caterpillars and beetles, are fairly large and easy to spot, as is the damage they cause. Some plants are more likely to suffer from abiotic disorders rather than plant diseases. There are thousands of genera and species of pathogenic microorganism which are associated, in one way or another, with plant and its produce. The pathogens that are principally involved in plant disease are fungi, bacteria nematodes, and virus. Correct identification of both the host plant and the causal agent of a disease or pest damage will enable a plant grower to choose the most effective management practices that will prevent further damage to crop plants without affecting harmless or beneficial organisms. Chapter 7, summarizes the losses caused due to abiotic and biotic stress, their detection, and management.

7.1 INTRODUCTION

Losses in quality and yield commonly occur every year due to the activities of plant diseases; yet accurate record of such losses and the diseases are not documented especially in developing countries due to poor management of agricultural produce (Alao, 2000; Opadokun, 2006). In developed countries such as the United States of America, despite all the measures and strategies of control being implemented, the losses of agricultural produce to plant diseases were estimated from 8% to 23%, by insects 4% to 21%, and by weeds 8% to 13%. While losses due to plant diseases in Canada, another developed country, range between 15.5%, 12.5%, and 10.5%, respectively. In the year 1990, it was estimated that these losses, if not checked, would reduce returns to the vegetable industry by $172.7, $138.2, and $115.2 million, respectively.

Losses of agricultural produce due to plant diseases occur due to one or more factors which could be either: (1) losses of plant and its produce due to attack by pathogens, (2) losses of plant and its produce due to attack by insects and causing damaging to the plant, and (3) losses of plant and its produce due to unfavorable environmental conditions, which are referred to as abiotic disorders. Although these plant diseases are caused by different factors, the symptoms for the diseases are somehow very similar. Therefore, the best way of identification a specific factor causing the plant diseases is by eliminating chances of other factors one by one; so first if eliminating chances for insect damage, then abiotic factors, and then pathogens (Yahaya and Alao, 2008; FAO, 2009; Monica et al., 2017a).

7.1.1   DISORDER BY INSECT PEST

The simplicity of finding insect damage very easily is the main reason why insect damage is considered first before other factors in plant disease identification. Some insects which are fairly large like caterpillars and beetles are easy to spot and also the damage caused by them. Likewise arthropods, mites, and other smaller insects can be easily spotted with a hand lens. The damage by insect is so conspicuous —creating holes in areas such as leaves or fruit, leaves with a ragged appearance due to insects chewing along leaf margins, or deformed leaves (Kim et al., 1975; Monica et al., 2017).

The presence of *sooty mold*, which is associated with sucking insects, and *frass*, the dark colored droppings is a sign that insects such as aphids and mealy bugs are in the plant. Therefore, the presence of sooty mold is identified in a plant such plant should be carefully inspected for the presence of insects especially aphids which are disease vectors (Kim et al., 1975; Monica et al., 2017).

Therefore, care should be taken to make good observations to confirm the evidence of involvement of insect in the plant disease. However, were all the necessary investigation end in a negative result then the next factor that should be considered for elimination is the "abiotic" as causative agent for the plant disease (Kim et al., 1975; Monica et al., 2017).

7.1.2 DISORDER BY ABIOTIC FACTORS

Eliminating chances of environmental factors is usually carried out before searching for plant pathogens. This is important because the successful development and spread of the pathogens certain environmental conditions are required. Therefore, the ability to determine an abiotic disorder will play a key role in disease diagnosis. Some plants suffer from abiotic disorders more than the plant diseases; therefore, investigation on the abiotic factors first may make the process of identification disease easy and less time-consuming (Opadokun, 1996; Alao, 2000; Yahaya and Alao, 2008).

The word "abiotic" stands for without life; it is referred to as the non-biological factor usually related with the environmental conditions, which has an impact on the survival of plant. The environmental factors which affect survival and development of plants consists of moisture, temperature, regime of light, soil pH, air quality, and nutrition. However, when one or more of these factors become adversely affected either at a higher or lower level than the normal required amount for a particular plant species, growth of the plant might be adversely affected and may result in development of disease. In addition the abiotic disorders could be a result of human activities such as application of fertilizer and pesticide (Opadokun, 1996; Alao, 2000; Yahaya and Alao, 2008).

The sign of damage within the environmental unit is one of the most typical features indicating the abiotic disorder. Plant diseases caused by environmental disorders are most likely affect plants in the environmental unit uniformly as compared with the diseases caused by attack of insect which usually occur in clumps or hot spots within the unit (Opadokun, 1996; Alao, 2000; Yahaya and Alao, 2008).

However, it is important to define two popular terms in abiotic disorders: the situation where the plant tissue lost the green pigment chlorophyll which results the plant tissue turning into various shades or yellow, the term is referred to as *chlorosis*. While dead tissue is called necrotic tissue, necrotic leaf tissue usually turns into brown or gray color (Alao, 2000; Opadokun, 2006; Yahaya and Alao, 2008).

7.1.3 DISORDER DUE TO TEMPERATURE CHANGES

Many plants are quite sensitive to temperature, especially the cold weather, particularly plants native to tropical parts of the world. For the cold sensitive plants, the damage may occur above 32°F (0°C). Formation of ice-crystals in between cells, due to freezing temperature, causes damage to the plants including rupture of cell membranes when contact occurs with the sharp edges of these crystals. With the continuous death to the number of adjacent cells (necrosis), the damage becomes noticeable to the naked eye. Therefore, the first and early symptoms of freeze injury are the areas of necrotic tissue, which are mainly at the margins and tips of leaves. Denaturation and coagulation occur when the temperature becomes excessively high, with consequences of drying out and death of the tissues. Usually, the symptoms of heat damage appear on the leaves as whitish and papery (Opadokun, 2006).

7.1.4 DISORDER DUE TO MOISTURE CHANGES

Well-being of the plant is adversely affected by lack of moisture, resulting in wilting. As lack of moisture increases signs of wilting become visible on the plant stem that will show more or less healthy looking vascular tissue in light colors, off-white in herbaceous plants and light brown in woody plants. Usually, while the stem tissue of plant wilted due to disease becomes dark colored. However, the symptoms for drought differ depending on the plant species, although the most common symptoms are similar in many plant species which include dehydration that is indicated by change of color or necrosis of margin and leaf tip (FAO, 2009; Monica et al., 2017). Furthermore, excessive availability of water to the plants may also be a problem to the normal plant development. For example, a disorder caused by too much supply of water to the plant which is called edema occurs on ornamental foliage plants with thick leaves. However, reduced rate of evaporation from plant containers and transpiration from the leaves surfaces may occur during period of prolonged rainy weather. Under this condition the plant leaves may likely engorge resulting in cell rupture with brownish lesions similar to that of leaf spots caused by living pathogens (Monica et al., 2017).

Lack of vigor and light green or pale yellow green patches may develop in leaves of plant that remain in water log soil for a longer period due to root oxygen deficiency (Monica et al., 2017). Likewise, a plant root starved with oxygen fails to function properly and finds it difficult to remove water and nutrients from the soil which may increase chances of diseases of root fungi (Monica et al., 2017).

7.1.5 DISORDER DUE TO CHANGES IN LIGHT

Different species of plants have different requirements for light. While certain plants require shade, others require a good amount of sunlight for their development. Less chlorophyll and leggy or spindly appearance is usually seen in sunlight-dependent plants when grown under a low-light regime. This happens when the length of the stem between internodes becomes longer than normal. Usually, this is the main challenge faced by plants that receive low light intensity. For example, potted plants of tomato, if grown during winter months, in containers, receive direct sunlight for a limited period and may likely show these symptoms. These plants will not only look thinner and taller than normal, also they will considerably droop under the weight of a normal fruit load (Agrios, 2005).

However, sunburn usually occurs in plants transplanted from a shady to a sunny location. The new leaves that develop after transplanting will be better adapted to the sun and will replace older leaves which develop necrosis in the center of the leaf (Opadokun 2016). Also, loss of shade coverage after wind storm may result in sunburn damage; however, certain plants will never look normal under the sun because they are not well-adapted to the sunny environment (Opadokun, 2006).

7.1.6 DISORDER DUE TO EXCESSIVE OR DEFICIENCY IN NUTRITION

Correct nutrition is a major component for normal plant growth, and the macronutrients include nitrogen (N), phosphorus (P), potassium (K), and magnesium (Mg); in addition to a lot of micronutrients, like iron (Fe), manganese (Mn), zinc (Zn), and boron (B), which must be supplied within a specific dosage in order to avoid either deficiency or excess of any nutrient (Agrios, 2005).

Although different plant species have different nutrient requirements, following is a list of the most common nutrient deficiencies, in order of relative importance, observed in Florida on general groups of plants grown in the landscape or gardens (Agrios, 2005).

- Turf: nitrogen and iron.
- Palms: boron, manganese, magnesium potassium, iron, and nitrogen.
- Broadleaf plants (vegetables and ornamentals; herbaceous and woody): magnesium, nitrogen, iron, boron, manganese, potassium, and zinc.

The plants with broadleaf suffering from low potassium levels usually show interveinal yellowing or necrosis of the oldest leaves. The low level of nitrogen is usually associated with plants that are uniformly lighter green than normal, especially on older leaves. Palms are especially susceptible to potassium deficiency, but the symptoms are different. A translucent orange, yellow, or necrotic spot is usually shown by the oldest leaves. Interveinal or marginal yellowing of the oldest leaves is normally associated with deficiency of magnesium and is normally found in the remaining part of green leaves (Agrios, 2005; Opadokun, 2006). Manganese, iron, copper, and zinc usually form insoluble salt in the soil. However, these minerals are not absorbed by roots were the condition is neutral or alkaline (pH 7.0 or high), resulting in development of deficiency symptoms which is shown by the yellowing of the newly developed leaves (Agrios, 2005). Also, deficiency of manganese is shown by the spotting of necrotic lesion, blotches, or streaking. The deficiency of iron, on the other hand, may be spotted as yellowing of the plant, which will be more severe on the newly developed leaves. While boron (B) deficiency appears as young leaves upward cupping of leaves, growing in a distorted condition.

However, excess supply of nutrients can be injurious to the plant. Toxicity from excess copper and boron are particularly noteworthy. For example, excess spray of boron and copper fertilizer has the capacity of causing necrotic leaf spots on leaves and may easily be identified or confused with fungal leaf spots or contact pesticide damage. In these types of situations, damage is more severe when the fertilizer accumulates into droplets in areas like tip of leaves. Generally, necrosis at the tip and margin of the oldest leaves result over application of boron fertilizer to the soil supporting growth of crop plants (Agrios, 2005).

However, plants may also suffer when the water-soluble fertilizer becomes too high and may results in burning of the root and wilting of young plants, which may look similar to damping-off caused by fungal pathogen. While the older plants may show necrosis at the margin and the leaf tip (Agrios, 2005).

7.1.7   DISORDER DUE TO PESTICIDE PHYTOTOXICITY

Phytotoxicity is the main problem in plants associated with application of pesticides. The methods of pesticide application on the plants vary and pesticides may be either directly sprayed on the plants, drop as granules at the lawn or base of the plant, or just drift from the application in the nearby plantations.

Leaf distortion, shoestring appearance, and growth-regulator-type injury on leaves can occur on the surfaces of leaves following absorption of some growth regulators, for example herbicides such as 2,4-D and dicamba.

Many other disease symptoms are associated with injury from other pesticides. These are: marginal yellowing and browning (necrosis) of leaves bleached (white) spots, stunted growth, interveinal or veinal yellowing, and stem and branch dieback. However, the interaction between hosts and pesticides is unique. But generally, systemic pesticides applied as foliar spray and contact normally which result in localized necrotic spot anywhere the spray landed on the surface of the plant, but damage is more severe when the pesticides accumulate in areas such as tip of leaf.

7.1.8   PLANT DISEASE CAUSED BY PATHOGENS

Thousands of genera and species of pathogenic microorganism are known, which are associated with plant diseases (Waller, 2002; Agrios, 2005). There are four pathogens that are principally associated with plants: fungi, bacteria, nematodes, and virus.

However, it should be noted that before a pathogen can infect a plant, the pathogen must enter the plant, obtain nutrients from it, and also neutralize its defensive mechanisms by secreting chemical substances that affect certain metabolic reactions of the host plant. However, this penetration and inversion is aided by, or in some cases, entirely occurs by the mechanical forces exerted by certain pathogens on plant cell walls (Agrios, 2005).

7.1.9 FUNGI

Pathogenic fungi constitute about 85% of plant diseases (Kuku et al., 2001; Agrios, 2005). Fungi are achlorophylous and their cell walls are composed of chitin and other polysaccharides instead of cellulose. Species of fungi are identified by the use of microscopic spores which they produce (Table 7.1). Spores, which are the asexual reproductive structures in fungi that can be carried for hundred miles by wind and splashing water current, insect or by human activity (Agrios, 2005). However, the soil fungi can move from one plant to another plant through growing from infested plant debris in the soil or by growing in intermingled roots. While some fungi— *Rhizoctonia* can survive for a long period in the soil even in the absence of host. Although fungi are known to enter into the plant through natural openings such as stomata or wounds, fungi may also enter through the plant cuticle (Waller, 2002; Agrios, 2005; Monica et al., 2017).

7.1.10 BACTERIA

Bacteria are very small one-celled microorganisms that can only be seen with aid of a powerful microscope (Agrios, 2005). Many plant pathogenic bacteria do not produce spores; however, bacteria can survive in the soil along with decaying plant materials for a considerable long period of time (Alao, 2000).

Bacteria are mainly dependent on the outside agents and disperse from one plant to another plant. The main way of dispersal is the splashing water (irrigation, wind-driven water) that is followed by dispersal by human contact. However, the simple way of spreading bacteria is by touching infected plant with the hand or equipment then touching a healthy plant (Agrios, 2005). However, bacteria are not known to penetrate plant cuticle but can enter through the plant wound or natural opening to trigger disease. There are particular sub-groups of bacteria which need an insect host for dispersal and entry into the plant host (Table 7.1).

TABLE 7.1 Some Important Plants (Food Crops) and Their Principal Diseases.

Plants	Fungal	Viral	Nematode	Bacterial
Tomato (*Lycopersicon esculentum*)	*Botrytis cinerea Aspergillus* spp. *Penicillium* spp.			
Pepper (*Capsicum annum*)	*Botrytis cinerea Aspergillus* spp. *Alternaria alternate*			
Lettuce (*Lactuca sativa*)	*Botrytis cinerea Aspergillus* spp.			
Pea (*Pisum sativum*)		Mosaic virus, *Ascochyta* spp.		
Carrot (*Daucus carota*)	*Alternaria dauci*			
Beans (*Phaseolus* spp.)				Xanthomonas campestris, Psedomonas syringe
Millet				
Common millet (*Panicum miliaceum*)				Downy mildew: *Sclerospora graminicola*
Finger millet (*Eleusine coracana*)	Blast: *Pyricularia setariae* Leaf blight: *Cochliobolus nodulosus*			
Maize (*Zea mays*)	Northern corn leaf blight: *Helminthosporium turcicum* (*Setosphaeria turcica*) Southern corn leaf blight: *H. maydis* (*Cochliobolus heterostrophus*) Rust: *Puccinia* spp.	Chlorotic dwarf: maize chlorotic dwarf machlovirus Streak: *Maize streak geminivirus* Yellow dwarf: Barley yellow dwarf luteovirus	Stewart's wilt: *Erwinia stewartii* Corn stunt disease: *Spiroplasma kunkelii*	Downy mildew: *Sclerospora* spp. and others

TABLE 7.1 *(Continued)*

Plants	Fungal	Viral	Nematode	Bacterial
	Smut: *Ustilago zeae*			
	Stalk and ear rots: *Gibberella zeae, Diplodia spp.* and others			
Sweet potato (*Lpomoea batatas*)	Scab: *Sphaceloma batatas* (*Elsino batatas*)	Feathery mottle: sweet potato feathery mottle potyvirus	Root-knot nematode: *Meloidogyne* spp.	Soil rot: *Streptomyces ipomoea*
	Fusarium wilt: *Fusarium oxysporum*			
	Black rot: *Ceratocystis fimbriata*			Little leaf: sweet potato little leaf phytoplasma
	Java black rot: *Botryodiplodia theobromae*			

Source: Adapted from Narayanasamy et al. (2006).

7.1.11 NEMATODES

The eelworms or the plant-parasitic nematodes are small worms usually less than 1 mm long, which thrive in the soil. Nematodes are broadly divided into two groups: ectoparasitic nematodes that attack the plant externally, and endoparasitic nematodes that live, at least for part of their life cycle, inside the host tissues. Parasitic nematodes have mouth part through which saliva is injected into the tissues of the host plant, result in the induction of greater damage most especially either tissue necrosis or the proliferation of the giant cells leading to galls. Nematodes such as *Xiphonema, Longidorus,* and *Trchodorus* while causing little direct damage, at the same time they may result in transmission of viral diseases to the host plant.

Thousands of nematodes species are free-living in the soil, feeding on fungi, bacteria, and other microbes. Many plant-parasitic nematodes feed on a very narrow spectrum of hosts, and very few species are regarded as agricultural pests. However, for example, Canada has relatively few nematodes in both—field and greenhouse vegetable crops which are of major economic importance (Webster, 1972).

Endoparasitic nematodes—These nematodes that live inside the host body. For example, Northern root-knot *Meloidogyne hapla* Chitwood penetrate feed and multiply within root tissues; some even penetrate the leaves, stems, and bulbs of almost all types of plants in both field and greenhouse. While other species of nematodes such as Southern root nematodes *Meloidogyne inconnita* (Kofoid and White) do not occur in the field but can survive once introduce into the greenhouse. While the root-lesion nematodes such as *Pratylenchus penetrans* (Cobb) Filip. and Stek. usually infect many of the major vegetable crops grown in the both field and greenhouse (Alao, 2000).

Ectoparasitic nematodes—These nematodes that live and feed outside the host body especially root tissues like cortex and epidermis, but these are the only possible were the nematodes have long stylet. While other species of ectoparasitic nematodes such as dagger nematodes *Xiphinema* spp. are more prevalent in vegetable growing fields and greenhouses but are usually identified from soil samples, however, they rarely cause any serious disease to the host plant. The dagger and needle nematodes usually prefer hosts with woody roots and are mainly associated with crop plants such as raspberry, grapes, roses, and strawberry than with vegetable crops, which are generally more soft-rooted (Nickle, 1984).

The damage caused by the plant-parasitic nematodes is somehow not easy to differentiate from those diseases caused by abiotic factors or other pathogens. The presence of unfavorable soil conditions such as insufficient decomposition of organic plant residues, poor fertility, frost heaving, and extreme moisture may lead to branching the tips of some young roots of plants, likewise a proliferation of secondary roots may indicate sign of attack by the soil nematodes.

Nematode diseases are normally assessed by visual examination of the infected plant tissue. However, most nematode diseases can only be identified after soil sampling and extraction; but both procedures may be expensive and time-consuming.

The plant-parasitic nematodes do not normally spread very rapidly; therefore, a minor infestation may not lead to evident symptoms or reduced productivity (Table 7.2).

TABLE 7.2   Host Ranges of Economically Important Nematode Pests on Crop Plants.

Plants	RKN	RLN	PCN	SBN	SRN	SCN
Bean	-	-			-	
Beet, chard, and spinach	-	-				-
Carrot	-	-				
Celery and celeriac	-	-				
Crucifers	-	-			-	-
Cucubits	-	-				
Ginseng	-	-t				
Greenhouse cucumber	-	-				
Greenhouse lettuce	-	-				
Greenhouse pepper	-	-				
Greenhouse tomato	-	-				
Lettuce, chicory, and endive	-	-				
Maize (sweet corn) - -			-		-	
Onion and other allium crops	-	-			-	
Parsnip	-					
Pea	-	-			-	

TABLE 7.2 *(Continued)*

Plants	RKN	RLN	PCN	SBN	SRN	SCN
Potato	-	-	-		-	
Rhubarb	-	-				-
Tomato, eggplant, and pepper	-	-	-		-	

PCN, potato cyst nematodes (Globodera spp.); RKN, root-knot nematode (*Meloidogyne hapla Chitwood*); RLN, root lesion nematode (*Pratylenchus penetrans* (Cobb) Filip. & Stek.); SBN, stem and bulb nematode (*Ditylenchus dipsasci* (Kuhn) Filip.); SCN, sugarbeet cyst nematode (*Heterodera schachi* Schmidt); SRN, stubby-root nematodes (*Paratrichodorus* and *Trichodorus* spp.).

Source: Adapted from Narayanasamy (2002).

7.1.12 VIRUS

The viruses are regarded as pathogen with features of both living and non-living. They are the smallest pathogens ever known to man and therefore, virus can only be seen with the aid of an electron microscope. The virus is generally made up of genetic material (RNA or DNA), which are usually wrapped in a coat of protein. They require a living host in order to survive and reproduce. Viruses are usually spread from diseased to healthy plants by insects, but can also be spread nematodes, fungi, and even humans (Agrios, 2005).

7.1.13 MANAGEMENT OF PLANT DISEASE

7.1.13.1 DISEASE IDENTIFICATION

Accurate identification of the disease causing, an agent play a vital role in the design of disease management practices that will prevent further damage to crop plants without affecting harmless or beneficial organisms (Agrios, 2006; Narayanasamy, 2006; Riley, 2012). Therefore, due to improper identification of diseases all control measures could be a waste of time and money and may lead to further plant losses (Pimentel et al., 1991; Narayanasamy, 2006). Usually, similar symptoms can be produced as a response to disease caused by different causal agents. However, the use of symptoms only in disease identification is inadequate; therefore, other methods have to be incorporated for an accurate and successful identification of the causal agents (Riley, 2002; Narayanasamy, 2006).

The most important tool for an accurate and successful identification of plant disease causal agent is the power of observation; in addition to asking so many questions to eliminate or identify the possible cause(s) of the problem and the need to take into consideration various environmental and cultural factors. The observation and question raised could lead (1) to identify the disease type and causative agent; (2) to narrow the problem down to specific possibilities which may involve further study before arriving at a final diagnosis; and (3) to baffle the problem completely (Agrios, 2005).

7.1.13.2 CORRECT PLANT IDENTIFICATION

For proper and successful identification of the plant affected, the first step is to note—both the scientific and common names of the plant. However, it should be noted that common names should not be relied upon because some distinctly different plant species may have similar common name; in addition common names used in one area may differ in another area or may be used for a completely different plant species (Alao, 2000; Narayanasamy, 2006). For example, the common name "vinca" has been used to describe two crop plants belonging to two different genera—*Vinca*, which is a perennial plant and *Catharanthus*, which is an annual plant. Also "monkey grass" is used to describe *Liriope* and *Ophiopogon* (mondo grass). Another example in forestry is the use of common name that can course confusion is "cedar", it is used to denote crop plants such as eastern red cedar (*Juniperus*), Western red cedar (*Thuja*), Port Orford cedar (*Chamaecyparis*), incense cedar (*Libocedrus*), and Atlas cedar (*Cedrus*). Therefore, the use of common names for identification of disease-causing agent is inadequate (Pimentel et al., 1991).

Apart from knowing the common and scientific names of infected plant, it is also necessary to know the specific variety or cultivar infected by the disease-causing agent. Usually, a great variation in susceptibility to a specific disease occurs within different cultivars of a plant species. For example, it should be taken into consideration the susceptibility of wheat to wheat stem rust caused by infection of *P. graminis* f. sp. *tritici*, it is clear that not all wheat cultivars are susceptible to all races of *P. graminis*. So, the major control measures for this disease development is based on planting cultivars of wheat each year that are resistant to the pathogen races and are predicted to be present during the growing season.

Ability to know the cultivar and its susceptibility to various diseases may assist to narrow down the possible disease-causing agents under consideration (Alao, 2000; Agrios, 2005). For example, tomato cultivars having "Better Boy" the genetic background is generally resistant to root-knot nematodes, however, those with the genetic background of the variety "Rutgers" are more susceptible; therefore, understanding the genetic background of a cultivar can be very helpful in identification of disease causal agents. Following the confirmation of the identity of plant species affected by the disease, the plant pathologist may be able to consult a list of plant diseases that infect the plant species which may lead in knowing the pathogenic agent and rule out other possibilities (Pimentel et al., 1991).

7.1.13.3 RECOGNIZE HEALTHY PLANT APPEARANCE

Knowledge of the normal appearance of healthy plant and its special growth habits, colors, and growth rates are important for a successful management of plant disease, otherwise it will be difficult to understand when something is wrong (Alao, 2000; Agrios, 2005). In many instances, ornamental shrubs are developed and marketed for the ornamental value of such bright-colored new growth. Therefore, were an individual did not understand the coloration is the normal appearance of the plant, he might otherwise think that the plant is diseased (Agrios, 2005; Opadokun, 2006; Yahaya and Alao, 2008). It is also important to understand that appearance can change as cultivar change. Naturally, some plant cultivars usually have yellow to pale green leaves (e.g., coleus varieties, new host cultivars, and herbs like golden oregano) which are at first glance may appear to have symptoms of root stress, soil pH disorder, or problem of under-fertilization (Wallace, 2002).

As soon as the normal appearance of the specific plant is known then comparisons can be made between the healthy and problem-facing plants (Agrios, 2005). Some of the characters that should be used for comparison include shape, coloration, overall plant size, leaf shape, bark, stem or trunk, texture and coloration, root distribution, and coloration. In addition, it is also very important to take note of all the normal events of the plant; typical example is the leaf drop, which may occur in a healthy plant. For example, some holly species usually drop leaves in the spring (Yahaya and Alao, 2008; Agrios, 2005).

While noting the condition of healthy plant, it is equally important to take note on the changes occurred in the part of the affected plant for any sign, for example are there sign of symptoms on the flowers, or fruit, leaves, stem or roots? Is the entire plant involved or only some part? Finding correct answers to the above questions may assist in correct identification of the plant disease.

7.1.14   IDENTIFICATION OF SIGN OF SYMPTOMS

7.1.14.1   IDENTIFY CHARACTERISTIC SYMPTOMS

Diseases spread involved progression of symptoms which may vary significantly from species to species. Therefore, describing characteristics of symptoms shown by the host plant may not necessarily be easy (Agrios, 2005). However, progression of symptoms is one of the most important characteristics associated with disorders caused by biotic agents. Primary and secondary symptoms may result from diseases, for example, toppling over of the tree or wind throw is a secondary symptom; however, decayed roots on a tree are a primary symptom. During the later stages of a disease development, secondary invaders may likely obscure the original disease symptoms so that symptoms observed at the later stages of the disease development may not be similar to the symptoms which developed in response to the infection by the original pathogen (Pimentel et al. 1991).

Therefore, it is very important to look for a progression of disease symptoms in plants. For example, symptoms observed due to improper herbicide usage may resemble spots which might be present due to infectious agent. Therefore, the changes observed may be with the herbicide injury, the symptoms may suddenly appear without any noticeable symptoms progression. Under such conditions, the spots may join the spray patterns of the herbicide. Some herbicides like 2,4-D, may result in distortion of the leaf and may be confused with other disease such as viral disease, but when new leaves are formed they automatically be symptoms free, showing complete absence of progression of symptoms (Opadokun, 2006).

7.1.14.2   IDENTIFYING SYMPTOM VARIABILITY

Variability of the expressed symptoms by the diseased plants may generally result to an improper diagnosis; such variations may occur from a couple of factors. Sometimes more than one disease and pathogen may be present infected plants (Agrios, 2005). The symptoms that are associated with these infected plants may significantly differ from the symptoms that may be expressed in response to infection by different pathogens when each separately acting. Therefore, the disease symptoms that may be exhibited by multiple pathogens infecting a plant could be either more or less severe than if the plant was infected with just one of the pathogens (Pimentel et al., 1991; Alao, 2000; Agrios, 2005).

7.1.14.3   *IDENTIFY SIGNS OF BIOTIC CAUSAL AGENTS*

The total observable evidence indicating that the presence of disease is the sign of plant disease-causing agent in a particular plant. This sign may consist of either the mycelia of a fungal agent, fungal spores, or spore-producing bodies (Agrios, 2005; FAO, 2009). While sign for insects as disease-causing agent may include the actual insect, insect egg, mite webbing, and insect frass. These signs are more specific to the disease-causing agents than the symptoms, and play important role in the identification of disease-causing agent and subsequent diagnosis of the disease. They use simple farm tools such as knife and a hand lens are highly recommended or may be very valuable in the field for the diagnostician. For example, for observation of mycelia mats of root rot fungi like *Armillaria* spp. this requires cutting into the bark of orna-mental plants and trees at the soil surface. While bacterial ooze may be seen by cutting into stems before placing them in clean water. For the other pathogens such as powdery mildews, the disease they course is usually identified and diagnosed by the observation of the gray to white mycelia and also by observation of conidia on the surface of leaves and flowers (Agrios, 2005).

Disease signs on plants can sometimes be overlooked were careful observations have not been conducted, this is because disease signs are not readily visible especially when taking a quick ride of the disease plant by only taking observation through the windshield of a moving truck; under such situations, the disease may not necessarily be visible to the naked eye (Agrios, 2005; Narayanasamy, 2006). Therefore, the use of dissecting or compound microscopes are recommended for the observation of specific spores and spore structures, which may results in further identification of possible disease-causing agents.

7.1.14.4   IDENTIFY PLANT PART AFFECTED: ARE SYMPTOMS ASSOCIATED WITH SPECIFIC PLANT PARTS?

During plant disease diagnosis, it is necessary to establish whether the symptoms observed are related to the specific plant parts. For example, disruption of the vascular system which may be indicated by browning of the vascular system is observed in a wilted plant, are necrotic lesions seen only on younger leaves or are the roots of the plants abnormal including rots, decreased feeder roots, etc.? Some diseases symptoms are mainly seen on some specific part of the plant and this observation is very valuable in diagnosis of plant disease (Agrios, 2005).

7.1.15   OBSERVATION OF DISEASE PATTERNS AND DISTRIBUTION OF SYMPTOMS

The first thing, a plant disease diagnostician should take into consideration is how the diseased plants are spread over the affected field area. He should find out whether the diseases are uniformly distributed across an area or are they localized? Also, it is important to determine the distribution pattern, for example, the diagnostician should find out whether the disease occurs only on the roadways or driveways or in low spots of a field, along the edges of the greenhouse near open windows or along a planted row, or is it occurring in the field and infecting plants at random? This pattern of distribution is important especially in determining at the possibility of non-infectious problems, like the improper use of herbicide or use of other various soil factors. Usually, diseases which are infectious generally occur over time with series of progression of symptoms. However, uniform pattern of distribution of an individual plant and uniform damage patterns over a large area only associated with abiotic agents not biotic agents (Agrios, 2005; Opadokun, 2006). Diseases caused by biotic agents generally observed when they are causing problems on a low percentage of plants at least at the beginning of the disease, unless there are extenuating situations, like the use of infected seeds. Even under that condition, hardly

100% infection will be seen. However, were disease occurs in 100% of the plants in an area, it is more likely the infection might have results from factors like deficiencies or toxicities of the soil, toxic chemicals such as improper use of pesticides, growth regulators, or air pollutants, like ozone or due to unfavorable climatic conditions; cold, temperatures, and drought.

Therefore, establish how the progression of symptoms on plants in the affected area is, were the all the symptoms appeared at the same time without any further development is an indication of episodic event like temperature changes or probably improper use of chemical. However, were the symptoms start in an area then spread to other areas with changes in the disease severity over time will indicate the presence of biotic agent (Narayanasamy, 2002; Agrios, 2005; FAO, 2009).

7.1.15.1 IDENTIFICATION OF HOST SPECIFICITY

Knowing whether the disease is occurring on only one plant species or on a different plant species, is important in plant disease management. If different plant species are affected, this will suggest the possibility of a non-infectious disease-causing agent, which may be associated to the cultural or environmental problems. For example, root rot of *Phytophthora* and *Pythium* can cause problems on many different plant species; however, because more than one plant species is infected does not remove the possibilities of infectious disease agents. In a situation where more than one plant species are involved, it is important to find out whether the plants related closely and the possibilities of both plants been infected by similar pathogen (Waller, 2002; Agrios, 2005).

7.1.16 GROWING ENVIRONMENT AND CULTURAL PRACTICE

Taking into consideration the cultural activities of an area around the infected plants are important for design of plant disease management strategies. Sometimes, diseases may occur in crop plants not due to anything wrong by the grower but could be due to what his/her neighbor has done in his/her area. Therefore, information pertaining to the growing environment by which the affected plant has been exposed is very vital in plant disease management (Narayanasamy, 2006). It is also equally important to keep proper records of changes in the environmental conditions. The environmental factors that should be considered and recorded properly are: rainfall, hail, lightning, prolonged drought, extreme temperatures (freezing and heat), temperature inversions (possible air pollutant and pesticide drift damage), and prevailing winds. All these abiotic factors may be important to the disease, site factors like soil pH, soil type, possible drainage problems should be checked and evaluated (Waller, 2002; Agrios, 2005).

For a successful management of plant disease maintenance and cultural activities are very essential. It is therefore important to keep a record for consultation about information on the pesticides or other chemicals that have been applied in an area, time of application, and the dosage applied. It is also essential to keep record of all the equipments used during application of pesticides or chemicals in the area of infected plants and all the possible observations that were made before and after the application (Narayanasamy, 2002; Opadokun, 2006).

KEYWORDS

- **abiotic disorders**
- **crop plants**
- **insect pest**
- **nematodes**
- **pathogens**
- **plant diseases**

REFERENCES

Alao, S. E. L. In *The Importance of post-harvest loss prevention*. Paper presented *at Graduation Ceremony of School of Torage Technology*. Nigerian Stored Product and Research Institute Kano, 2000; pp 1–10.

Agrios, G. H. *Plant Pathology*; Academic Press: London, 2005.

Elliott, M.; Pernezny, K.; Palmateer, A.; and Havranek, N. Guidelines to Identification and Management of Plant Disease Problems: Part I. Eliminating Insect Damage and Abiotic Disorders, 2017a.

Elliott, M.; Pernezny, K.; Palmateer, A.; and Havranek, N. Guidelines for Identification and Management of Plant Disease Problems: Part II. Diagnosing Plant Diseases Caused by Fungi Bacteria and Virus, 2017b.

Food and Agricultural Organization. Corporate Document Repository Report, 2009; pp 1–10.

Kim, S. H.; Forer, L. B.; Longnecker, J. L. Recovery of Plant Pathogens from Commercial Peat-Products. *Proc. Am. Phytopathol. Soc.* **1975**, *2*, 124.

Kuku, F. O.; Akano, D. A.; Olarewaju, T. O.; Oyeniran, J. O. *Mould Deterioration on Vegetables.* NSPRI Technical Report. No. 8, 1985 Annual Report. 2001, 89–94.

Lopez-Garcia, R.; Park, D. L; Phillips, T. D. Integrated Mycotoxins Management System. Internet Resource: File: 11A Mycotoxins Management Systems Foods, Nutrition and Agriculture, 2004. www.fao.org/document/show_cdr.asp. [Accessed Dec. 30, 2017]

Narayanasamy, P. *Microbial Plant Pathogens and Crop Disease Management*; Science Publishers Inc.: Enfield, New Hampshire, USA, 2002.

Narayanasamy, P. *Post-Harvest Pathogens and Diseases Management*; John Wiley and Sons, Inc.; Hoboken, New Jersey, United States, 2006.

Nickle, W. R.; Ed. In Plant and Insect Nematodes; Dekker; New York, NY, 1984; pp 925.

Nickle, W. R.; Ed. In Manual of Agricultural Nematology; Dekker: New York, NY, 1991; pp 1035.

Opadokun, J.S. Reduction of Post Harvest Losses in Fruits and Vegetables. Lectures Delivered at AERLS/Nigerian Stored Product Research Institute. *Joint National Crop Protection Workshop, IAR, Zaria 1987.* 2006; pp 3–26.

Pimentel, D.; McLaughlin, L.; Zepp, A.; Lakitan, B.; Kraus, T.; Kleinman, P.; Vancini, F.; Roach, W. J.; Graap, E.; Keeton, W. S.; Selig, G. Environmental and Economic Impacts of Reducing US Agricultural Pesticide Use. In *Handbook of Pest Management in Agriculture;* Pimentel, D., Ed.; CRC Press: Boca Raton, Florida, 1991; Vol. 2, pp 679–720, 773.

Riley, M. B.; Williamson, M. R.; Maloy, O. Plant Disease Diagnosis. *Plant Health Instr.* American Phytopathological Society, **2002** (Online Early Access). DOI: 10.1094/ PHI-I-2002-1021-01.

Waller, J. M. Postharvest Diseases. In *Plant Pathologist Pocketbook;* Waller, J. M., Leanne, M., Waller, S. J., Eds.; CAB International: UK, 2002; pp 33–54.

Yahaya, S. M.; Alao, S. E. An Assessment of Postharvest Losses of Tomato *Lycopersicon esculentum* and *Pepper capsicum* Annum in Selected Irrigation Areas of Kano State, Nigeria. *Int. J. Res. Biosci.* **2008**, *2006*, 53–56.

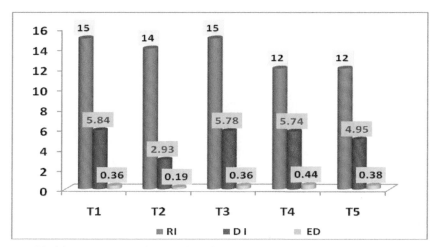

T1, Rhizobium + PSB (Local); T2, Rhizobium + PGPR (L) + PSB (L); T3, Rhizobium + PSB + PGPR (N); T4, Rhizobium + PGPR; and T5, Control; DI, diversity index, E_D, equitablility; and RI, richness index.

FIGURE 5.1   Diversity and evenness of fungi under different treatments.

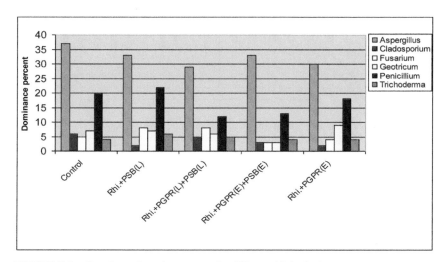

FIGURE 5.2   Dominant fungal genera under different biological treatments.

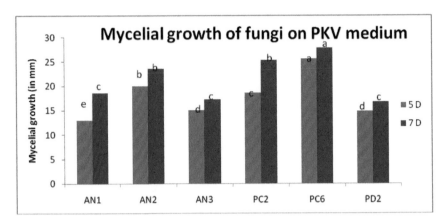

FIGURE 5.3 Mycelial growth of *Aspergillus* and *Penicillium* spp. on PKV medium on fifth and seventh day of incubation.

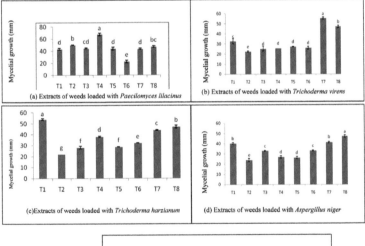

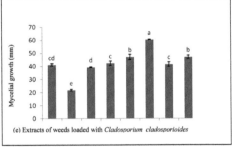

FIGURE 5.4 Effect of extracts of weeds loaded with beneficial fungi on mycelial growth of *Fusarium oxysporum* f. sp. *lycopersicae* (FOL).

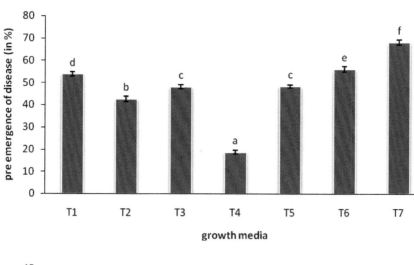

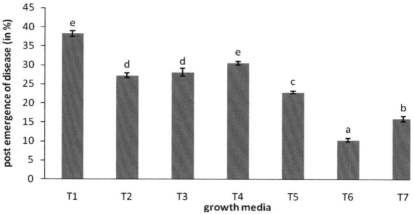

T1, no fungal inoculation, T2, AN+ TV+ PL; T3, AN+TV+PL+CC; T4, AN+TV+PL+ HG; T5, AN+TV+PL+ HG+CL; T6, AN+TV+PL+ CC+PP+FS; T7, all fungal inoculants except CC.

FIGURE 5.5   Treatments effect on disease indices of plant pathogenic fungal consortium on tomato seedlings.

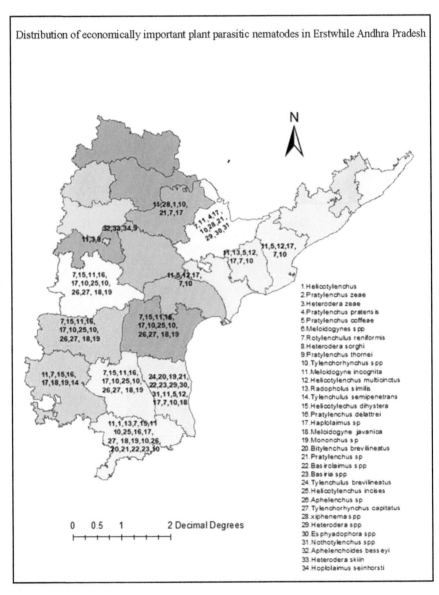

Distribution of economically important plant parasitic nematodes in Erstwhile Andhra Pradesh

1.Helicotylenchus
2.Pratylenchus zeae
3.Heterodera zeae
4.Pratylenchus pratensis
5.Pratylenchus coffeae
6.Meloidogynes spp
7.Rotylenchulus reniformis
8.Heterodera sorghi
9.Pratylenchus thornei
10.Tylenchorhynchus spp
11.Meloidogyne incognita
12.Helicotylenchus multicinctus
13.Radopholus similis
14.Tylenchulus semipenetrans
15.Helicotylechus dihystera
16.Pratylenchus delattrei
17.Haplolaimus sp
18.Meloidogyne javanica
19.Mononchus sp
20.Bitylenchus brevilineatus
21.Pratylenchus sp
22.Basirolaimus spp
23.Basiria spp
24.Tylenchulus brevilineatus
25.Helicotylenchus incises
26.Aphelenchus sp
27.Tylenchorhynchus capitatus
28.xiphenema spp
29.Heterodera spp
30.Esphyadophora spp
31.Nothotylenchus spp
32.Aphelenchoides besseyi
33.Heterodera skiin
34.Hoplolaimus seinhorsti

MAP 9.1

FIGURE 9.1   Guava orchard showing infestation of *Meloidogyne entrolobii.*

FIGURE 9.2   Guava seedlings and roots showing galls and drying symptoms.

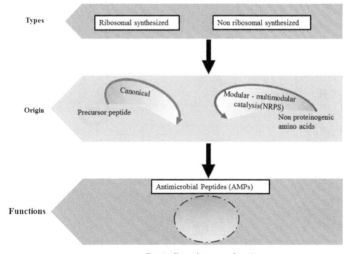

FIGURE 11.1   Antimicrobial peptides of ribosomal origin are synthesized using canonical pathway, whereas nonribosomal origin in general includes peptaibols which are synthesized through modular pathway. Still irrespective of their origin, both AMPs either target cell membrane or intracellular pathways.

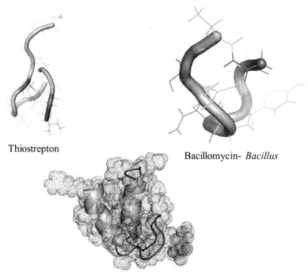

Thiostrepton

Bacillomycin- *Bacillus*

S-linked glycopeptide sublancin 168

FIGURE 11.2a   Ribosomal-derived antimicrobial peptides from bacterial source.

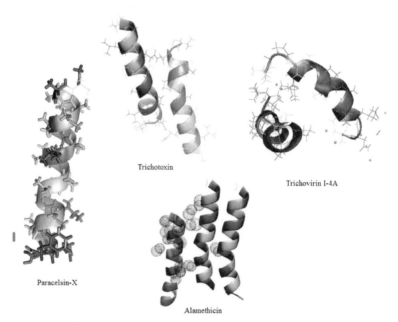

Trichotoxin

Trichovirin I-4A

Paracelsin-X

Alamethicin

FIGURE 11.2b   Peptaibol of *Trichoderma* representing nonribosomal-derived antimicrobial peptides from *Trichoderma reesei* Aib as light blue and other sky blue.

```
Trichodecenin_TD_I      ------------------Z--decenoylGlyGlyLeuAibGlyI------------leL
Trichorozin_I           AcAibAsnIleLeuAibProI-----------leLeuAibProValOH-----------
Harzianin_HB_I          AcAibAsnLeuIleAibProI-----------ValLeuAibProLeuOH-----------
Pseudokinin_KLIII       AcAibAsnIleIleAibProL-----------euLeuAibProNH--------------
NA_VII                  AcAibAlaAl---aAibIvaGlnAibAibAib---------SerLeuAibOCH-------
Paracelsin_A            AcAibAlaAibAlaAibAlaGlnAibValAibGlyAibAibProValAibAibGlnGlnP
Saturnisporin_SA_I      AcAibAlaAibAlaAibAlaGlnAibLeuAibGlyAibAibProValAibAibGlnGlnP
Atroviridin_A           AcAibProAibAlaAibAlaGlnAibValAibGlyLeuAibProValAibAibGlnGlnP
Alamethicin_F-30        AcAibProAibAlaAibAlaGlnAibValAibGlyLeuAibProValAibAibGluGlnP
```

```
Trichodecenin_TD_I      euOH
Trichorozin_I           ----
Harzianin_HB_I          ----
Pseudokinin_KLIII       ----
NA_VII                  ----
Paracelsin_A            heOH
Saturnisporin_SA_I      heOH
Atroviridin_A           heOH
Alamethicin_F-30        heOH
```

FIGURE 11.3   Multiple alignment of nonribosomal-derived antimicrobial peptides (AMPs) such as peptaibols of *Trichoderma* origin.

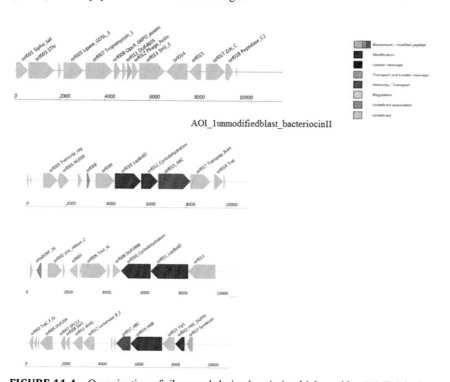

FIGURE 11.4   Organization of ribosomal-derived antimicrobial peptides (AMPs) in the genome of *Bacillus* species mostly used as biocontrol agents; green color indicates AOI type of antimicrobial peptides.

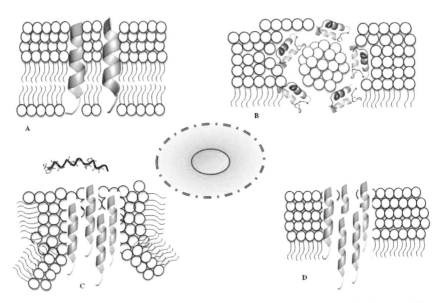

FIGURE 11.5 Schematic representation of mode of action of antimicrobial peptides A-Torroidal, B-Carpet-like model high concentrations of peptide molecules disrupt the membrane in a detergent-like manner breaking the lipid bilayer into set of separate micelles. C- Barrel stave in which hydrophobic regions of AMPs align with the tails of the lipids and the hydrophilic residues form the inner surface of the forming pore and D- Toroidal pore model in which peptides aggregates and hydrophilic heads of the lipids are electrostatically dragged by charged residues of AMPs.

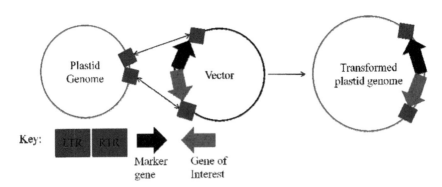

FIGURE 13.2 Foreign gene integration in chloroplast genome through homologous recombination.

PART II
Nematode Diseases and Management

CHAPTER 8

Role of Biological Agents for the Management of Plant Parasitic Nematodes

AAMIR RAINA, MOHAMMAD DANISH*, SAMIULLAH KHAN, and HISAMUDDIN SHEIKH

Department of Botany, Aligarh Muslim University, Aligarh, 202002, Uttar Pradesh, India

Corresponding author. E-mail: danish.botanica@gmail.com

ABSTRACT

The crop loss incurred by the infestation of a wide range of nematodes is valued in billions of dollars and has posed a serious threat to global food security. The nematode affects economically important crops all over the world and incurs a significant decline in the crop yield. Management of nematodes is given consideration and different methods have been developed. Intriguing evidence suggests that the biological method to control nematode infestation is the best and viable approach. Control of nematodes by the use of biocontrol agents offers a promising substitute to the use of chemical agents and chemical-associated harmful effects. The microbes such as fungi and bacteria have the ability to reduce the nematode population and have played a vital role in decreasing the crop yield loss caused by nematodes. The development and subsequent implementation of biocontrol agents are very unpredictable and quite hectic on a large scale. The implementation of an ideal biocontrol agent requires an elaborated understanding of the mechanisms of infectivity of the antagonist against nematode populations, and meticulous exploration of the interactions among biocontrol isolates, nematode, soil microbiota, plant microenvironment, and ecological implication must be developed.

8.1   INTRODUCTION

Plant parasitic nematodes devastate the crops across the globe and pose a serious threat to the overall crop production. In addition to the significant reduction in yield, they also impair the quality of crops. Keeping this in view the management of these small creatures holds great importance. Biocontrol is a novel, eco-friendly approach relying on soil microorganisms and, therefore, provide a feasible and sustainable perspective in many agrosystems for management of soil pests such as root-knot nematodes (RKNs, *Meloidogyne* spp.). The use of microbes as biological nematicide has previously been used for decades. The fact that microorganisms including bacteria, fungi, mites, and predatory nematodes are the best-known biocontrol agents of parasitic nematodes is well established. However, there is a need for proper formulation of some important strains of bacteria and fungi and their commercialization.

8.1.1   PLANT PARASITIC NEMATODES

The majority of plant-parasitic nematodes (PPNs) are either ecto- or endoparasites feeding on or inside the plant roots while a small proportion of nematodes are ariel feeders. They play a key role in damaging the economically important crops and have the capability to serve as vectors of viruses and cause a wide spectrum of diseases in association with other pathogens. Nematodes also facilitate emanate infestation by secondary pathogens namely, bacteria and fungi.[1] Ngangbam and Devi[2] advocate that nematodes cause devastation in almost all the crops across the globe. They live in and around rhizosphere and have the ability to live even in the absence of hosts. The figures for yield loss caused due to PPN in several important crops are presented in Table 8.1.

Classically, nematode management has relied on synthetic chemical nematicides; however, these nematicides have proven unreliable for small-holder systems as they are very costly and unavailable to farmers.[3,4] In addition to this, chemical nematicides have posed a threat to the environment, hence have been withdrawn from markets or their use severely restricted.[4] Other management techniques, such as soil solarization and fumigation with soil disinfectants namely, dazomet, metham sodium, and formaldehyde at recommended doses, flooding, use of resistant cultivars and use of cover crops have been practiced but have their individual limitations.

TABLE 8.1 Yield Loss Caused Due to Nematode Damage Across Different Countries.

Crop	Yield loss in percentage (%)							
	Brazil	Greece	Philippines	Russian Federation	Egypt	USA	India	China
Banana	30	1–4	26.4–45.4	–	10–25	–	7.9–34.6	10–25
Cassava	10	–	–	–	–	–	–	–
Coconut	10–35	–	–	–	–	–	–	–
Guava	10–35	–	–	–	0–25	–	–	–
Maize	20	1–3	–	–	5–15	0–7	17–29	–
Millet	15	–	–	–	–	–	–	–
Groundnut	10–20	–	–	–	15–50	1.5–5	21.6	5–20
Potato	20–35	2–10	2–70	0–20	0–20	1.2–12	–	5–10
Sorghum	10–40	–	–	0–10	0–10	–	–	–
Sugarcane	10–15	–	–	10–25	10–30	–	–	–
Soybean	15–25	–	35.2–50	33–58	10–25	8	–	5–25
Tobacco	10–15	7–15	–	–	20–85	0.9	–	5–30
Citrus	10–30	2–5	–	–	15–65	–	6.8–17.5	–
Coffee	10–20	–	–	–	–	–	–	–
Cotton	8–18	–	–	–	5–25	1.6	18–32	–
Tomato	20–30	8–20	37.5–96.1	30–50	–	2–3	27.2	5–35
Grape	20–30	7–10	–	–	15–75	12–20	11–28	–

TABLE 8.1 *(Continued)*

Crop	Yield loss in percentage (%)							
	Brazil	Greece	Philippines	Russian Federation	Egypt	USA	India	China
Melons	20	6–15	14.3–60.5	–	0–45	10	20	5–30
Okra	35	5–15	–	–	10–60	5.7	14.1	–
Papaya	5–15	–	–	–	0–25	–	–	–
Pepper	5–10	5–15	–	–	0–35	7	–	–
Sugar beet	–	4–8	–	10–25	10–25	5	–	–
Wheat	–	1–4	–	5–10	0–15	–	32.4–66.6	5–30
Rice	–	–	25.5–61.6	75	0–25	4	10.5	5–15

Source: Askary, T. H.; Martinelli, P. R. P., Eds. In *Biocontrol Agents of Phytonematodes*, CABI: Wallingford, UK, 2015.

For example, solarization is expensive, can be technically challenging for farmers, and negatively impacts on the beneficial soil microorganisms.[5,6] Flooding is not suitable for all locations and is dependent on the type of crop and nematode species[7]; resistant cultivars are often highly specific to nematode species and are not readily available to farmers in developing countries[7–9] and cover crops that are considered most effective such as rattlebox (*Crotalaria spectabilis*) and castor (*Ricinus communis*) may be toxic to livestock and are mostly species specific.[10]

8.1.2 IMPACT OF PARASITIC NEMATODES ON AGRICULTURE ECONOMY

Agriculture intensification, poor agronomic practices such as the implementation of monoculture, use of chemical fertilizer, and inconsistent irrigation are the main factors that led to increase in parasitic nematodes and decrease in global crop productivity with a resultant global cost of >$120 billion p.a.[11,12] As per the reports of the Karssen and Moens,[13] Moens et al.,[14] more than 100 RKN species are infecting thousands of plant species and remains a serious threat to the sustainable food production across the globe. Up to now, there are about 3400 species of nematodes that are plant parasitic. The recent reports advocate 250 species from 43 genera reflect phytosanitary risk yet there are may be many species of phytosanitary importance.[15] The most important genera in terms of juvenile diffusion include *Globodera, Heterodera,* and *Meloidogyne* of the family Heteroderidae.

8.1.3 CYST CEREAL NEMATODES

The wheat and barley production has incurred huge losses due to cyst cereal nematodes such as *Heterodera avenae* and is posing a major threat to these crops in many regions of the world. It has been observed that cereal cyst nematode cause significant yield losses especially under rainfed conditions and less irrigated regions of Australia, China, Pakistan, and the United States. Yield loss due to cereal cyst nematode (CCN) are 15%–20% on wheat in Pakistan,[16] 17%–77% on barley and 40%–92% on wheat in Saudi Arabia,[17] 20% on barley and 23%–50% on wheat in Australia,[18] 42% on rainfed wheat in Turkey.[19] There has been a reduction of about 50% on barley yield due to *H. latipons* in Cyprus[20] in China about

33% yield loss in cucumber have been reported by.[21] The nematodes have also caused a huge reduction in yield of about 87%.[22] There are several obstacles in estimating the perfect crop loss caused due to nematodes. The current statistics reflect massive economic loss on the yield of about US $9 million in India, £3 million in Europe, and about AUS $72 million in Australia has incurred due to the CCN infestation.[19]

8.2 MANAGEMENT OF NEMATODES

With regard to population explosion and food security, management of parasitic nematode should be given utmost importance with novel strategies. Because it is too difficult to eliminate the nematodes from the soil, the overall goal of proper management is to reduce the nematode populace to least possible level. As per the reports of Food and Agriculture Organization,[23] it has estimated that more than 800 million people face starvation. Plant diseases play a critical role in decreasing the agricultural productivity and is considered as the biggest threat to sustainable food production across the globe.[24] As per the reports of Nicol et al.,[19] about 12% of world total food production is lost due to PPN. This is a huge quantity to be ignored and hence the need of the hour is to keep it as low as possible. The best viable approach to attain the objective is through the management of nematodes. In past management of nematodes have been principally carried out through the use of chemicals; however, this method is having a wide spectrum of limitation and, therefore, it is mandatory to substitute this with some suitable approach. Chemical nematicides are costly, unavailable, and pose a harmful effect on the environment.[25]

8.2.1 *BIOLOGICAL CONTROL OF NEMATODES*

The apprehension over the deleterious effects of chemical nematicides on the environment and human health has facilitated the need for safer, eco-friendly control measures. The biological control of nematode growth and populace offers a promising substitute for the vigorous use of chemical nematicides. Biological control relying on soil microorganisms may offer feasible and sustainable perspectives in many agrosystems for management of nematodes. This is considered as an appropriate to chemical nematicides as it is cheap, readily available, and eco-friendly; in addition to this, it plays a crucial role in facilitating the sustainability in agricultural production.[26] The reduction in root galling expressed in terms of Root-Knot Index form the basis for the assessment of efficiency of biocontrol agents in the management of PPN.[27] Biocontrol agents also have the ability to improve the plant health and the hence overall productivity. Several organisms have been shown possess antagonistic activity against PPN[28,29] among these fungi have proved to best in terms of efficiency and efficacy.

8.2.2 NEMATOPHAGOUS FUNGI

Nematophagous fungi are a class of microfungi that have the ability to capture, kill, and digest nematodes. Numerous fungal strains have been assessed for their antagonistic action against nematodes and it has been proven that fungi possess some striking features to act as best biocontrol agent of nematodes, for example, enzymes like collagenase, chitinases, and serine protease have the ability to rupture the cuticle of adult nematode or eggshells, thereby causing significant mortality at very early stage.[30] The cuticle of adult nematode is mainly composed of proteins, keratin, collagen, and fibers while as the chitin forms the major portion of nematode egg. Therefore, the degradation of collagen through the fungal collagenase is believed to be essential for the control of nematode population.[30] The fungi employ the association of mechanical activity along with the hydrolytic enzymes to penetrate the nematode cuticle. The enzymes degrade the main constituent (protein and chitin) of nematode cuticle and egg shells.

8.2.3 MECHANISM OF INFECTION

The nematophagous fungi involve a wide array of hydrolytic enzymes in infecting the nematodes and nematode eggs, these are discussed in next section.

8.2.3.1 SUBTILASES

Spatafora et al.[31] have reported that Pezizomycotina contains the filamentous, sporocarp-producing nematophagous fungi is the largest subphylum of Ascomycota. As per the reports of Yang et al,[32] fungal extracellular enzymes have been recognized as a strong tool to penetrate and infect the nematodes. Li et al,[33] are of the opinion that subtilisin-like serine proteases, amongst the extracellular enzymes, have been exalted as the essential enzyme with central roles in penetrating and colonizing their nematode host. In 1990, the first serine protease from *V. suchlasporium* was isolated, purified, characterized, and tested positive for the activity of degrading certain cyst nematode proteins.[30] Importantly, intriguing evidence emerged suggesting that subtilisin-like serine proteases isolated from nematode-trapping fungi regulate important processes of penetration, degradation, and digestion of nematode cuticles.[34-39] Several kinds of serine protease including PII and Aoz1 from *Arthrobotrys oligospora*,[40] pSP-3 from *Paecilomyces lilacinus*[41] and VCP1 from *Paecilomyces chlamydosporia*[42] have been isolated, purified and cloned. These enzymes disrupt the interrupt the physiological integrity of the nematode cuticle that facilitates penetration and colonization.

8.2.3.2 CHITINASES

The potential of fungi as biocontrol agent has proved a great promise to sustain the agriculture and reduce the crop loss incurred by the wide spectrum of nematodes. The degradation of chitin of nematode egg shell through the action of fungal extracellular enzymes has been supported by several workers.[43] Chitinases, a prerequisite for hyphal growth participate in infection of mycoparasites and nematopathogenic fungi are a class of inducible enzyme that catalyzes chitin, the main constituent of nematode cuticles.[44] Tikhonov et al.[45] identified and purified CHI43, the first chitinase with nematicidal activity.

As per the reports of Huang et al.,[30] the activity of chitinase CHI43 obtained from nematophagous fungi of *V. chlamydosporium* and *V. suchlasporium* increased with time when cultured in medium with containing colloidal chitin as the main source of C and N. Chitinase of CHI43 or/ and serine protease of P32 treated eggs of *G. pallida* revealed the prominent effect of CHI43 on degradation of nematode eggshell.[30] Scars and slight peelings were recorded on the egg surfaces when CHI43 or P32 were applied individually, however, a combination of CHI43 and P32 caused more serious damage. Thus, overall experimentation concluded the involvement of chitinase in the disruption of the cuticle of nematode eggshells. Furthermore, it was also revealed that a combination of different hydrolytic enzymes, including chitinase and serine protease, proved to be more efficient in keeping the nematode population density low. Recently, RKN (*Meloidogyne javanica*) was kept under control by the exogenous application of *Trichoderma harzianum* BI[46] and concluded that different concentrations (102–108 spores/mL) of *T. harzianum* BI reduced nematode infection proportionally with the increase in chitinase activity in comparison to control. *Trichoderma* spp. has been studied by different workers as a biological control agent against nematode diseases of crops.[47,48] Several scientists are of the opinion that *T. harzianum* have the potential to act as an effective bioagent for the management of the citrus nematode.[49,50] Windham et al.[51] treated the soil with *T. harzianum* and *T. koningii* preparation and reported a huge reduction in egg production in the RKN *Meloidogyne arenaria*. Seifullah and Thomas[52] studied and confirmed by the low-temperature scanning electron microscopy, the parasitism of *Globodera rostochiensis* by the exogenous application of *T. harzianum*. Rao et al.,[49] and Sharon et al.,[50] advocated that *T. harzianum* isolates can decrease the *M. javanica* infection to a greater extent. They further studied the antagonistic effect in detail and have proposed the mechanism of action in the following two ways:

1. Boost in the activity and/ or pool of chitinase and protease enzymes result in direct parasitism of nematode eggs and larva through the breakdown of chitin and proteins.[50,53]
2. Indirect parasitism involves the development of systemic resistance. Chitinase and protease which reflect antifungal activities appear to participate in the *Meloidogyne javanica Trichoderma* spp. interaction.[50]

Sahebani and Hadavi[46] have shown that *T. harzianum* BI can be used as an efficient biocontrol agent against *M. javanica*. In addition to this, their experimentation also demonstrated that inoculation of tomato seedlings with *T. harzianum* can notably mitigate the population of this nematode and hence disease severity.

8.2.3.3 COLLAGENASE

Collagenases are the group of calcium and zinc-dependent enzymes that have the ability to hydrolyze collagen in their native triple helix and denatured form. Collagenases are becoming increasingly important commercially. Collagenases have the ability to damage the nematode cuticle as the collagen forms the main constituent of nematode cuticle. Bedoya et al.[54] observed the collagenolytic activity of proteolytic enzymes in 10 different isolates of *Paracoccidioides brasiliensis*. Their results indicated that about 70% and 80% of the isolates secrete collagenolytic enzymes. These results further support the view that *P. brasiliensis* has the capability to secrete collagenolytic enzymes and can be used in the control wide of nematode attack. Tosi et al.[55] have demonstrated the antagonistic activity of collagenolytic enzymes secreted by a nematophagous Antarctic fungus *Arthrobotrys tortor* and a wide range of species of the genus *Arthrobotrys* against *Caenorhabditis elegans*. The results showed a threefold increase in the collagenase secreted by the *A. tortor* in comparison to other species and hence it was concluded that this fungus offers a promising strategy for the nematode management. Several workers have advocated that the production of collagenolytic proteases from nematode-trapping fungi can be exploited for commercial purposes.[40] Nematode-trapping fungi, potent biological control agents are unique in capturing the plant parasitic nematodes by employing several trapping devices. These fungi harbor a wide array of trapping devices including adhesive networks, adhesive columns, constricting rings, adhesive knobs, and nonconstricting rings to capture, kill, and digest nematodes. A striking example nematode-trapping fungi is *A. oligospora*, which employs special hype of 3D networks along with the secretion of extracellular enzymes to capture, penetrate, and immobilize nematodes.[30] Schenck et al.[56] have demonstrated the production of collagenase production in Nematode-trapping fungi such as *Arthrobotrys amerospora*.

8.2.4   *VOLATILE ORGANIC COMPOUNDS WITH NEMATICIDAL ACTIVITIES FROM FUNGAL ISOLATES*

Some of workers have also demonstrated that the volatile organic compounds (VOCs) produced by fungi such as *Aspergillus candidus, Penicillium brevicompactum, Penicillium clavigerum, Penicillium cyclopium, Emericella nidulans, Penicillium crustosum, Penicillium expansum, Tritirachium oryzae,* and *Penicillium glabrum* have been shown to harbor great antagonistic activity against plant pathogens in general and nematodes in particular.[57,58] Freire et al.,[59] while working with *Fusarium oxysporum*, identified VOCs substances with very strong immobility and mortality to *M. incognita* and reduced infectivity. They further showed that *Fusarium oxysporum* and *F. solani* isolates also led to 88%–96% mortality to *Meloidogyne incognita* second-stage juveniles. Riga et al.[60] evaluated the VOCs produced by the fungus *Muscodor albus* and recorded that in vitro second stage juveniles (J2) ranging from 82% to 95% in *Paratrichodorus allius, Pratylenchus penetrans,* and *Meloidogyne chitwoodi.*

8.2.5   NEMATOPHAGOUS BACTERIA

Numerous bacterial isolates and most prominent among them are *Pseudomonas* spp. and *Pasteuria spp.*as they possess strong nematicidal activities against RKN.[61-63] A very high number of *Bacillus* isolates have also been identified and found to have nematicidal properties against *M. javanica* in vitro[64] and in vivo[65-67]. Very little is known about the mechanism involved, however, as per the reports of Adam et al.,[68] *Bacillus subtilis* isolates play a crucial role in alleviating gall formation and also induced systemic resistance in tomato plants, thereby reducing nematode infectivity. Padgham and Sikora,[69] also report similar results and advocate that an isolate of *Bacillus megaterium* to inhibit or reduce root penetration and migration of *M. graminicola* to the root zone of rice plants.

In the last few decades, there has been a mounting interest in the exploration of bacterial antagonists of nematodes.[70] Oosterdorp and Sikora,[71] have demonstrated that nematode invasion of roots can be reduced by *Rhizobacteria* treatment to seeds before sowing. During the putrefaction of organic matter in the soil, several bacteria release metabolic byproducts, enzymes, and toxins which may make these organisms as essential natural antagonists of nematodes.

8.3   PASTEURIA PENETRANS

Hewlett et al.[72] have recently, reported another group of nematode antagonists, *Pasteuria penetrans*, cosmopolitan in distribution in agricultural soils feed as obligate parasites of nematodes.[73] Several workers[70,74,75] have advocated that *Pasteuria* species, potential economical and eco-friendly biological control agents can considerably control plant-parasitic nematodes affecting wide range of plants such as egg-plant, wheat, tobacco, tomato, soybean, bean, pepper, peanut, rye, cucumber, chickpea, grape, mung, and okra.

8.3.1   MECHANISM OF INFECTION

Pasteuria penetrans is a potent biocontrol agent for the RKN *Meloidogyne* spp. A four-step proposed mechanism is as follows:

1. Attachment of *Pasteuria* spores to the nematode cuticles.
2. Spore germination inside the roots.
3. Formation and proliferation of microcolonies inside the female nematode
4. Finally, the release of endospores after the disruption of the genital system of the female nematode.[76,77]

8.3.2   RHIZOBACTERIA

Some free-living soil bacteria produce a good amount of nematicidal compounds and can be commercialized. Several researchers have at randomly screened rhizosphere bacteria for nematicidal property about 8% were shown to possess activity. Isolates *of Agrobacterium spp*, *Bacillus spp*, and *Pseudomonas spp*, are known for their antagonistic activity against soil-borne bacterial and fungal pathogens also have the potential to eradicate a wide range of nematodes. Li et al.[78] studied *Rhizobacteria* for inhibition of the RKN and soil-borne fungal pathogens, they further report that isolates such as *Brevibacillus brevis* or *Bacillus subtilis* have proven to have strong nematicidal activity by causing the severe mortality of J2 larvae of *Meloidogyne spp.* to a greater extent. Similarly, Insunza et al.[79] have shown that 16 bacteria isolates out of 44 reduced 50%–100% nematode population densities such as *Paratrichodorus pachydermous* and *Trichodorus primitives*. Ali et al.[80] have reported that soil drenching with *B. subtilis* and *P. aeruginosa* significantly declined the root-rot, root-knot infection, and nematode population in *Vigna mungo*.

8.4 PREDACIOUS NEMATODES

Cobb[81] was the first who gave the idea of using predatory nematode for management of plant-parasitic nematode. Later, Steiner and Heinly[82] reported the use of *Clarkus papillatus* for controlling *Meloidogyne* spp. and other plant-parasitic nematodes in sugar beet fields. Over the past few years, interest in using potent predators such as mononchids, dorylaimids, aphelenchids, and diplogasterids for nematode control has risen. Among them dorylaimida predators have greater efficiency and are highly effective bioagents because of their short life cycles, chemotaxis sense and resistance to adverse conditions.[83] The most beneficial and promising aspect of the dorylaimid, nygolaimid, and diplogasterid predators is that maintenance of their populations is very easy and their resistance to unfavorable environmental conditions as they will remain sufficient in soil even in the absence of prey nematodes.[84] They have the ability to reduce population density of plant parasitic nematodes in virtually all soils and also produce nutrients in the available form to plant, which improves plant health to withstand nematode infectivity. Lal et al.[85] found that an increase in the population of predatory nematode incidentally reduces the population of RKN. Significant decrease in the population densities of *M. incognita* and potato cyst nematode *Globodera rostochiensis* was due to the presence of a predatory nematode, *P. punctatus*.[86] The opinion of Khan and kim[83] stated that dorylaimids are highly effective biocontrol predators because their population densities can be easily enhanced by the addition of organic nutrient.

8.5 SUMMARY

Over the past two decades, a vast study has been undertaken to identify, screen, and evaluate the biocontrol potential of a variety of microorganisms for the management of nematode. But only a few commercial biocontrol products from the beneficial microorganism, such as bacteria, fungi, and other organisms with nematicidal activity, have been commercialised in the agriculture sector. The development and subsequent implementation of bioagents are quite hectic on a large scale.[87] The most important about the development of a commercial biocontrol agent is that it must be capable of targeting host in a laboratory test. In order to develop effective biocontrol strategies, a comprehensive understanding of the mechanisms and vast exploration of the interactions among nematodes, such as soil microbiota, plant microenvironment, and ecological implication must be understood. In the past few years, it has been very well emphasized and reviewed on the interactions between the microorganism and nematode; plant and environment[88–91], including integrated pest management (IPM), are very effective and eco-friendly methods to decline the pest infectivity. The IPM aims at the synergistic impact of biocontrol and other methods, such as biofertilizers, soil nutrient amendments, cultivation of resistant plant varieties, and crop rotation at appropriate times, so that they improve plant health and by increasing rhizospheric colonization and enhance antagonistic activity against nematode.[89,92,93] To achieve the overall motive, we need to have accurate knowledge about biology, ecology, and mechanisms of infection of antagonistic organisms. Detailed insight about the underlying mechanism understanding on the molecular level of the various biocontrol agents not only will lead to a proper and effective nematode management decision, but also could pave a way to the development of novel biocontrol strategies for the management of plant-parasitic nematodes. Advances in molecular biology have widened the horizon of our knowledge and now we are able to explore the underlying molecular mechanisms of infection and signaling pathway that enable the defense system. This may give knowledge about mass production of biocontrol agents that can be used commercially.

KEYWORDS

- biocontrol
- biological agents
- nematode
- root-knot nematodes
- yield loss

REFERENCES

1. Powell, N. T. Interactions Between Nematodes and Fungi in Disease Complexes. *Ann. Rev. Phytopathol.* **1971,** *9* (1), 253–274.

2. Ngangbam, A. K.; Devi, N. B. An Approach to the Parasitism Genes of the Root-Knot Nematode. *Int. J. Phytopathol.* **2012**, *1* (1), 81–87.

3. Gaur, H. S.; Perry, R. N. The Biology and Control of the Plant-Parasitic Nematode *Rotylenchulus reniformis*. *Agric. Zool. Rev.* **1991**, *4*, 177–212.

4. Renčo, M.; Kováčik, P. Response of Plant Parasitic and Free-Living Soil Nematodes to Composted Animal Manure Soil Amendments. *J. Nematol.* **2012**, *44* (4), 329–336.

5. Gaur, H. S.; Perry, R. N. The Use of Soil Solarization for Control of Plant-Parasitic Nematodes. *Nematological Abstracts*, **1991**, *60* (4), 153–167.

6. Kaşkavalci, G. Effects of Soil Solarization and Organic Amendment Treatments for Controlling *Meloidogyne incognita* in Tomato Cultivars in Western Anatolia. *Turk. J. Agric. For.* **2007**, *31* (3), 159–167.

7. Sikora, R. A.; J. Bridge, J. L. Starr. "Management Practices: An Overview of Integrated Nematode Management Technologies." Plant Parasitic Nematodes in Subtropical and Tropical Agriculture, 2nd ed. CAB International: Wallingford, 2005; pp 793–825.

8. Roberts, P. A. Current Status of the Availability, Development, and Use of Host Plant Resistance to Nematodes. *J. Nematol.* **1992**, *24* (2), 213–227.

9. Hockland, S.; Niere, B.; Grenier, E.; Blok, V.; Phillips, M.; Den Nijs, L.; Anthoine, G.; Pickup, J.; Viaene, N. An Evaluation of the Implications of Virulence in Non-European Populations of *Globodera pallida* and *G. rostochiensis* for Potato Cultivation in Europe. *Nematology* **2012**, *14* (1), 1–13.

10. McSorley, R. Host Suitability of Potential Cover Crops for Root-Knot Nematodes. *J. Nematol.* **1999**, *31* (4S), 619–623.

11. Karuri, H. W.; Olago, D.; Neilson, R.; Njeri, E.; Opere, A.; Ndegwa, P. Plant-Parasitic Nematode Assemblages Associated with Sweet Potato in Kenya and Their Relationship with Environmental Variables. *Trop. Plant Pathol.* **2017**, *42* (1), 1–12.

12. Wachira, P. M.; Kimenju, J. W.; Okoth, S. A.; Mibey, R. K. Stimulation of Nematode-Destroying Fungi by Organic Amendments Applied in Management of Plant Parasitic Nematode. *Asian J. Plant Sci.* **2009**, *8* (2), 153–159.

13. Karssen, G.; Moens, M; Root-Knot Nematodes. In *Plant Nematology*, Perry, R. N.; Moens, M., Eds.; CABI Publishing: Wallingford, UK, 2006; pp 59–90.

14. Moens, M.; Perry, R. N.; Starr, J. L. Meloidogyne Species: a Diverse Group of Novel and Important Plant Parasites. In *Root-knot Nematodes;* Roland N. Perry, Maurice Moens, James L. Starr, Eds.; CABI International Publisher: Oxfordshire, UK, 2009; Vol. 1, pp 1–13.

15. Singh, S. K.; Hodda, M.; Ash, G. J. Plant-Parasitic Nematodes of Potential Phytosanitary Importance, Their Main Hosts and Reported Yield Losses. *Eppo Bull.* **2013**, *43* (2), 334–374.

16. Nicol, J. M.; Rivoal, R.; Trethowan, R. M.; Van Ginkel, M.; Mergoum, M.; Singh, R. P. CIMMYT's Approach to Identify and Use Resistance to Nematodes and Soil-Borne Fungi, in Developing Superior Wheat Germplasm. In *Wheat in a Global Environment*; Springer: Netherlands, 2001; pp 381–389.

17. Ibrahim, A. A.; Al-Hazmi, A. S.; Al-Yahya, F. A.; Alderfasi, A. A. Damage Potential and Reproduction of *Heterodera avenae* on Wheat and Barley under Saudi Field Conditions. *Nematology* **1999**, *1* (6), 625–630.

18. Abidou, H.; El-Ahmed, A.; Nicol, J. M.; Bolat, N.; Rivoal, R.; Yahyaoui, A. Occurrence and Distribution of Species of the Heterodera avenae Group in Syria and Turkey. *Nematologia Mediterranea* **2005**, *33* (2), 195–201.

19. Nicol, J. M.; Turner, S. J.; Coyne, D. L.; Den Nijs, L.; Hockland, S.; Maafi, Z. T. Current Nematode Threats to World Agriculture. In *Genomics and Molecular Genetics of Plant-Nematode Interactions*; Springer: Netherlands, 2011; pp 21–43

20. Nicol, J. M.; Rivoal, R. Global Knowledge and Its Application for the Integrated Control and Management of Nematodes on Wheat. In *Integrated Management and Biocontrol of Vegetable and Grain Crops Nematodes*; Springer: Netherlands, 2008; pp 251–294.

21. Kayani, M. Z.; Mukhtar, T.; Hussain, M. A. Effects of Southern Root Knot Nematode Population Densities and Plant Age on Growth and Yield Parameters of Cucumber. *Crop Protect.* **2017**, *92*, 207–212.

22. Jones, J. T.; Haegeman, A.; Danchin, E. G.; Gaur, H. S.; Helder, J.; Jones, M. G.; Kikuchi, T.; Manzanilla-López, R.; Palomares-Rius, J. E.; Wesemael, W. M.; Perry, R. N. Top 10 Plant-Parasitic Nematodes in Molecular Plant Pathology. *Mol. Plant Pathol.* **2013**, *14* (9), 946–961.

23. FAO. World Food and Agriculture. Food and Agriculture Organization of the United Nations: Rome, 2013.

24. Strange, R. N.; Scott, P. R. Plant Disease: A Threat to Global Food Security. Ann. Rev. Phytopathol. **2005**, *43,* 83–116

25. Quesada-Moraga, E.; Herrero, N.; Zabalgogeazcoa, Í. Entomopathogenic and Nematophagous Fungal Endophytes. In *Advances in Endophytic Research.* Springer: India, 2014; pp 85–99

26. Atandi, J. G.; Haukeland, S.; Kariuki, G. M.; Coyne, D. L.; Karanja, E. N.; Musyoka, M. W.; Fiaboe, K. K.; Bautze, D.; Adamtey, N. Organic Farming Provides Improved Management of Plant Parasitic Nematodes in Maize and Bean Cropping Systems. *Agric., Ecosys. Environ.* **2017**, *247*, 265–272.

27. Khan, T.; Shadab, S.; Afroz, R.; Aziz, M. A.; Farooqui, M. Study of Suppressive Effect of Biological Agent Fungus, Natural Organic Compound and Carbofuran on Root Knot Nematode of Tomato (*Lycopersicon esculentum*). *J. Microbiol. Biotechnol. Res.* **2017**, *1* (1), 7–11.

28. Costa, S. R.; van der Putten, W. H.; Kerry, B. R. Microbial Ecology and Nematode Control in Natural Ecosystems. In *Biological Control of Plant-Parasitic Nematodes*; Springer: Netherlands, 2011; pp 39–64.

29. Stirling, G. R. Biological Control of Plant-Parasitic Nematodes: An Ecological Perspective, a Review of Progress and Opportunities for Further Research. In Biological Control of Plant-Parasitic Nematodes; Springer: Netherlands, 2011; pp 1–38.

30. Huang, X.; Zhao, N.; Zhang, K. Extracellular Enzymes Serving as Virulence Factors in Nematophagous Fungi Involved in Infection of the Host. *Res. Microbiol.* **2004**, *155* (10), 811–816.

31. Spatafora, J. W.; Sung, G. H.; Johnson, D.; Hesse, C.; O'Rourke, B.; Serdani, M.; Spotts, R.; Lutzoni, F.; Hofstetter, V.; Miadlikowska, J. Reeb, V. A Five-Gene Phylogeny of Pezizomycotina. *Mycologia* **2006**, *98* (6), 1018–1028.

32. Yang, J.; Tian, B.; Liang, L. and Zhang, K. Q.; 2007. Extracellular Enzymes and the Pathogenesis of Nematophagous Fungi. *Appl. Microbiol. Biotechnol.* **2007**, *75* (1), 21–31.

33. Li, J.; Yu, L.; Yang, J.; Dong, L.; Tian, B.; Yu, Z.; Liang, L.; Zhang, Y.; Wang, X.; Zhang, K. New Insights into the Evolution of Subtilisin-Like Serine

Protease Genes in Pezizomycotina. *BMC Evolut. Biol.* **2010,** *10* (1), https://doi. org/10.1186/1471-2148-10-68.

34. Wang, B.; Liu, X.; Wu, W.; Liu, X.; Li, S. Purification, Characterization, and Gene Cloning of an Alkaline Serine Protease from a Highly Virulent Strain of the Nematode-Endoparasitic Fungus *Hirsutella rhossiliensis. Microbiol. Res.* **2009,** *164* (6), 665–673.

35. Wang, R. B.; Yang, J. K.; Lin, C.; Zhang, Y.; Zhang, K. Q. Purification and Characterization of an Extracellular Serine Protease from the Nematode-Trapping Fungus *Dactylella shizishanna. Lett. Appl. Microbiol.* **2006,** *42* (6), 589–594.

36. Yang, J.; Huang, X.; Tian, B.; Wang, M.; Niu, Q.; Zhang, K. Isolation and Characterization of a Serine Protease from the Nematophagous Fungus, *Lecanicillium psalliotae*, Displaying Nematicidal Activity. *Biotechnol. Lett.* **2005,** *27* (15), 1123–1128.

37. Yang, J.; Li, J.; Liang, L.; Tian, B.; Zhang, Y.; Cheng, C.; Zhang, K. Q. Cloning and Characterization of an Extracellular Serine Protease From the Nematode-Trapping Fungus *Arthrobotrys conoides. Arch. Microbiol.* **2007a,** *188* (2), 167–174.

38. Yang, J.; Liang, L.; Zhang, Y.; Li, J.; Zhang, L.; Ye, F.; Gan, Z.; Zhang, K.Q. Purification and Cloning of a Novel Serine Protease from the Nematode-Trapping Fungus *Dactylellina varietas* and Its Potential Roles in Infection Against Nematodes. *Appl. Microbiol. Biotechnol.* **2007b,** *75* (3), 557–565.

39. Zhang, Y.; Liu, X.; Wang, M. Cloning, Expression, and Characterization of Two Novel Cuticle-Degrading Serine Proteases from the Entomopathogenic Fungus Cordyceps sinensis. *Res. Microbiol.* **2008,** *159* (6), 462–469.

40. Minglian, Z.; Minghe, M.; Keqin, Z. Characterization of a Neutral Serine Protease and its Full-Length cDNA from the Nematode-Trapping Fungus *Arthrobotrys oligospora. Mycologia* **2004,** *96* (1), 16–22.

41. Bonants, P. J.; Fitters, P. F.; Thijs, H.; den Belder, E.; Waalwijk, C.; Henfling, J. W. D. A Basic Serine Protease from *Paecilomyces lilacinus* with Biological Activity Against *Meloidogyne hapla* Eggs. *Microbiology* **1995,** *141* (4), 775–784.

42. Morton, C. O.; Hirsch, P. R.; Peberdy, J. P.; Kerry, B. R. Cloning of and Genetic Variation in Protease VCP1 from the Nematophagous Fungus *Pochonia chlamydosporia. Mycol. Res.* **2003,** *107* (1), 38–46.

43. Seidl, V. Chitinases of Filamentous Fungi: A Large Group of Diverse Proteins with Multiple Physiological Functions. *Fungal Biol. Rev.* **2008,** *22* (1), 36–42.

44. Takaya, N.; Yamazaki, D.; Horiuchi, H.; Ohta, A.; Takagi, M. Intracellular Chitinase Gene from *Rhizopus oligosporus*: Molecular Cloning and Characterization. *Microbiology* **1998,** *144* (9), 2647–2654.

45. Tikhonov, V. E.; Lopez-Llorca, L. V.; Salinas, J.; Jansson, H. B. Purification and Characterization of Chitinases from the Nematophagous Fungi *Verticillium chlamydosporium* and *V. suchlasporium. Fungal Genet. Biol.* **2002,** *35* (1), 67–78.

46. Sahebani, N.; Hadavi, N. Biological Control of the Root-Knot Nematode *Meloidogyne javanica* by *Trichoderma harzianum. Soil Biol. Biochem.* **2008,** *40* (8), 2016–2020.

47. Chérif, M.; Benhamou, N. Cytochemical Aspects of Chitin Breakdown during the Parasitic Action of a *Trichoderma* sp. on *Fusarium oxysporum* f. sp. *radicislycopersici. Phytopathology* **1990,** *80* (12), 1406–1414.

48. Chet, I. *Trichoderma*-Application, Mode of Action and Potential as Biocontrol Agent of Soilborn Plant Pathogenic Fungi. In *Innovative Approaches to Plant Disease Control;* Chet, I., Ed.; John Wiley & Sons: New York, 1987; pp 137–160.

49. Rao, M. S.; Reddy, P. P.; Nagesh, M. Evaluation of Plant Based Formulations of *Trichoderma harzianum* for the Management of *Meloidogyne incognita* on Egg Plant. *Nematologia Mediterranea* **1998**, *26* (1), 59–62.

50. Sharon, E.; Bar-Eyal, M.; Chet, I.; Herrera-Estrella, A.; Kleifeld, O.; Spiegel, Y. Biological Control of the Root-Knot Nematode *Meloidogyne javanica* by Trichoderma *harzianum*. *Phytopathology* **2001**, *91* (7), 687–693.

51. Windham, G. L.; Windham, M. T.; Williams, W. P. Effects of Trichoderma spp. on Maize Growth and *Meloidogyne arenaria* Reproduction. *Plant Dis.* **1989**, *73* (6), 493–495.

52. Seifullah, P.; Thomas, B. J. Studies on the Parasitism of *Globodera rostochiensis* by *Trichoderma horzianum* Using Low Temperature Scanning Electron Microscopy. *Afro-Asian J. Nematol.* **1996**, *6*, 117–122

53. Suarez, B.; Rey, M.; Castillo, P.; Monte, E.; Llobell, A. Isolation and Characterization of PRA1, a Trypsin-Like Protease from the Biocontrol Agent *Trichoderma harzianum* CECT 2413 Displaying Nematicidal Activity. *Appl. Microbiol. Biotechnol.* **2004**, *65* (1), 46–55.

54. Bedoya-Escobar, V. I.; Naranjo-Mesa, M. S.; Restrepo-Moreno, A. Detection of Proteolytic Enzymes Released by the Dimorphic Fungus Paracoccidioides brasiliensis. *J. Med. Vet. Mycol.* **1993**, *31* (4), 299–304.

55. Tosi, S.; Annovazzi, L.; Tosi, I.; Iadarola, P.; Caretta, G. Collagenase Production in an Antarctic Strain of Arthrobotrys Tortor Jarowaja. *Mycopathologia* **2002**, *153* (3), 157–162.

56. Schenck, S.; Chase, T.; Rosenzweig, W. D.; Pramer, D. Collagenase Production by Nematode-Trapping Fungi. *Appl. Environ. Microbiol.* **1980**, *40* (3), 567–570.

57. Schalchli, H.; Tortella, G.R.; Rubilar, O.; Parra, L.; Hormazabal, E.; Quiroz, A. Fungal Volatiles: An Environmentally Friendly Tool to Control Pathogenic Microorganisms in Plants. *Crit. Rev. Biotechnol.* **2016**, *36* (1), 144–152.

58. Fischer, G.; Schwalbe, R.; Möller, M.; Ostrowski, R.; Dott, W. Species-Specific Production of Microbial Volatile Organic Compounds (MVOC) by Airborne Fungi from a Compost Facility. *Chemosphere* **1999**, *39* (5), 795–810.

59. Freire, E. S.; Campos, V. P.; Pinho, R. S. C.; Oliveira, D. F.; Faria, M. R.; Pohlit, A. M.; Noberto, N. P.; Rezende, E. L.; Pfenning, L. H.; Silva, J. R. C. Volatile Substances Produced by *Fusarium oxysporum* from Coffee Rhizosphere and Other Microbes Affect *Meloidogyne incognita* and *Arthrobotrys conoides*. *J. Nematol.* **2012**, *44* (4), 321–328.

60. Riga, E.; Lacey, L. A.; Guerra, N. Muscodor Albus, a Potential Biocontrol Agent Against Plant-Parasitic Nematodes of Economically Important Vegetable Crops in Washington State, USA. *Biol. Control* **2008**, *45* (3), 380–385.

61. Cho, M. R.; Na, S. Y.; Yiem, M. S. Biological Control of *Meloidogyne arenaria* by *Pasteuria penetrans*. *J. Asia-Pacific Entomol.* **2000**, *3* (2), 71–76.

62. Timper, P.; Kone, D.; Yin, J.; Ji, P.; Gardener, B. B. M. Evaluation of an Antibiotic-Producing Strain of *Pseudomonas fluorescens* for Suppression of Plant-Parasitic Nematodes. *J. Nematol.* **2009**, *41* (3), 234 p.

63. Bagheri, N.; Ahmadzadeh, M.; Heydari, R. Effects of *Pseudomonas fluorescens* Strain UTPF5 on the Mobility, Mortality and Hatching of Root-Knot Nematode *Meloidogyne javanica*. *Arch. Phytopathol. Plant Protect.* **2014**, *47* (6), 744–752.

64. Ashoub, A. H.; Amara, M. T. Biocontrol Activity of Some Bacterial Genera Against Root-Knot Nematode, *Meloidogyne incognita. J. Am. Sci.* **2010,** *6* (10), 321–328.

65. Oliveira, D. F.; Campos, V. P.; Amaral, D. R.; Nunes, A. S.; Pantaleão, J. A.; Costa, D. A. Selection of Rhizobacteria Able to Produce Metabolites Active Against *Meloidogyne exigua. Eur. J. Plant Pathol.* **2007,** *119* (4), 477–479.

66. Wei, L.; Shao, Y.; Wan, J.; Feng, H.; Zhu, H.; Huang, H.; Zhou, Y. Isolation and Characterization of a Rhizobacterial Antagonist of Root-Knot Nematodes. *PloS One* **2014,** *9* (1), e85988.

67. Xiong, J.; Zhou, Q.; Luo, H.; Xia, L.; Li, L.; Sun, M.; Yu, Z. Systemic Nematicidal Activity and Biocontrol Efficacy of *Bacillus firmus* Against the Root-Knot Nematode *Meloidogyne incognita.* World *J. Microbiol. Biotechnol.* **2015,** *31* (4), 661–667.

68. Adam, M.; Heuer, H.; Hallmann, J. Bacterial Antagonists of Fungal Pathogens Also Control Root-Knot Nematodes by Induced Systemic Resistance of Tomato Plants. *PloS One.* **2014,** *9* (2), e90402.

69. Padgham, J. L.; Sikora, R. A. Biological Control Potential and Modes of Action of *Bacillus megaterium* Against *Meloidogyne graminicola* on Rice. *Crop Protect.* **2007,** *26* (7), 971–977.

70. Dickson, D. W.; Oostendorp, M.; Giblin-Davis, R. M.; Mitchell, D. J. Control of Plant-Parasitic Nematodes by Biological Antagonists. *Pest Manag. Subtrop., Biol. Control - FL Perspect.* **1994,** 575–601

71. Oostendorp, M.; Sikora, R. A.; Seed Treatment with Antagonistic Rhizobacteria for the Suppression of *Heterodera schachtii* Early Root Infection of Sugar Beet. *Rev. Nematol.* **1989,** *12* (1), 77–83.

72. Sayre, R. M.; Starr, M. P. Bacterial Diseases as Antagonists of Nematodes. In *Diseases of Nematodes,* Poinar, G. O, Jansson, H. B., Eds.; CRC Press: Boca Raton, FL, 1988; Vol. 1, pp 70–101.

73. Chen, Z. X.; Dickson, D. W.; McSorley, R.; Mitchell, D. J.; Hewlett, T. E. Suppression of Meloidogyne Arenaria Race 1 by Soil Application of Endospores of *Pasteuria penetrans. J. Nematol.* **1996,** *28* (2), 159–168.

74. Hewlett, T. E.; Cox, R.; Dickson, D. W.; Dunn, R. A. Occurrence of *Pasteuria* spp. in Florida. *J. Nematol.* **1994,** *26* (4S), 616–619.

75. Weibelzahl-Fulton, E.; Dickson, D. W.; Whitty, E. B. Suppression of *Meloidogyne incognita* and *M. javanica* by *Pasteuria penetrans* in Field Soil. *J. Nematol.* **1996,** *28* (1), 43–49.

76. Mankau, R.; Imbriani, J. L.; Bell, A. H. SEM Observations on Nematode Cuticle Penetration by *Bacillus penetrans. J. Nematol.*1976, *8* (2), 179–181.

77. Sayre, R. M.; Wergin, W. P. Bacterial Parasite of a Plant Nematode: Morphology and Ultrastructure. *J. Bacteriol.* **1977,** *129* (2), 1091–1101.

78. Li, B.; Xie, G. L.; Soad, A.; Coosemans, J. Suppression of *Meloidogyne javanica* by Antagonistic and Plant Growth-Promoting Rhizobacteria. *J. Zhejiang Univ. Sci. B* **2005,** *6* (6), 496–501.

79. Insunza, V.; Alström, S.; Eriksson, K. B. Root Bacteria from Nematicidal Plants and Their Biocontrol Potential Against Trichodorid Nematodes in Potato. *Plant Soil* **2002,** *241* (2), 271–278.

80. Ali Siddiqui, I.; Ehetshamul-Haque, S.; Shahid Shaukat, S. Use of Rhizobacteria in the Control of Root Rot–Root Knot Disease Complex of Mungbean. *J. Phytopathol.* **2001**, *149* (6), 337–346.

81. Cobb, N.A. A genus of Free-Living Predatory Nematodes: Contributions to a Science of Nematology VI: (With 75 Illustration in the Text). *Soil Sci.* 1917, *3* (5), 431–486.

82. Steiner, G.; Heinly, H. The Possibility of Control of *Heterodera radicicola* and Other Plant-Injurious Nemas by Means of Predatory Nemas, Especially by *Mononchus papillatus* Bastian. *J. Washington Acad. Sci.* **1922**, *12* (16), 367–386.

83. Khan, Z.; Kim, Y. H. A Review on the Role of Predatory Soil Nematodes in the Biological Control of Plant Parasitic Nematodes. *Appl. Soil Ecol.* **2007**, *35* (2), 370–379.

84. Webster, J. M. Nematode and Biological Control. In *Economic Nematology*; Webster, J. M. Ed.; Academic Press: New York, 1972; p 563.

85. Lal, A.; Sanwal, K.C.; Mathur, V. K. Changes in the Nematode Population of Undisturbed Land with the Introduction of Land Development Practices and Cropping Sequences. *Indian J. Nematol.* **1983**, *13* (2), 133–140.

86. Small, R. W. The Effects of Predatory Nematodes on Populations of Plant Parasitic Nematodes in Pots. *Nematologica* **1979**, *25* (1), 94–103.

87. Meyer, S. L. United States Department of Agriculture–Agricultural Research Service Research Programs on Microbes for Management of Plant-Parasitic Nematodes. *Pest Manag. Sci.* **2003**, *59* (6–7), 665–670.

88. Kerry, B. R. Rhizosphere Interactions and the Exploitation of Microbial Agents for the Biological Control of Plant-Parasitic Nematodes. *Ann. Rev. Phytopathol.* **2000**, *38* (1), 423–441.

89. Barker, K.R. Perspectives on Plant and Soil Nematology. Ann. Rev. Phytopathol. **2003**, *41* (1), 1–25.

90. Davies, K. G. Interactions between Nematodes and Microorganisms: Bridging Ecological and Molecular Approaches. *Adv. Appl. Microbiol.* **2005**, *57*, 53–78.

91. Dong, L. Q. Zhang, K. Q. Microbial Control of Plant-Parasitic Nematodes: A Five-Party Interaction. *Plant Soil* **2006**, *288* (1–2), 31–45.

92. Barker, K. R.; Koenning, S. R. Developing Sustainable Systems for Nematode Management. *Ann. Rev. Phytopathol.***1998**, *36* (1), 165–205.

93. Meyer, S. L.; Roberts, D. P. Combinations of Biocontrol Agents for Management of Plant-Parasitic Nematodes and Soilborne Plant-Pathogenic Fungi. *J. Nematol.* **2002**, *34* (1), 1–8.

CHAPTER 9

Principles and Concepts of Integrated Nematode Management in Major Crops

B. S. SUNANDA[1,*] and RAVULAPENTA SATHISH[2]

[1]*Assistant Scientific Officer (Nematology) and Centre In-charge, AICRP (Nematode)*

[2]*Senior Research Fellow (Entomology), National Institute of Plant Health Management (NIPHM), Hyderabad 500030, India*

Corresponding author. E-mail: patilsunanda722@gmail.com

ABSTRACT

Plant parasitic nematodes are hidden enemies of crops. The symptoms of the damage they cause are not easily recognizable. Their microscopic size further reduces the chances of being recognized as the causal organisms of any damage. Some nematodes predispose plants to other pathogens, while other nematodes act as various vectors. On a worldwide basis, these worms are estimated to cause crop yield losses of over US$ 78 billion. These losses are estimated to be over 14% in the developing countries and about 9% in the developed countries. For the management the use of disease or pathogen free planting materials is most recommended method of nematode exclusion. However, in case where there is no availability of such material, treatment of infected plant material is essential before planting. In current chapter we are summarizing the loss caused by nematode, historical development of nematode study, and their management strategies.

9.1 INTRODUCTION

Erstwhile, the Andhra Pradesh state (which includes the present states of Andhra Pradesh and Telangana) is considered as seed hub of South

India with more than 400 seed companies and seedling nurseries around Hyderabad. The state produces major cereals, millets, pulses, vegetables, fruits, and commercial crops like cotton, tobacco, and sugarcane. In addition, protected cultivation in polyhouses is catching up by mostly growing vegetables and cut flowers. Nematodes which are often least considered for management have become serious threats for majority of horticultural and field crops both under open field and protected cultivation. Though efforts on identification of plant-parasitic nematodes of Andhra Pradesh are very less but have started more than five decades ago. Progress on management of nematode diseases in Andhra Pradesh is limited to a few crops. In 2014, the state was bifurcated into Andhra Pradesh and Telangana. In the erstwhile Andhra Pradesh, universities such as Osmania University, Hyderabad and Acharya NG Ranga Agricultural University (now PJTSAU, Rajendranagar, Bapatla, and Tirupathi campuses), Hyderabad, national institutes like Directorate of Rice Research (now IIRR), Directorate of Oilseeds Research (IIOR), National Institute of Plant Health Management (NIPHM), and National Bureau of Plant Genetic Resources, (NBPGR) Regional Station and international institute, International Crops Research Institute for Semi-Arid Tropics conducted nematological activities.

9.1.1 AGRO CLIMATIC ZONES AND MAJOR CROPS OF ANDHRA PRADESH

The agrarian state has a geographical area of 1,62,760 sq km spread over in 13 districts. The state is a part of the peninsular shield and has three physiographic divisions. There are five different types of soils to cultivate a wide range of crops. Red soils occupy about 65% of the state, while black soils account for about 25%. The alluvial soils of riverine and marine deposits occupy about 5% of the area. The others are coastal sands, laterite soils, and so on. Out of 162.76 lakh ha, 62.35 lakh ha (38.3 %) area is put to net cultivation including fisheries during 2014–15. The state has 36.63 lakh ha forest (22.5%) and 22.59 lakh ha (13.9%) fallow lands. Based on the amount and distribution of rainfall, the state has been divided into six Agro-climatic zones (Table 9.1) The rainfall of the state ranges between 500 mm in the scarce zone and 1400 and above, in high altitude tribal areas. Among the food crop grown, rice dominates in the state followed by sugarcane, groundnut, cotton, and mesta. Other principal crops grown are soghum, bajra, maize, ragi, horse gram, black gram, green gram, red gram, cowpea, sesame, sunflower, and tobacco are popular. Horticulture sector also contributes to the state GDP. The major fruits cultivated are mango, banana, acid lime, orange, amla, plantations, guava, papaya, sapota, pomegranate, custard apple, musk melon, watermelon, and cashew nut. With regard to vegetables, tomato, brinjal, chili, onion, tapioca, cucurbits, beans, cabbage, cauliflower, and drumstick are the prominent ones. Commercial crops such as sugarcane, tobacco, turmeric, ginger, curry leaf, coconut, betel leaf, arecanut, tamarind, and coffee are also under cultivation. Flowers such as jasmine, rose, crossandra, marigold, and chrysanthemum provide income as well as employment.

TABLE 9.1   Agro-climatic Zones of Andhra Pradesh.

Sl. No.	Agro climatic Zone	Districts
1	North coastal	Srikakulam, Vizayanagaram, and Visakhapatnam
2	Godavari zone	East Godavari and West Godavari
3	Krishna	Krishna, Guntur, and Prakasam
4	Southern	Chittoor, YSR Kadapa, and S.P.S.Nellore
5	Scarce rainfall	Kurnool and Ananthapur
6	High altitude and tribal	Srikakulam, Visakhapatnam and East Godavari

Source: Department of Agriculture, Andhra Pradesh (2014–2015).

The state has a total population of 49.83 million (Census, 2011), with density of 308 per sq km and literacy level of 67.41%. With these potentials, the state has gained first rank in terms of fish production, fish exports, and egg production.

9.1.2   AGRO CLIMATIC ZONES AND MAJOR CROPS OF TELANGANA

The state is bestowed with diverse tropical and sub-tropical climatic conditions. The region has an area of 114.84 lakh ha and population of 352.87 lakhs as per 2011 census. The Krishna and Godavari rivers flow through the state from West to East. Based on climatic parameters that is, rainfall, soils, and cropping pattern and so on, the state (10 districts) is divided into four agro-climatic zones (Table 9.2). The agricultural planning for each zone is supported with the research and recommendations of Regional Agricultural Research Stations of Professor Jaya Shankar, Telangana State

Agriculture University (PJTSAU) set up within each zone. More than 51% of the total geographical area is under cultivation and 23.89% under forest. Agriculture is dominated by rain-fed cultivation.

TABLE 9.2   Agro Climatic Zones of Telangana.

Sl. no.	Name of the zone	Districts
1	Northern Telangana zone	Karimnagar, Nizamabad, and Adilabad
2	Central Telangana zone	Warangal, Khammam, and Medak
3	Southern Telangana zone	Mahbubnagar, Nalgonda, and Rangareddy (+ Hyderabad)
4	High altitude and tribal areas zone	High altitude and tribal areas of Khammam and Adilabad districts

Source: Department of Agriculture Telangana (2014–2015).

In Telangana state crops grown in both *Kharif* and *Rabi* seasons put together cover an area of 53.15 lakh ha. The important crops grown are rice 14.15 lakh ha, maize 6.91 lakh ha, pulses 3.11 lakh ha, groundnut 0.12 lakh ha, cotton 16.93 lakh ha, chilies 0.73 lakh ha, and sugarcane 0.72 lakh ha. 75% of area was sown in *kharif* and the remaining area of 25% was cultivated in *Rabi* season during 2014–15. Horticulture sector in Telangana has emerged as a potential player in the economy. Enterprising and progressing farming community is willing to adopt new technologies that is, green houses, mulching, drip automation and so on. Horticulture sector contributes approximately 5.16% GSDP (Rs. 18,703 crores) of the state. Telangana stands 1st in turmeric production. In vegetables, it stands 11th in area and 13th in production. In case of fruits, the state stands 3rd in area and 8th in production. Vegetables constitute 71% of the total horticulture cropped area followed by spices and flowers. Export potential is very high for mango, banana and vegetables, and flowers. International Airport has Potential to become major export hub.

9.2 HISTORICAL DEVELOPMENTS OF NEMATOLOGY IN ANDHRA PRADESH AND TELANGANA STATES

Research on nematodes in united Andhra Pradesh dates back to 1960s when V. M. Das first published his work on soil nematodes of Andhra Pradesh (Das, 1960). Root-knot nematode problem on tobacco led to creation of the post of nematologist by ICAR in Central Tobacco Research

Institute and early work on tobacco nematode management is classical from Andhra Pradesh. Several entomologists were trained in Nematology from the Agricultural University completing their Ph.Ds from the Division of Nematology, Indian Agricultural Research Institute, New Delhi in late 60s and early 70s. However, nematology work worth mentioning happened after the report of Kalahasti Malady from South zone of Andhra Pradesh and the classical work was done including development of resistant varieties of groundnut is a landmark in the history of Nematology. Nematology could not be established as a separate Division in the University of Agriculture till now. A nematologist position filled in Central Plant Protection and Training Institute, Hyderabad (now known as National Institute of Plant Health Management) started training in Nematology and also reported the widespread occurrence of white tip nematode on paddy in the region. A new regional station was started in 1985 by the National Bureau of Plant Genetic Resources of ICAR. Nematologist from NBPGR Regional Station, Hyderabad started operating for quarantine processing of global germplasm exchange and evaluation of germplasm for nematode resistance. In 1986, a nematologist post was filled in International Crops Research Institute for Semi-Arid Tropics and the work on survey to detect nematode problems, screening germplasm of groundnut, pigeon pea, and chickpea started systematically including the research on nematode management. Later, a nematologist post was filled in Directorate of Rice Research (now known as Indian Institute of Rice Research), Hyderabad. Rice nematode problems detection, germplasm screening, and their management received greater attention from the region. Indian Institute of Oilseeds Research filled its Nematologist position recently and the nematode problems of oilseeds started receiving attention. Some of the significant events in the history of nematology research in the erstwhile Andhra Pradesh are given below:

• 1960—First Nematology Thesis on Soil Nematodes in India from Erstwhile Andhra Pradesh.
• 1971—Pigeon pea cyst nematode (*Heterodera cajani*) reported on pigeon pea in Andhra Pradesh.
• 1971—Distribution of (*Heterodera avenae, H. zeae, H. cajani, and Anguinatritici*) in India. *Indian J. Nematol.* 1, 106–111.

- 1975–76—Kalahasti Malady, A Serious Disease Caused by Nematode, *Tylenchorhynchus brevilineatus* in Groundnut was Recorded from Nellore and Chittoor districts.
- 1984—*A Nematode Disease of Peanut Caused by Tylenchorhynchus brevilineatus. Plant Dis.*, 68 (6), pp 526–529.
- 1979—Widespread Damage to Paddy due to White-tip Nematode, *A. besseyi* in Ranga Reddy District Around Hyderabad.
- 1979—A Serious Outbreak of White Tip Nematode Disease *Aphelenchoides besseyi* in Rice Crops at Hyderabad. *Indian J. Plant Prot.* 7, 218–219.
- 1986—*Meloidogyne javanica* Recognised as Pest on Acid Lime/ Citrus Orchards
- 1986—Occurrence of *Meloidogyne javanica* on Citurs in Andhra Pradesh (India). *Int. Nematol. Network Newsl.* 3, 9–10.
- 1978—Reported Banana Burrowing Nematode (*Radopholus similis*) in Major Banana Growing Areas of Andhra Pradesh.
- 1978—Occurrence and Distribution of *Radopholus similis* (Cobb, 1983) Thorne, 1949 in South India. *Indian J. Nematol.* 8, 49–58.
- 1991—Systematic Survey Conducted Root Knot Nematode (*Meloidogyne incognita*) Occurrence and Damage in Vegetables.
- 2008—Diversity and Community Structure of Major Plant Parasitic Nematodes in Selected Districts of Andhra Pradesh, India, *Indian J. Nematol.* 38, 68–74.
- 1994—Citrus Nematode *Tylenchulus semipenetrans* predominantly, Followed by Reniform Nematode *Rotylenchulus reniformis*, Spiral and Lesion Nematodes Recorded on Citrus.
- 1994—Occurrence and Distribution of *Tylenchulus semipenetrans* in Andhra Pradesh. *Indian J. Nematol.* 24, 106–111.
- 1995—First Report of Rice Root-knot Nematode, *Meloidogyne graminicola*.
- 1991—Reported Root-knot Nematode, (*Meloidogyne javanica*) Race-3 is an Important Nematode Parasite of Groundnut.
- 1991—Occurrence of *Meloidogyne jauanica* on Groundnut in Andhra Pradesh, India. *Indian J. Nematol.*, 21, 166.
- 1995—Host Races of *Meloidogyne jauanica,* with Preliminary Evidence that the "Groundnut race" is Widely Distributed in India. *Int. Arachis Newsl.,* 15: 43–44.
- 2001—Ufra Nematode Reported from Andhra

- 2001—Ufra Nematode, (*Ditylenchus angustus*) is Seed Borne Crop Protection 21(1), 75–76.
- 2005—Record of Ufra Nematode, *Ditylenchus angustus* on Rice in Andhra Pradesh, India. *Oryza* 42 (3), 242–243.
- 2001—25% of Maize Growing Area Infested with Maize Cyst Nematode, *Heterodera zeae*
- 2006—Incidence of *Meloidogyne incognita* on Pomegranate was Reported from Anathapur District of Andhra Pradesh.
- 2007—A New Report of Root-knot Nematode, *Meloidogyne incognita* on Pomegranate, (*punica granatum*) from Andhra Pradesh. *Indian J. Nematol.* 2007, 37 (2), 201.
- 2006—*M. incognita, Pratylenchus coffeae* and *Radopholus similis* on Banana from Andhra Pradesh.
- 2006—Community Structure of Plant Parasitic Nematodes in Banana Plantations of Andhra Pradesh, India. *Indian J. Nematol.* 36, 209–212.
- 2014—Establishment of Separate Full fledge Nematology Laboratory at National Institute of Plant Health Management (NIPHM) Rajendranagar, Hyderabad.
- 2016—First Center of All India Coordinated Research Project on Nematodes and cropping systems, sanctioned by division of Nematology, ICAR – IARI, New Delhi NIPHM Hyderabad as voluntary Center of AICRP (Nematodes) vide F.No. CS 4-4/2014-PP, September 7, 2016.

9.3 DIVERSITY OF PHYTO-NEMATODES IN ANDHRA PRADESH

In view of diverse ecosystems and a large number of crops cultivated, huge nematode diversity is reported in the erstwhile Andhra Pradesh. An account of diversity of plant-parasitic nematode species and known races is given below:

9.3.1 SEDENTARY ENDO PARASITES

Root knot nematodes, *M. incognita, M. javanica, M. arenaria*, and *M. graminicola* are reported. *M incognita* is a pest on pulses, oilseeds, cotton, vegetables, tuber crops, betel vine, and among the fruits, banana, pomegranate, muskmelon, and grapes. *M. incognita* is reported on several other hosts including corkwood tree. Prevalence of *M. incognita* race-2 was confirmed in tomato, chili, and brinjal ecosystems. *M. javanica* is reported to be pest on groundnut, vegetables including gherkin and banana. Race-3 of *M. javanica* on groundnut is prevalent in the south and scarce rainfall zones of Andhra Pradesh. *M. arenaria* on groundnut and *M. graminicola* mostly on upland rice are known to occur. Root-knot nematodes are a major problem in protected cultivation on vegetables and horticultural nurseries known from Telangana. Recently, *M. enterolobii* has become a menace for guava nurseries and cultivation in Telangana. Cyst nematodes, *Heterodera cajani* on pigeon pea and other pulses and *H. zeae* on maize are major pests while other cyst nematodes reported are *H. sorghi* on sorghum, *H. raskii*, and *Bilobodera* spon grasses are reported. *H. cajani* is also reported on castor from Telangana.

9.3.2   SEMI-ENDOPARASITES

Reniform nematode is pest on maize, pulses particularly pigeon pea, oilseeds particularly castor, sunflower and groundnut, cotton, turmeric, vegetables, banana, grapes, acid lime, and sweet orange. Renifrom nematode is highly prevalent on grapes and castor in Telangana, while it occurs in the south zone of Andhra Pradesh on citrus species while on cotton it occurs in both AP and Telangana. *Tylenchulus semipenetrans* on acid lime and sweet orange was predominant among the plant parasitic nematode community having the highest absolute frequency, absolute density, and prominence value that were recorded maximum level in clay soils followed by clay loam, sandy loam, and laterite soils. Ufra nematode, *Ditylenchus angustus* has been found infecting irrigated rice in Godavari delta of Andhra Pradesh.

9.3.3   MIGRATORY ENDOPARASITES

The Burrowing nematode, *Radopholus similis* is reported as pest on banana from coastal ecosystem and also known to infest several vegetables. Lesion nematodes reported on rice, maize, pulses, groundnut, cotton, vegetables, acid lime, sweet orange, and banana. *Pratylenchus penetrans, P. pratensis, P. zeae, P. coffeae, P. thornei*, and *P.* delattrei are predominant lesion nematode species. Lesion nematodes are important on rice, maize, cotton, and vegetables both in Andhra Pradesh and Telangana. In Andhra Pradesh, lesion nematodes are pests mainly on banana, groundnut, and citrus. *P. zeae* on maize, *P. coffeae* on banana, and *P. thornei* on pulses deserve attention for further studies. *Hirschmannella oryzae* is a pest commonly encountered in irrigated rice ecosystems.

9.3.4 ECTOPARASITES

Stunt, spiral, lance, dagger, and stubby root nematodes and several other ectoparasites are reported both from Telangana and Andhra Pradesh. Most important disease that attracted attention is Kalahasti malady of groundnut attributed to *Tylenchorhynchus brevilineatus* from the south zone of AP. Later in one of the surveys in ground affected with Kalahasti malady in south zone of AP, *M. arenaria* was also detected. *T. capitatus, T. digitatus*, and several other unknown species are reported to be associated with several crops. Among the spiral nematodes, *Helicotylenchus multicinctus* is the most important on banana, although it is reported on several other crops such as maize, pulses, oilseeds, and vegetables. Other species reported are *H. dihystera, H. imperialis, H. incises*, and so on. Several other ectoparasites are recorded in the rhizosphere of different crops that include *Macroposthonia ornata*, some species of *Ditylenchus, Basirolaimus, Xiphinema*, and *Trichodorus*.

A large number of nonplant parasites (predators, parasites, entomopathogens, and other soil nematodes) are reported from Andhra Pradesh and Telangana which are not covered as it is beyond scope of this article.

9.4 MAJOR NEMATODE PROBLEMS OF THE STATE

9.4.1 NEMATODE DISEASE OF RICE

Rice being a staple food crop is being infected by about 300 nematode species belonging to 35 genera. Among them, nematode species from ten genera are economically important in rice production of different agroecological conditions. Three are reported from Andhra Pradesh they are *Ditylenchus angustus* responsible for *Ufra* disease, *Meloidogyne graminicola* responsible for root knot-disease and *Pratylenchus* spp. causing lesion disease (Prasad et al., 2012).

TABLE 9.3 Plant-parasitic Nematodes Reported from Different Districts of Erstwhile Andhra Pradesh State (Includes Present Telangana and Andhra Pradesh States)

District	Nematode	Crop	Source
Ananthapur	*Meloidogyne incognita*	Pomegranate	Sudheer et al. (2007)
Chittoor	*Meloidogyne incognita* race-2	Tomato, chilli, brinjal, okra, French bean, gherkin, onion, bitter gourd, pulses, banana	Naidu et al. (2007)
	Helicotylenchus sp.	Tomato, chilli, gherkin, groundnut, pulses, paddy, sorghum, mango, marigold, crossandra, chrysanthemum	
	Radopholus similis	Tomato, chilli, bitter gourd, and banana	
	Rotylenchulus reniformis	Tomato, chilli, brinjal, onion, groundnut, and pulses	
	Tylenchorhynchus brevilineatus	Groundnut, pulses, and paddy	
Kurnool, Kadapa, Prakasam, Mahaboobnagar, and Ananthapur	*Rotylenchulus reniformis, Helicotylenchus dihystera, Meloidogyne incognita, Pratylenchus delattrei, Hoplolaimus* sp., *Xiphinema* sp., *Tylenchorhynchus* sp., *Helicotylenchus incises, Tylenchus* sp., *Hirschmanniella oryzae, Aphelenchus* sp., *Tylenchorhynchus capitatus, Criconema* sp., *Meloidogyne javanica, Mononchus* sp.	Tomato, brinjal, okra, chilli, onion, cabbage, ridge gourd, bitter gourd, snake gourd, cucurbits, beans, and cluster bean	Prasad Rao et al. (2007)
Khammam	*Rotylenchulus reniformis, Meloidogyne incognita, Pratylenchus pratensis, Haplolaimus* spp., *Tylenchorhychus* spp., *Xiphenema* spp., *Pratylenchus* spp.	Cotton	Murali and Vanita Das (2014).
Nellore	*Tylenchorhychus brevilineatus*	Groundnut	Reddy et al. (1984)

TABLE 9.3 *(Continued)*

District	Nematode	Crop	Source
Chittore and Nellore	*Bitylenchus brevilineatus, Pratylenchus* spp., *Basirolaimus* spp., *Tylenchorhynchus* spp. *Basiria* spp., *Ecphyadophora* spp., *Nothotylenchus* spp., *Sakia* spp., and *Tenunemelus* sp. *Heterodera* spp. *Meloidogyne* spp. *Rotylenchulus Reniformis*	Groundnut	Mani and Ratnakumar (1990).
Anantapur	*Tylenchulus semipenetrans,*	Citrus	Mani (1994)
Chittoor	*Rotylenchulus reniformis,*		
Cuddapha	*Tylenchorhynchus* spp.,		
Godawari East	*Helicotylenchus* spp.,		
Godawari West	*Neorylenchus* spp.,		
Guntur	*Pratylenchus* spp.,		
Karimnagar	*Basirolaimus* spp.,		
Khammam	*Xiphinema* spp.,		
Karnool	*Meloidogyne* spp		
Nellore			
Prakasam			
Warangal			
Hyderabad	*Aphelenchoides besseyi*	Rice	Savitri et al. (1998)
Hyderabad	*Heterodera raskii* n. sp.	Bulb grass (*Cyperusbulbosus*)	Basnet and Jayaprakash (1984).
Rangareddy	*Meloidogyne incognita* race 2	Tomato, chilli, and brinjal	Kiranbabu et al. (2011)

TABLE 9.3 *(Continued)*

District	Nematode	Crop	Source
West Godavari, East Godavari, Guntur, Nellore	*Meloidogyne incognita*	Banana	
West Godavari, East Godavari, Guntur, Nellore	*Pratylenchus coffeae*	Banana	
West Godavari	*Radopholus similis*	Banana	
West Godavari, East Godavari, Guntur, Nellore	*Helicotylenchus multicinctus*	Banana	Sundararaju (2006)
West Godavari, East Godavari, Guntur, Nellore	*Hoplolaimu* ssp.	Banana	
West Godavari, East Godavari, Guntur, Nellore	*Rotylenchulus reniformis*	Banana	
West Godavari, East Godavari, Guntur, Nellore	*Tylenchorhynchus* sp.	Banana	
Nellore	*Radopholus similis*	Banana	
	M. javanica	Banana	
East Godavari	*Rotylenchu* ssp.	Banana	
Hyderabad	*Hoplolaimus seinhorsti*	Cauliflower	Luc (1957)

TABLE 9.3 *(Continued)*

District	Nematode	Crop	Source
Hyderabad	*Pratylenchus thornei*	Maize	Singh and Khan (1981)
AP	*Heterodera sorghi*	Sorghum	Sharma and Sharma (1988)
AP	*Heterodera zeae*	Maize	Kaushal et al. (2007)
Warangal	*Rotylenchus reniformis*	Cotton	Vindhyarani and
Warangal	*Heterodera* sp., *Helicotylenchus* sp.	Maize	Raghuramulu (2011)
Warangal	*Meloidogne incognita, Xiphinema* sp., *Helicotylenchus* sp., *Tylenchorhychus* sp., *Pratylenchu* ssp.	Okra	
Warangal	*Meloidogne incognita, Helicotylenchus* sp., *Helicotylenchus* sp., *Tylenchorhychus* sp., *Pratylenchus* sp.	Tomato	
Warangal	*Rotylenchus reniformis*	Chilli	
Warangal	*Meloidogne incognita, Pratylenchus* sp., *Hoplolaimus* sp.	Brinjal	

9.4.1.1 CROP LOSSES

Estimation of crop losses due to these nematodes in Andhra Pradesh is not exactly assessed. However, 5% to 100% losses are reported from other epidemic states in case of *Ufra* disease, 16% to 32 % in root-knot disease and 13% to 33 % in lesion nematode incidence.

MAP 9.1   **(See color insert.)**

9.4.1.2 SYMPTOMS

Infection of *D. angustus* causes mosaic or chlorotic discoloration in emerging leaves. Yellowish or pale green splash-patterns on affected leaves and leaf sheaths are noticed. The appearance of brown to dark brown spots on leaves and leaf sheaths is common. At the reproductive stage of crop, nematodes reach the space between the inner sides of imbricate whorl of leaf sheaths to feed on the ear primordia and developing ear heads. As a result, ear heads emerge in a twisted and crinkled manner with empty spikelet or do not emerge at all. The combined symptoms are called as *Ufra*.

TABLE 9.4 Crop Wise Distribution of Plant Parasitic Nematodes in Andhra Pradesh and Telangana States.

Sl. no.	Crop	Nematode reported
1	Maize	*Helicotylenchus*
		Pratylenchus zeae
		Heterodera zeae
2	Pulses	*Pratylenchus coffeae*
		Pratylenchus
		*Meloidogyne*s spp
		Rotylenchulus reniformis
		Heterodera cajani
		Pratylenchus thornei
		Helicotylenchus spp.
3	Oilseed crops	*Meloidogyne* spp.
		Rotylenchulus reniformis
		Tylenchorhynchus spp.
		Helicotylenchus spp.
4	Vegetable crops	*Meloidogyne* spp.
		Rotylenchulus reniformis
		Tylenchorhynchus spp
		Helicotylenchus spp.
5	Fiber crops	*Meloidogyne incognita*
		Rotylenchulus reniformis
		Helicotylenchus spp.
		Pratylenchus spp.

TABLE 9.4   *(Continued)*

Sl. no.	Crop	Nematode reported
6	Banana	*Helicotylenchus multicinctus*
		Pratylenchus spp.
		Meloidogyne incognita
		Rotylenchulus reniformis
		Radopholus similis
		Pratylenchus spp.
7	Citrus	*Tylenchulus semipenetrans*
8	Grape	*Meloidogyne incognita*
		Rotylenchulus reniformis
9	Betelvine	*Meloidogyne incognita*
10	Tuber crops	*Meloidogyne incognita*

Source: Economically Important Plant Parasitic Nematodes Distribution ATLAS–2010, ICAR, New Delhi.

Root-knot nematode affected plants show depletion in vigor, stunted growth, chlorotic, and curled leaves in nurseries and main field. The nematode infection is characterized by the formation of small galls near the tips of the roots. Excessive branching of affected roots occurs.

In case of lesion nematode infestation, plant show chlorosis of leaves, stunted, and smothered growth in patches, swollen with water-soaked lesions which develop into black necrotic lesions on the root surface are noticed. Often infected roots decay, when such plants are pulled infected root portions and associated population remain in the soil.

9.4.1.3   MANAGEMENT

Ufra disease

9.4.1.3.1   Cultural Management

Burning of infested stubbles, preventing flood water from the river, completely drying fields when they are fallowed, plowing to destroy loci of infection in stubbles and rotation with non-host crops are the best practices. Growing a non-host crop under crop rotation is also recommended.

9.4.1.3.2   Host Plant Resistance

Cultivars such as IR63142-J8-B-2-1, Rayada 16-06, CN 540, NC 493, TCA 55, Brazil-65, Rayada 16-05, Rayada 16-06, Rayada 16-07, Rayada 16-08, Rayada 16- 011, Rayada 16-013, Ba Tuc, AR 9, IR 13437-20-P1, and IR 17643-4 are found tolerant to resistant to Ufra.

9.4.1.3.3   Chemical management

It is managed by application of carbofuran @ 0.75 kg a.i. /ha at transplanting and 1.5 kg a.i. /ha in main field. Two foliar sprays with carbosulfan 40 EC at 0.2% followed by two sprays of triazophos 40 EC at 0.2% is also effective.

9.4.2   ROOT-KNOT NEMATODE

9.4.2.1   CULTURAL MANAGEMENT

Soil application of FYM, Crop rotation with non-host crops which is, sweet potato, cowpea, sesamum, castor, sunflower, soybean, turnip, and cauliflower inhibit nematode development. In situ green manuring with marigold and burning of 15 cm deep rice hulls are also useful in reducing nematode population.

9.4.2.2   BIOLOGICAL MANAGEMENT

Application of *Pseudomonas flourescens* @ 20 g/m^2 was found to be effective in reducing the nematode numbers. Isolates of *Trichoderma* are also the potential biological control agents of *M. graminicola* in rice.

9.4.2.3   CHEMICAL MANAGEMENT

Carbofuran, phorate, isazophos, cartap, carbosulfan, or quinalphos when given as soil application @ 1 kg a.i. /ha significantly reduce the root galling.

9.4.3   PIGEON PEA CYST NEMATODE

Heterodera cajani was found to cause 16%–34% of yield losses and in maize *Rotylenchulus reniformis* was reported to cause 8% losses. Three chickpea cultivars (N 31, N 59, and ICCC 42) and a promising chickpea breeding line (ICCV 90043) have been identified as tolerant to the root-knot nematode, and two promising short-duration pigeon pea breeding lines (ICPL 83024 and ICPL 85045) and selections from medium-duration lines (ICPLs 8357, 85068, 85073, 89050, 89051, and 90097) have been identified as tolerant to the reniform nematode (Sharma, 1997).

9.4.4   KALAHASTI MALADY OF GROUNDNUT

During 1975–76, a severe disease of peanut characterized by reduction in pod size and brownish discoloration of pod surface was noticed near Kalahasti village of Andhra Pradesh and since then it is popularly known as Kalahasti malady. This disease was reportedly caused by the nematode *Tylenchorhynchus breveliniatus* (Reddy et al., 1984)

9.4.4.1   CROP LOSSES

Estimation of crop losses due to Kalahasti malady was limited to visual estimation and was accounted to cause 40%–60% yield loss.

9.4.4.2   SYMPTOMS

Affected plants have small, brownish yellow lesions on pegs, pod stalks, and developing young pods. The margins of the lesions will be slightly elevated because of the proliferation of host cells around the lesion. Pod stalks were reduced in length, kernel becomes discolored. Affected plants will be stunted and dark green in color.

9.4.4.3   MANAGEMENT

Field application of aldicarb 10G or carbofuron 3G @ 6 kg a.i. /ha reduced the nematode population considerably (Reddy et al., 1984).

9.4.5 ROOT-KNOT NEMATODE IN POMEGRANATE

Pomegranate, a commercial fruit crop of the state was found infested by *Meloidogyne incognita* in Ananthapur district during the survey conducted in *kharif* (2006).

9.4.5.1 CROP LOSSES

Infested plants were stunted resulting in economic yield losses and more than 5-year-old plants were worst affected with small and shriveled fruits unfit for marketing.

9.4.5.2 DAMAGE SYMPTOMS

Infested plants exhibited yellowing of foliage resulting in stunted plant growth with less number of fruits or undersize fruits or no fruits which might be due to the nematode-induced nutritional deficiency. In severe cases, galls were predominantly found on entire root system. The young galls were white in color turned to light brown and hardy when they became old. The intensity of root-knot nematode damage increased with increase in age of the plant. Five-year old plants were severely affected by root-knot nematode (Sudheer et al., 2007).

9.4.5.3 MANAGEMENT

There are technologies officially communicated or recommended for management of this nematode in pomegranate. However, farmers are tackling the problem on their own by adopting crop rotation and application of neem cake.

9.4.6 LESION NEMATODE

9.4.6.1 CULTURAL MANAGEMENT

Crop rotation with *Phaseolus radiates* decreases the root-lesion nematode, application of neem or mahua, mustard, karanj, pongamia, groundnut, or cotton cakes will reduce the nematode population. Growing greengram or blackgram as inter crop or in rotation with rice also helps in reducing the populations of lesion nematodes

9.4.6.2   CHEMICAL MANAGEMENT

Application of carbofuran or phorate @ 1 kg a.i./ha soil in the affected crops reduces the nematode injury and avert losses in grain yield up to 48.5%.

9.5   TECHNOLOGIES DEVELOPED

Over the past four years, National Institute of Plant Health Management has been advocating ad-hoc recommendations for management of root-knot nematode which is most commonly noticed. Use of FYM and neem cake fortified with *Paecilomyces lilacinus, Trichoderma viride,* and *Pseudomonas fluorescens* is tested, found promising and recommended for poly houses especially for hybrid vegetables where pesticide residue is a major concern. Crop rotation with non-hosts has also been found in field conditions for vegetables such as tomato, chili, and brinjal to reduce root-knot nematode population. In case of guava, *Meloidogyne enterolobii* is of immediate concern of everyone stakeholder. Strict quarantine measures are advised to nursery units engaged in importing of seedlings. Field management of this nematode by integrated approach using chemicals and biocontrol agents is at final stage and after validation, it will be transferred to other farmers during next season. Besides this, Indian Institute of Rice Research (IIRR) is developing technologies for integrated nematode management in rice and also working on identification of resistant cultivars in collaboration with AICRP (Nematodes). In case of pigeon pea and chickpea ICRISAT has developed resistant lines but very limited literature is available about their practical exploitation.

9.6   EMERGING NEMATODE PROBLEMS

9.6.1   GUAVA DECLINE

Guava decline, a deadly disease due to root-knot nematodes (*Meloidogyne* sp.) and *Fusarium* wilt disease complex is spreading at an alarming rate in

Andhra Pradesh and Telangana states. The disease was first time recorded from Rangareddy, Sangareddy, Medak, and Nalgonda districts of Telangana, and East and West Godavari districts of Andhra Pradesh by National Institute of Plant Health Management (NIPHM) in 2015.

9.6.1.1 CROP LOSSES

In infected fields, incidence of this disease varied from 40% to 80% and in some nurseries, 90%–100% incidence was noticed. Nurseries growers had no other option but to destroy an entire lot of saplings in order to prevent its further spread and produce new healthy saplings. On an average 30%–60% yield loss is caused due to this disease and in severe cases, up to 100% loss is observed in guava orchards (Figs 9.1 and 9.2).

FIGURE 9.1   (See color insert.) Guava orchard showing infestation of *Meloidogyne entrolobii.*

FIGURE 9.2   (See color insert.) Guava seedlings and roots showing galls and drying symptoms.

9.6.1.2 SYMPTOMS

It causes chlorosis, stunted growth, wilting, extensive root galling, partial to complete rotting of roots which ultimately cause death of plants both in nurseries and orchards. The symptoms start with yellowing of plants followed by withering, giving the tree barren look and roots revealing a dirty root appearance with beaded knots. Plants become flaccid, broken reveal a hollow twig. Wilted leaves, leaf dropping, drying of branches, and decline in productivity are also witnessed. Underground symptoms included root galls and partially rotted roots. Roots are completely galled of varying size with a dirty appearance and many of them were compound galls. In association with the wilt causing fungi, there is extensive rotting of roots and death of plants within months.

9.6.1.3 MANAGEMENT

9.6.1.3.1 Cultural Practices

- Movement of nematode infected root stocks across the states should be strictly restricted.
- Use of nematode-free saplings for planting.
- Removal and destruction of nematode infected saplings or trees.
- Maintaining the orchard free from weeds and alternate hosts.

9.6.1.3.2 Nematode Management in Nurseries

Treatment of soil mixture used for raising guava rootstocks:

- A ton of soil mixture has to be mixed with 50–100 kg of neem cake or pongamia cake enriched with the bio-pesticides such as 1 kg *Paecilomyces lilacinus*, 1 kg *Pseudomonas fluorescens*, and 1 kg *Trichoderma harzianum*.
- 5 kg of Carbofuran/phorate can also be added to one ton of soil mixture.

9.6.1.3.3 Integrated Management Practices

- Apply 3–4 kg of bio-pesticide enriched vermicompost/farm yard manure (FYM)/compost per plant at an interval of 3–4 months.
- Mix 20 kg of bio-pesticide enriched neem cake/pongamia cake in 200 L water, leave it for two days. This can be used for drenching @ 2–3 L/plant or filter it thoroughly and use it for sending along with the drip, once at an interval of 15–20 days.
- Farmers have been advised to apply carbofuron 3G @ 100 g/plant mixed with 1 kg sand to facilitate uniform application around the trunk of each plant. After 15 days, apply neem cake @ 500 g/plant fortified with *Trichoderma viridi, Pseudomonas fluroscens,* and *Paecilomyces lilicinous* bio-control agents (100:2:2:1).
- Neem cake will act as bio nematicide and bio-control agent will restrict the growth of the parasitic nematodes by way of parasitism, competition, and suppression. For edible and fleshy fruit crops application of neem cake fortified with bio-control agents and the use of neem oil for drenching are safe.
- Use FYM 15 kg/plant as source of nutrition fortified with *Trichoderma* and *Pseudomonas* (100:1:1) to enhance the impact of bio-control agents and encourage early recovery.
- It is advised to apply neem cake 200 g/pit fortified with bioagents at the time of planting as a precautionary measure. As a precautionary measure, farmers should avoid infested seedlings for planting.
- For nursery operators, proper soil solarisation is recommended to cure their potting soil for nematodes and other microorganisms. The chemical soil fumigants available in the market shall be used under the supervision of plant protection officials.

Both in Telangana and Andhra Pradesh, new nematode problems are emerging every year. Due to poor knowledge of plant-parasitic nematode diseases and their symptoms among the growers, often nematode diseases go unnoticed. Adoption of modern cultivation technologies such as polyhouse cultivation, greenhouse, and shade house cultivation has attracted the attention toward nematode diseases. Because nematodes have become number one enemies of all these protected cultivation practices. In both the states, flower and vegetable crops under protected cultivation are suffering from severe nematode diseases. In open field also, vegetables such as tomato, brinjal, chili, okra, capsicum, and so on are suffering from various kind of nematode infestations. Field crops such as pigeon pea, rice, and groundnut are no longer free from nematodes. The synergistic effect of nematode and wilt causing *Fusarium* spp. in pigeon pea has become a complex challenge for breeders and pathologists.

Horticulture is the major revenue generating sector in both the states and is very lucrative. Many are engaged in nursery entrepreneurship importing and marketing of various fruits, flowers, and hybrid seedlings. However, nematodes are hindering this market, especially quarantine nematodes. Recently NIPHM has identified Root-knot nematode *Meloidogyne enterolobaii* based on perennial pattern in guava fields causing severe wilt and death of affected trees. The source was traced to the guava seedlings imported from Taiwan and Bangladesh. This quarantine nematode has caused complete loss of affected guava orchards. Recently the Project Coordinator, All India Coordinated Research Project (AICRP), Nematodes and his team have visited the affected orchards and prepared an action plan to mitigate this problem in coordination with NIPHM.

9.7   MAJOR EXTENSION ACTIVITIES

Management of nematode disease was a major challenge among the farmers of both Telangana and Andhra Pradesh. In order to educate farmers and disseminate the technologies for effective management, various extension activities were implemented regularly by nematology section of NIPHM across both the states.

- About 12 on campus and 7 off-campus trainings were organized to farmers for creating awareness about nematode diseases, their symptoms, economic importance, and management.
- Over the past four years, more than 300 farmers' fields were visited for diagnosis of nematode diseases and recommended management practices.
- Demonstrations were organized for nematode management in vegetables using organic amendments mixed with biopesticides.
- 19 exclusive training programs on nematodes were organized to the ICAR/SAU/KVK scientists, agriculture, horticulture, and extension officials from state Department of Agriculture on detection and diagnosis of nematode disease in fields and poly houses.

- Participated in exhibitions, kisan melas, and workshops to show-case the technologies of eco-friendly management of plant-parasitic nematodes.
- Information on nematodes was published in pamphlets, bulletins, and folders in local language and distributed among the farmers.
- A one-day workshop was organized to horticultural officers and nursery growers at Kadiyam, Rajmandary in Andhra Pradesh in collaboration with Nursery Growers Association to sensitize about the importance of quarantine nematodes in guava and its management.
- Stake holders' meetings are also organized especially of protected cultivation growers and progressive horticultural farmers to get the feedback of technologies transferred and revalidation of adopted technologies.
- Analysis of soil for nematode infestation and issuing test report for the establishment of new polyhouses growers.
- Providing consultancy service for the farmers about nematode management in horticulture crops.

9.8 SUCCESS STORIES OF TECHNOLOGY ADOPTION AND NEMATODE MANAGEMENT

Response to technology adoption in nematode management is very good in case of guava cultivation. However, the result is very quick and encouraging in protected cultivation. Numbers of poly houses around Hyderabad and guava growers have been adopted nematode control measures suggested by NIPHM and are successfully operating.

KEYWORDS

- **nematode**
- **horticultural crops**
- **economic loss**
- **symptomology**
- **management stratgies**

REFERENCES

Basnet, C. P.; Jayaprakash, A. *Heterodera raskiin.* sp. (Heteroderidae: Tylenchina), a Cyst Nematode on Grass from Hyderabad. *J. Nematol.* **1984,** *16* (3), 213–216.

Das, V. M. Studies on Nematode Parasites of Plants in Hyderabad. *Zeitschriftt fur Prasitenkunde* **1960,** *19,* 553–605.

Joshi, P. K.; Singh, N. P.; Singh, N. N.; Gerpacio, R. V.; Pingali, P. L. *Maize in India: Production Systems, Constraints and Research Priorities*; D.F., CIMMYT: Mexico, 2005.

Kaushal, K. K.; Srivastava, A. N.; Pankaj, Chawla, G.; Singh, K. Cyst forming nematodes in India: A Review. *Indian J. Nematol.* **2007,** *37,* 1–7.

Kiranbabu, T.; Raghuprakash, K. R.; Varaprasad, K. S.; Sivaramakrishnan, S.; Anuradha, G. Identification of Prevalent Race of Root Knot Nematode in Ranga Reddy District of Andhra Pradesh. *Pest Manag. Horticult. Ecosyst.* **2011,** *17* (2), 156–158.

Koshy, P. K.; Swarup, G. Distribution of *Heterodera avenae, H. zeae, H. cajani and Anguina tritici* in India. *Indian J. Nematol.* **1971,** *1,* 106–111.

Mani, A. Occurrence of *Meloidogyne javanica* on citurs (India) *Int. Nematol. Netwk. News* **1986,** *3,* 9–10.

Mani, A.; Ratna Kumar. Plant Parasitic Nematodes Associated with Groundnut in Andhra Pradesh. *Indian J. Nematol.* **1990,** *20* (1), 44–48.

Mani, A. Occurrence and Distribution of *Tylenchulus semipenetrans* in Andhra Pradesh. **1994,** *24* (2), 106–111.

Murali, A.; Vanita Das, V. Biodiversity of Plant Parasitic Nematodes Associated with Cotton in Khammam District of Andhra Pradesh, India. *Inter. J. Pharm.* **2014,** *5* (10), 795–797.

Naidu H.; Harinath, P.; Haritha, V.; John Sudheer, M. Community Analysis of Plant Parasitic Nematodes in Chittoor district of Andhra Pradesh. *Ind. J. Nematol.* **2007,** *37* (2), 207–211.

Prasad, J. S.; Somasekhar, N.; Varaprasad, K. S. Status of Rice Nematode Research in India, 2012; pp 1–7, http://www.rkmp.co.in.

Prasada Rao, G.M.V.; John Sudheer, M.; Priya, P. Community Analysis of Plant Parasitic Nematodes Associated with Vegetable Crops in Selected Districts of Andhra Pradesh. *Ind. J. Nematol.* **2007,** *37* (2), 221–213.

Reddy, D. D. R.; Subrahmanyam, P.; Sankara Reddy, G. H.; Raja Reddy, C.; Siva Rao, D. V. A Nematode Disease of Peanut Caused by *Tylenchorhynchus brevilineatus. Plant Dis.* **1984,** *68,* 526–529.

Savitri, H.; Wahab, T.; Sattar, M. A.; Reddy, B. M.; Wahab, T. Prevalence of White Tip Nematode (*Aphelenchoides besseyi* Christie) in Rice Samples of Andhra Pradesh. *J. Res.* **1998,** *26,* 74–76.

Sharma, R.; Prasad, J. S. First Record of *Meloidogyne graminicola* on Rice in Andhra Pradesh. *Oryza* **1995,** *32,* 59.

Sharma, S. B. In *Diagnosis of Key Nematode Pests of Chickpea and Pigeonpea and Their Management.* Proceedings of a Regional Training Course, 25-30 Nov 1996, ICRISAT, Patancheru, India. Patancheru 502 324, Andhra Pradesh, India: International Crops Research Institute for the Semi-Arid Tropics, 1997, p 112.

Sharma, S. B.; Sharma, R. Occurrence of the Sorghum Cyst Nematode, f-Ieteroderasorghi in Andhra Pradesh. *Ind. J. Nematol.* **1988,** *18,* 329.

Singh, D. B; Khan, E. Morphological Variations in Populations of *Pratylenchus thornei* Sher and Allen, 1953. *Ind. J. Nematol.* **1981,** *11,* 53–60.

Sudheer; Mohan, J.; Kalaiarasan, P.; Senthamarai, M. Report of Root-Knot Nematode, *Meloidogyne incognita* on Pomegranate, *Punicagranatum* L. from Andhra Pradesh. *Ind. J. Nematol.* **2007,** *3* (2), 201–202.

Sundararaju, P. Community Structure of Plant Parasitic Nematodes in Banana Plantations of Andhra Pradesh, India. *Indian J. Nematol.* **2006,** *36* (2), 209–212.

Vindhya Rani, P.; Raghu Ramulu, G. Distribution of Plant–Parasitic Nematodes in Selected Agro-ecosystems. *Asian J. Animal Sci.* **2011,** *6* (2), 203–205.

PART III
Biocontrol

CHAPTER 10

Lichens: A Novel Group of Natural Biopesticidal Sources

VINAYAKA S. KANIVEBAGILU* and ARCHANA R. MESTA

Department of Botany, Kumadvathi First Grade College, Shimoga Road, Shikaripura 577427, Shimoga, Karnataka, India

Corresponding author. E-mail: ks.vinayaka@gmail.com

ABSTRACT

Lichens are commonly used as spices and they have been used in traditional medicine from age-old days in various parts of India and world. The lichens have unique secondary metabolites when compared with higher plants and have found their uses in many fields such as medicine, dyes, cosmetics, deodorants, preservatives, and also in biopesticides. Large quantities of agricultural products are wasted every year due to insects, pests, and weeds. Various chemicals have been used to overcome this problem. The synthetic pesticides have an adverse effect on the health of human beings and other creatures and also lead to different types of environmental pollution. There are about 800 known metabolites from lichens such as salazinic acid, usnic acid, gyrophoric acid, stictic acid, and so on. Among these secondary metabolites, most are found to have pesticidal effect. Barbatic acid and barbatolic acids show antimicrobial activity. Diffractaic acid and evernic acids are known for their fungicidal effects. Hence, the present study aims at the biological control of pests and insects by utilizing the lichen substances as a potential source.

10.1 INTRODUCTION

Lichens are one of the important floral communities and they play a key role in ecological succession. Lichens are the simplest form of plants consisting of a symbiotic phenotype of nutritionally specialized fungi in association with green algae or cyanobacterium. The phycobiont help in food synthesis through photosynthesis mechanism; mycobiont gives protection to the algae. Lichens were well known as the first colonizer of xeric succession on the earth[1] and are distributed in all forms of environments from cold desert to evergreen forests of the tropical regions. There are a wide range of habitats throughout the world which dominate about 8% of terrestrial ecosystem. About 20,000 lichens are reported from all over the world.

Higher level of defense chemicals are produced in the organisms which grow slowly in the low resource habitats.[2] Lichens are very slow growing organisms; they are well known for their active metabolites, and mostly these compounds are synthesized by mycobiont of lichen.[3] The metabolites produced by lichens are unique in nature. The lichen metabolites are weak acids as they help in the breakdown of sedimentary rocks and provide a loose substrate for other nonvascular and vascular plants. These secondary metabolites play an ecological role like antiallergens, antiherbivory in lichens.[4] Lichen substances such as usnic acid, barbatic acid, fumaric acid, and protocetraric acids act as allergens and diffractic acid, barbatic acid, and isousnic acid affect the growth of higher plants.[5,6]

10.2   IMPORTANCE OF LICHENS

These lichen substances have a wide range of uses. These are used in human and animal food, medicines, preparation of dyes, perfume industries, and also in the pollution monitoring problems. Lichens are capable of curing disease like blood and heart problems, brochiolities, antiobisity, piles, dyspepsia, scabies, digestive problems, and so on.[7,8,9]

Medicinal uses of lichens have been practiced since centuries in countries like America, Europe, China, and India.[10] In recent years, researchers have focused on the application aspects of lichens. The pesticidal effect of lichen makes them aone important biological insecticide.[11,12]

10.3   HISTORY AND DEVELOPMENT OF BIOPESTICIDES

Food is the basic necessity of all living organism including human beings. According to Food and Agricultural Organization (FAO), in 2001 the population of the world was estimated to be 6.134 billion. The population of the world in 2016 was 7.4 billion and this population is found to be increasing year by year especially in developing countries. This increasing population requires additional agricultural production every year. To fulfill the requirement of additional agricultural products, the biodiverse agricultural land should not be converted into monoculture agricultural land. Instead, the scope toward the improvement in crop productivity should be encouraged. About 40–42% of the produced crops are destroyed by the pests, insects, diseases, and weeds.[13,14] To overcome this problem many chemical fertilizers, pesticides, and insecticides have been used. The use of these synthetic pesticides and insecticides causes the environmental pollution such as water, air, and soil pollution; they also affect the health of humans and other organisms. This has made the researchers to think about the natural products from plants and other organisms as a potential source of pesticides and insecticides, which supports an eco-chemical method in the pest control.[15] The uses of higher plants as antifungal, anti-microbial, insecticidal, and pesticidal material have been well documented in different parts of the world.[16,17] In traditional methods plant parts like leaves and roots were used for protective storage of crops.[18,19] Hence these are best source for the development of biochemical-based pesticides. Till today, about 6000 species of angiosperm belonging to 235 families have been recorded for anti-insect property.[20]

10.4 POTENTIAL UTILIZATION OF LICHEN METABOLITES

The utilization of lichen as traditional medicine has been known since the time of very first civilization.[21] Because of the presence of different secondary metabolites, lichens have been used for medicines, perfumes, cosmetics, dyes, food, and as pesticides. Many of the lichen substances exhibit antimicrobial, antioxidant, cytotoxic properties, and reported as potential source of pharmaceutically useful chemical.[22] These lichen secondary metabolites have importance in ecological key roles for protecting against biotic (herbivore, competition) and abiotic (UV light) activities.[23] The lichen having unusual chemical substances have potential bioactive sources used for development of novel biopesticides[23] (Table 10.1). More than 50–60% of lichens showed with antibacterial or antibacterial or antiviral properties.[24] Many researchers have been carried out to prove the antifungal and antimicrobial activity of the lichens.

TABLE 10.1   Activity of Lichen Metabolites.

Sl. no.	Compound name	Chemical class	Use or potential use
1.	(–)-16⟨-hydroxykaurane	Diterpene	Cytotoxic
2.	(+) and (–) isousnic acid	Usnic acid	Antimicrobial fungicidal
3.	(+) and (–) usnic acid	Usnic acid	Antimicrobial fungicidal herbicidal
4.	16-*O*-acetylleucotylic acid	Triterpene	Antimicrobial
5.	7-acetoxy-22-hydro-hopane	Triterpene	Antimicrobial
6.	Alectosarmentin	Dibenzofuran	Antimicrobial
7.	Atranorin	*para*-depside	Fungitoxic
8.	Barbatic acid	*Para-depside*	PSII inhibitor
9.	Didymic acid	*Dibenzofuran*	Antimicrobial
10.	Diffractaic acid	*Para- depside*	Fungitoxic
11.	Durvilldiol	*Triterpene*	Antimicrobial
12.	Durvillonol	*Triterpene*	Antimicrobial
13.	Entothein	Dibenzopyranone	Bactericidal
14.	Epanorin	Pulvinic acid	Antifeedant
15.	Eulecanorol	Triterpene	Antimicrobial
16.	Fallacinal	Anthraquinone	Antimicrobial
17.	Friedelin	Triterpene	Antimicrobial
18.	Gyrophoric acid	Para-depside	PSII inhibitor
19.	Haemathamnolic acid	meta- depside	PSII inhibitor
20.	Hiascic acid	para- depside	Fungitoxic
21.	Lecanoric acid	para- depside	Fungitoxic PSII inhibitor
22.	Leprapinic acid	Pulvinic acids	Antibacterial
23.	Leucotylic acid	Triterpene	Antimicrobial
24.	Leucotylin	Triterpene	Antimicrobial
25.	Phlebic acid	Triterpene	Antimicrobial
26.	Pinastric acid	Pulvinic acids	Atiherbivory
27.	Polyporic acid	Terphenylquinone	Antibacterial
28.	Pulvinic dilactone	Pulvinic acids	Antifeedant
29.	Pyxinic acid	Triterpene	Antimicrobial
30.	Retigeranic acid	Triterpene	Antimicrobial
31.	Retigerdiol	Triterpene	Antimicrobial
32.	Rhizocarpic acid	Pulvinic acids	Antifeedant
33.	Rugulosin	Anthraquinone	Antimicrobial

TABLE 10.1 *(Continued)*

Sl. no.	Compound name	Chemical class	Use or potential use
34.	Stictaurin	Pulvinic acids	Antifeedant
35.	Taraxene	Triterpene	Antimicrobial
36.	Thiophanic acid	Xanthones	Fungicidal
37.	Thiophaninic acid	Xanthones	Fungicidal
38.	Triterpene C	Triterpene	Antimicrobial
39.	Triterpene D	Triterpene	Antimicrobial
40.	Ursolic acid	Triterpene	Cytotoxic
41.	Vulpinic acid	Pulvinic acids	Anti-herbivory
42.	Zeorin	Triterpene	antimycobacterial

10.4.1 ANTIFUNGAL ACTIVITY

Antifungal and antimicrobial activity of extracts from *Usnea* species is experimentally proved.[25] Different concentrations of lichen extracts have variable degree of lichen activity.[26] *Everniastrum cirrhatum, Nephroma arcticum, Parmelia tinctorum, Ramalina farinacea, Telochistes flavicans,* and *Usnea undulata* show effective antifungal activity.[27,28,29]

Heterodermia leuocomela, E. cirrhatum, Leptogium sp., *Lobaria* sp., *Cladonia* species were effective against humans, plant, and mammalian pathogenic fungi.[30,31] *Letharia vulpine* and *Peltigera rufescens* shows toxicity against *Sitophilus granaries*.[32]

10.4.2 ANTIMICROBIAL ACTIVITY

Secondary metabolites like depsides and depsidones group of compounds such as pulvinic acid, usnic acid, aliphatic acids, and orcinol type are well known for their effective antimicrobial activity.[33] Leishmanicidal and antiprotozoal activity of secondary metabolites from lichens were first reported by Fournet.[34]

Evernia prunastri, Everniastrum cirrhratum, Platismatia glauca, Parmelia flaventior, Parmotrema pseodotinctorum, Parmelia saxatilis, P. kamstachandalis, Ramalina hossei, R. pollinaria, R. polymorpha, Umbilicaria nylanderiana, and *Usnea pictoides* were well known for their antimicrobial activity[35–42] (Table 10.1). *Cladonia delicate, C. glauca,*

C. borbonica, Parmelia conspersa, P. physodes, P. rudecta, Thamnolia vermicularis, Umbilicaria papulosa, Xanthoria parietiana show antibacterial activity.[43]

The extracts from *Rocella montagnei, Heterodermia burnetiae, R. hossei,* and *H. burnetiae* are known to have an inhibitory effect against diseases like dental caries, burn, and urinary tract infections and other infections which are caused by the microorganisms such as *Streptococcus aureus, Enterococcus faecalis, E. coli,* and *Klebsiella pneumonia.*[44] Usnic acid is reported for inhibitory effect against tuberculosis-causing bacteria[45] and *Staphylococcus cureus.*[46]

10.4.3   INSECTICIDAL ACTIVITY

Many lichen secondary metabolites are known for their insecticidal activity. Lichen substances such as usnic acid, selazinic acid, sekikaic acid, tannins, terpenoides, and steroids are well known for its insecticidal effect.[47]

The extracts of *Heterodermia leucomela, P. tinctorum, P. pseudotictorum, Pyxine consocians, Ramalina nervulosa, R. pacifica, R. hossei, R. conduplicans, Rocella montagnei, Usnea galbinifera,* and *U. longissima* are well known for their insecticidal, larvicidal, and antihelminthic activity.[40,47–51] Some special compounds like 4-*O*-methylcryptochlorophaeic acid, lichexanthone, 3,6-dimethyl-2-hydroxy-4-methoxybenzoic acid, and cabraleadiol monoacetate showed significant larvicidal activity.[51]

10.4.4   ANTIHERBIVORY

The growth of the plant is inhibited by many of the lichen substances.[51–53] These lichen substances act as antifeedants for insects and animals.[54–58] The secondary metabolites from lichens have the capacity to inhibit the seed germination in vascular plants and spore germination in mosses.[55]

10.5   CONCLUSION

The use of synthetic pesticide causes dangerous environmental pollution and lead to the death of nontargeted organism. Hence, the use of biopesticide has been encouraged as there is no harm to the environment and also to other organisms. The secondary metabolites produced by lichens were unique in nature and they have produced a diverse group of secondary metabolites like depsides, despidones, and pulvinic acid. The antimicrobial, antiinsecticidal, and antiherbivore properties of the lichen compounds have made the lichen as a potential source of biopesticide.

KEYWORDS

- **antiherbivorous**
- **antimicrobial**
- **biopesticides**
- **leishmanicidal**
- **lichen**
- ***Ramalina***
- **secondary metabolite**

REFERENCES

1. Taylor, T. N.; Hass, H.; Remy, W.; Kerp, H. The Oldest Fossil Lichen. *Nature* **1995**, *378*, 244–244.
2. Coley, P. D. Effect of Plant Growth Rate and Leaf Lifetime on the Amount and Type of Anti-herbivore Defense. *Oecologia* **1988**, *74*, 531–536.
3. Ahmadjian, V. *The Lichen Symbiosis*; Wiley: New York, 1993; p 250.
4. Lawrey, J. D. Lichen Secondary Compounds: Evidence for a Correspondence Between Antiherbivore and Antimicrobial Function. *Bryologist* **1989**, *92*, 326–328.
5. Follmann, G.; Nakagava, M. Keimhemmung von Angiospermensamen durch Flechtenstoffe. *Naturwissenschaften* **1963**, *50*, 696–697.
6. Huneck, S.; Scheiber, K.; Wachstumsregulatorische Eigenschaften von Flechten- und Moos- Inhaltsstoffen. *Phytochemistry* **1972**, *11*, 2429–2434.
7. Saklani, A.; Upreti, D. K. Folk Uses of Some Lichens in Sikkim. *J. Ethnopharmacol.* **1992**, *27*, 229–233.
8. Lal, B.; Upreti D. K. Ethnobotanical Notes on Three Indian Lichens. *Lichenologist* **1995**, *27*, 77–79.
9. Negi, H. R.; Kareem, A. Lichens: The Unsung Heroes. *Amrut* **1996**, *1*, 3–6.
10. Romagni, J. G., Dayan, F. E. Structural Diversity of Lichen Metabolites and Their Potential Use. In *Advances in Microbial Toxin Research and its Biotechnological Exploitation;* Upadhyay, R. K. Ed.; Kluwer Academic and Plenum Publishers: New York, 2002; pp 151–170.

11. Dayan, F. E.; Romagni, J. G. Structural Diversity of Lichen Metabolites and Their Potential for Use. In *Advances in Microbial Toxin Research and Its Biotechnological Exploitation*; R. Upadhyaya, Ed.; Kluwer Academic/Plenum Publishers, 2002.

12. Emsen, B.; Yildirim, E.; Aslan, A.; Anar, M.; Ercisli, S. Insecticidal Effect of the Extracts of *Cladonia foliacea* (Huds.) Willd. and *Flavoparmelia caperata* (L.) Hale Against Adults of the Grain Weevil, *Sitophilus granarius* (L.) (Coleoptera: Curculionidae). Egypt. *J. Biol. Pest Contr.* **2012**, *22*, 145–149.

13. Yildirim, E.; Emsen, B.; Aslan, A.; Bulak, Y.; Ercisli, S. Insecticidal Activity of Lichens Against the Maize Weevil, *Sitophilus zeamais* Motschulsky (Coleoptera: Curculionidae). *Egypt. J. Biol. Pest Contr.* **2012**, *22*, 151–156.

14. Agrios, N. G. *Plant Pathology*; Academic: San Diego, 1997.

15. Oerke, E. C.; Dehne, H. W.; Shoenbeck, F.; Weber, A. *Estimated Losses in Major Food and Cash Crops*; Elsevier: London, 1994.

16. Dubey, N. K., Shukla, R., Kumar, A., Singh, P., Prakash, B. Prospects of Botanical Pesticides in Sustainable Agriculture. *Curr. Sci.* **2010**, *98* (4), 479–480.

17. Dalziel, J. M. *The Useful Plants of West Tropical Africa*; Crown Agents for Overseas Governments: London, 1937.

18. Ayensu, S. *Medicinal Plants of West Africa*; Reference Publications: Algonae, 1978.

19. Tripathi, P.; Dubey, N. K. Exploration of Natural Products as an Alternative Strategy to Control Post Harvest Fungal Rotting of Fruits and Vegetable. *Postharvest Biol. Technol.* **2004**, *32*, 235–245.

20. Rajendran, S.; Sriranjini, V. Plant Products as Fumigants for Stored Product Insect Control. *J. Stored Prod. Res.* **2008**, *44*, 126–135.

21. Saxena, R. C. Botanical Pest Control. In *Critical Issues in Insect Pest Management*; G. S. Dhaliwal, Heinrichs, Eds.; Commonwealth Publisher: New Delhi, India, 1998; pp 155–179.

22. Nayaka, S.; Upreti, D. K.; Khare, R. Medicinal Lichens of India. In *Drugs from Plants*, 2010.

23. Molnár, K.; Farkas, E. Current Results on Biological Activities of Lichen Secondary Metabolites: A Review. *Z. Naturforsch.* **2010**, *65*, 157–173.

24. Franck, E.; Dayan.; Joanne, G.; Romagni. Lichens as a Potential Source of Pesticides; *Pest. Outlook* **2001**, 229–232.

25. Dayan, F. E.; Romagni, J. G. Lichens as a Potential Source of Pesticides; *Pest. Outlook.* **2001**, *6*, 229–232.

26. Crittenden, D.; Porter, N. Lichen Forming Fungi: Potential Sources of Novel Metabolites. *Trends Biotechnol.* **1991**, *9*, 409–414.

27. Prasad, C.; Manoharachary; Kunwar, I. K. Effect of Lichen Extracts on the Growth of Fungi. Ind. Bot. Soc. **1994**, *73*, 353–354.

28. Cansaran, D.; Kahya, D.; Yurdakulol, E.; Atakol, A. Identification and Quantification of Usnic Acid from the Lichen *Usnea* Species of Anatolia and Antimicrobial Activity. *Zeitschrift für Naturforschung* **2006**, *61* (11–12), 773–776.

29. Land, C. J.; Lundstrom, H. Inhibition of Fungal Growth by Water Extracts From the Lichen *Nephroma arcticum*. *Lichenologist* **1998**, *30*, 259–262.

30. Kekuda, P. T. R.; Vinayaka, K. S.; Swathi, D.; Suchitha, Y.; Venugopal, T. M.; Mallikarjun, N. Mineral Composition, Total Phenol Content and Antioxidant Activity

of a Macrolichen *Everniastrum cirrhatum* (Fr.) Hale (Parmeliaceae). *E-J. Chem.* **2011,** *8* (4), 1886–1894.

31. Saklani, A.; Jain, S. K. *Cross Cultural Ethnobotany of North India*; Deep: New Delhi, 1994; pp 31–32.

32. Shahi, S. K.; Shukla, A. C.; Dikshit, A.; Upreti, D. K. Use of Lichen as Antifungal Drug Against Superficial Fungal Infections. *Arom. Plant Sci.* **2000,** *22* (4A) and *23* (1A), 169–172.

33. Esimone, C. O.; Eck, G.; Duong, T. N. Potential Antirespiratory Syncytial Virus Lead Compounds from Aglaia Species. *Pharmazie 63*, 1–6.

34. Azenha, G.; Iturriaga, T.; Michelangeli F. I.; Rodriguez, E. Ethnolichenology, Biochemical Activity, and Biochemistry of Amazonian Lichen Species. *Cornell Univ. Undergrand Res. Prog. Biodivers.* **1998,** *1*, 8–14.

35. Fournet, A.; Ferreira, M. E.; Rojas de Arias, A.; Torres de Ortiz, S.; Inchausti, A.; Yaluff, G.; Quilhot, W.; Fernandez, E.; Hidalgo, M. E. Activity of Compounds Isolated from Chilean Lichens Against Experimental Cutaneous Leishmaniasis. *Comp. Biochem. Physiol. C.* **1997,** *116*, 51–54.

36. Honda, N. K.; Pavan, F. R.; Coelho, R. H. Antimycobacterial Activity of Lichen Substances. *Phytomedicine* **2010,** *17*, 328–332.

37. Kumar, P. S. V.; Kekuda, P. T. R.; Vinayaka, K. S.; Sudharshan, S. J.; Mallikarjun, N.; Swathi, D. Studies on Antibacterial, Anthelmintic and Antioxidant Activities of a Macrolichen *Parmotrema pseudotinctorum* (des. Abb.) Hale (Parmeliaceae) from Bhadra Wildlife Sanctuary, Karnataka. *Int. J. PharmTech Res.* **2010a,** *2* (2), 1207–1214.

38. Pavithra, G. M.; Vinayaka, K. S.; Rakesh, K. N.; Junaid, S.; Dileep, N.; Kekuda, P. T. R.; Siddiqua, S.; Naik, A. S. Antimicrobial and Antioxidant Activities of a Macrolichen *Usnea pictoides* G. Awasthi (Parmeliaceae). *J. Appl. Pharm. Sci.* **2013,** *3* (08), 154–160.

39. Swathi, D.; Suchitha, Y.; Kekuda, P. T. R.; Venugopal, T. M.; Vinayaka, K. S.; Mallikarjun, N.; Raghavendra, H. L. Antimicrobial, Anthelmintic and Insecticidal Activity of a Macrolichen *Everniastrum cirrhatum* (Fr.) Hale. *Int. J. Drug Dev. Res.* **2010,** *2* (4), 780–789.

40. Kekuda, T. R. P.; Vinayaka, K. S.; Kumar, S. V. P.; Sudharshan, S. J. Antioxidant and Antibacterial Activity of Lichen Extracts, Honey and Their Combination. *J. Pharm. Res.* **2009,** *2* (12), 1875–1879.

41. Vinayaka, K. S.; Krishnamurthy, Y. L.; Kekuda, P. T. R.; Kumar, P. S. V.; Sudharshan, S. J.; Chinmaya, A. Larvicidal and Wormicidal Efficacy of Methanolic Extracts of Five Macrolichens Collected from Bhadra Wildlife Sanctuary. *Biomedicine* **2009,** *29* (4), 327–331.

42. Gulluce, M.; Aslan, A.; Sokmen, M.; Sahin, F.; Adiguzel, A.; Agar, G.; Sokmen, A. Screening the Antioxidant and Antimicrobial Properties of the Lichens *Parmelia saxatilis, Platismatia glauca, Ramalina pollinaria, Ramalina polymorpha* and *Umbilicaria nylanderiana. Phytomedicine* **2006,** *13*, 515–521.

43. Mazid, M. A.; Hasan, C. M.; Rashid, M. A. Antibacterial Activity of *Parmelia kamstchandalis. Fitoterapia* **1999,** *70*, 615–617.

44. Florencio, B. Antibacterial Substances from Lichens. *Econ. Bot.* **1952,** *6* (4), 402–406.

45. Kekuda, P. T. R.; Vivek, M. N. et al. Biocontrol Potential of Parmotrema Species Against *Colletotrichum capsici* Isolated from Anthracnose of Chilli. *J. Biol. Sci. Opin.* **2014**, *2* (2), 166–169.

46. Vartia, K. O. Antibiotics in Lichens. In *The Lichens*; V. Ahmadjian, M. E. Hale, Jr., Eds.; Academic Press: New York, 1973; pp 547–561.

47. Asahina, Y.; Shibata, S. Chemistry of Lichen Substances. *Jpn. Soc. Prom. Sci. Ueno*, 1954.

48. Praveen Kumar, S. V.; Prashith Kekuda, T. R.; Vinayaka, K. S.; Swathi, D.; Chinmaya A. Insecticidal Efficacy of *Ramalina hossei* H. Magn and G. Awasthi and *Ramalina conduplicans* vain. Macrolichens from Bhadra Wildlife Sanctuary, Karnataka. *Biomedicine* **2010**, *30* (1), 100–102.

49. Kumar, P. S. V.; Kekuda, P. T. R.; Vinayaka, K. S.; Swathi, D.; Mallikarjun, N.; Nishanth, B. C. Studies on Proximate Composition, Antifungal and Anthelmintic Activity of a Macrolichen *Ramalina hossei* H. Magn & G. Awasthi. *Int. J. Biotechnol. Biochem.* **2010b**, *6* (2), 191–201.

50. Vinayaka, K. S.; Kumar, P. S. V.; Mallikarjun, N.; Kekuda, P. T. R. Studies on Insecticidal Activity and Nutritive Composition of a Macrolichen *Parmotrema pseudotinctorum* (des. Abb.) Hale (Parmeliaceae). *Drug Inv. Today* **2010**, *2* (2), 102–105.

51. Kumar, P. S. V.; Kekuda, P. T. R.; Vinayaka, K. S.; Sudharshan, S. J. Anthelmintic and Antioxidant Efficacy of two Macrolichens of Ramalinaceae. *Pharm. J.* **2009**, *1* (4), 238–242.

52. Boustie, Joel.; Grube, Martin. Lichen-A Promising Source of Bioactive Secondary Metabolites. *Plant Gen. Res.* **2005**, *3* (2), 273–287.

53. Ramaut, J. L.; Thonar, J. Inhibition de la Germination de Differentes Grains d' Angiosperms par *Evernia prunastri* (L.) *Ach. I. An. R. Soc. Esp. Fis. Quim.* **1972**, *68*, 575–595.

54. Ramaut, J. L., Thonar, J. Inhibition de la Germination de Differentes Grains d' Angiosperms par *Evernia prunastri* (L.) *Ach. II. An. R. Soc. Esp. Fis. Quim.* **1972b**, *68*, 597–607.

55. Lawrey, J. D. Vulppiinic Acid Pinastric Acids as Lichen Antiherbivore Compounds: Contrary Evidence. *Bryologist* **1983**, *86*, 365–369.

56. Lawrey, J. D. Biological Role of Lichen Substances. *Bryologist* **1986**, *89*, 111–122.

57. Ahad, A. M.; Goto, Y.; Kiuchi, F.; Tsuda, Y.; Kondo, K.; Sato, T. Nematocidal Principles in "Oakmoss Absolute" and Nematocidal Activity of 2,4-dihydroxybenzoates. *Chem. Pharm. Bull.* **1991**, *39*, 1043–1046.

58. Emmerich, R.; Giez, I.; Lange, O. L.; Proksch, P. Toxicity and Antifeedant Activity of Lichen Compounds Against the Polyphagous Herbivorous Insect Spodoptera Littoralis. *Phytochemistry* **1993**, *33*, 1389–1394.

59. Giez, I.; Lange, O. L.; Proksch, P. Growth-retarding Activity of Lichen Substances Against the Polyphagous Herbivorous Insect *Spodoptera littoralis. Biochem. Ecol.* **1994**, *22*, 113–120.

CHAPTER 11

Antimicrobial Peptides from Biocontrol Agents: Future Wave in Plant Disease Management

VIVEK SHARMA[1*] and RICHA SALWAN[2]

[1]University Centre for Research and Development, Chandigarh University, Gharuan 140413, Punjab, India

[2]Richa Salwan, College of Horticulture and Forestry, Neri, Hamirpur (HP) 177 001, India

*Corresponding author. E-mail: ankvivek@gmail.com

ABSTRACT

Microbes play an important role in affecting plant health. The nonpathogenic microbes such as *Bacillus* spp., Actinobacteria, Pseudomonads, *Rhizobium* and *Trichoderma* offer a sustainable, ecofriendly and economical solution for the management of various crop diseases compared with chemical-based management practices. Several mechanisms including production of cell wall hydrolases and bioactive secondary metabolites targeting the plant pathogens are considered to play important roles in the biocontrol activity of these microbes. In particular, the production of low-molecular weight peptides having ability to target the pathogen cell wall or membrane offer attractive choice for future plant disease management. These peptides are broadly categorized into ribosomal and nonribosomal peptides. The genomes of beneficial microbes harbor wide gene clusters of ribosomal-synthesized peptides in canonical fashion, whereas peptides of nonribosomal are modular in their organization. The posttranslational modified peptides of ribosomal origin include lanthipeptides, tail-cyclized peptides, sactipeptides, unmodified bacteriocins, and other large antimicrobial proteins. On the other side, the antimicrobial peptides (AMPs) synthesized by nonribosomal peptide synthetase enzyme complexes represent a diverse class of peptaibols. This chapter highlights the nature, mechanism, and the potential of these AMPs from beneficial microbes for the biological management of plant diseases.

11.1   INTRODUCTION

The global food security is a major concern worldwide and significant loss in crops by pathogenic microbes such as bacteria and fungi account for ~10% of the total loss at global level (Strange and Scott, 2005). The management of plant diseases is largely based on use of synthetic chemicals which causes considerable side effects to ecosystem in the form of pollutants (Makovitzki et al., 2007). The emergence of drug-resistant "ESKAPE" pathogens presents a challenging task and requires alternate measures for combating the pathogen (Gill et al., 2015). Furthermore, the genetic variation and continuous coevolution of plant pathogens often leads to the origin of highly virulent strains of pathogens which are often difficult to manage. Therefore, the development of highly effective and ecofriendly measures is the focus of current agriculture research. The, noncytotoxic membrane-targeting peptides having lipopolysaccharide (LPS)-binding affinities have emerged as promising candidates for plant disease management in agricultural applications (Datta et al., 2015).

Antimicrobial peptides (AMPs) of 5–100 amino acids have a broad spectrum of activity against various organism (Bahar and Ren, 2013). These AMPs are valuable resource for students, researchers, and have emerged as novel antibiotics for a long time. The history of AMPs started in 1939, when Dubos (1939a) showed the efficacy of *Bacillus* extract which was proved to protect mice from pneumococci infection (Dubos, 1939b). In subsequent studies, Hotchkiss and Dubos (1940) identified an AMP and named it as gramicidin. After that, the efforts on AMPs have gained extensive speed. The rapid-action and broad-spectrum antimicrobial activities of these low-molecular peptides have emerged as potential therapeutics for microbial infections caused by antibiotic-resistant strains of bacteria (Hancock, 1997; Brogden, 2005; Sahl, 2006; Datta et al., 2016). Despite a higher variability in their sequence, mass, charge, and structure, the AMPs are generally small peptides and constitute a class of molecules with wide distribution in nature. The next-generation AMPs have gained attention due to the antimicrobial activity against different microbes including bacteria, fungi, and viruses. Also, a number of organisms including plants, animals, and humans are known to produce AMPs in response to their defense against pathogens. The effectiveness of these molecules against pathogens which are resistant to conventional drugs offers several advantages (Aoki et al., 2012).

The development of agriculturally relevant transgenic plants expressing AMPs can prove a sustainable approach in combating plant diseases and also decreases concurrent pathogens ability to develop resistance (Wang et al., 2011). Besides targeting the membrane, some of AMPs can directly interact and inhibit important pathways related to DNA replication and protein synthesis (Brogden, 2005) even at low concentration. These AMPs contain an active site for intracellular target (Otvos et al., 2000; Kragol et al., 2001) or inhibition of cell wall synthesis such as nisin (Brumfitt et al., 2001). For majority of antifungal AMPs, their ability to target cell does not have clear relation with their structures, for example, antifungal peptides with α-helical (Jiang et al., 2008) or extended (Lee et al., 2003), and β-sheet structure (Barbault et al., 2003). The antifungal AMPs have been reported for causing death of fungi either by targeting cell wall (De Lucca et al., 1998, 1999) or by targeting intracellular processes (Lee et al., 1999). The ability of AMPs to efficiently bind fungal cells and then causing either the disruption of fungal membranes (Terras et al., 1992) or altering, membrane permeabilization (Van der 2010), or formation of pores in the membrane directly (Moerman et al., 2002). The studies also revealed that AMPs-mediated activity is less vulnerable to develop resistance than other methods (Datta et al., 2015). Furthermore, the promising AMPs are capable of killing bacteria under in vitro conditions at very low dose in the range from 0.25 to 4 μgmL^{-1}. The additional benefits of these peptides include their ability to destroy the target cells rapidly, broad spectrum activity, and effectiveness against already antibiotic-resistant pathogens.

The continuous and massive use of fungicides and pesticides in agriculture sectors have led to the development of resistance among pathogens and created several environmental and health risks. These peptides can be explored as an alternative to their chemical counterpart for the protection of plants/animals against diseases. Therefore, these low-molecular peptides offer exciting possibilities over conventional antibiotics. A number of *Bacillus* species have been explored for the production of AMPS such as fusaricidins, iturins, fengycins, polymixins, and agrastatins (Stein, 2005), and reported for their synergistic antifungal activity against a variety of phytopathogens. Under in vitro conditions, a strong inhibitory effect of lipopeptide against *F. oxysporum, Pythium ultimum, Rhizoctonia solani, Rhizopus* sp., and *B. cinerea* has been described (Ongena et al., 2005; Table 11.1). Similarly, the pretreatment of bean seedlings with *B. subtilis* M4 supernatant prior resulted in a decrease in disease symptoms (Breen et al., 2015). The antimicrobial potential of these AMPs has also been described in *planta*. The modes-of-action of certain AMPs such as fengycins and peptaibols resulted not only in the inhibition of the plant pathogen but also led to increase disease resistance through systemic resistance and the production of phenolic compounds (Ongena et al., 2005; Montesinos, 2007).

The results of the biotechnological research as well as genetic engineering related to AMPs have proven their potential over the period of time in reducing economic losses to various agricultural crops caused by pathogens. Therefore, this study provides a unifying view of both peptides of ribosomal and nonribosomal origin which is so far studied differently. In this book chapter, focus is given on understanding the functional and structural characteristics of both ribosomal and nonribosomal derived AMPs from biocontrol and plant growth-promoting bacteria (PGPB). A historical preview of their development, functional mechanisms, and recent developments in bioinformatics research in relation to AMPs are also discussed.

11.2   TYPES OF AMPs

The AMPs of microbial origin can be broadly classified into: (1) ribosomal represented by majority of bacteria and (2) nonribosomal of fungal origin depending upon their synthesis. Ribosomal peptides results from the cleavage of a precursor protein. The nonribosomal peptides synthesized through nonribosomal peptide synthetases (NRPS) are represented by modular complexes which are found in an operon in bacteria or clusters of gene in eukaryotes and synthesize one peptide per operon or cluster (Breen et al., 2015). A common general centric model of both theses peptides is their ability to interact with membrane.

TABLE 11.1 Antimicrobial Peptides Used Against Pathogens in Agricultural Crops.

Microbial origin	AMP types	Antifungal	Reference
Ribosomal	*Bacillus subtilis* B-3 was	Iturin	control of peach brown rot caused by *Monilinia fructicola*	Gueldner et al., 1988
	B. subtilis strain M4	Fengycin homologs	*F. oxysporum, Pythium ultimum, Rhizoctonia solani, Rhizopus sp.,* and *B. cinerea.*	Ongena et al., 2005
	Paenibacillus B2	Polymixin B	Antagonistic to *E. carotovora* and the root pathogenic fungus *F. solani* and *Fusarium acuminatum*	Selim et al., 2005
	Bacillus amyloliquefaciens strain PPCB004	Lipopeptides, fengycins, iturins, and surfactins, along with bacillomycin	*Alternaria citri, Botryosphaeria sp., Colletotrichum gloeosporioides, Fusicoccum aromaticum, Lasiodiplodia theobromae, Penicillium crustosum,* and *Phomopsis perse*	Arrebola et al., 2010
	B. amyloliquefaciens strain Q-426	Bacillomycin D, fengycins A, and B	*F. oxysporum* f. sp. *spinaciae*	Zhao et al., 2014
	B. amyloliquefaciens strain NJN-6	Bacillomycin D and of the macrolactin family, macrolactin A, 7-O-malonylmacrolactin A, and 7-O-succinyl macrolactin A	*Ralstonia solanacearum*	Yuan et al., 2011; 2012
	Bacillus atrophaeus strain CAB-1	Fengycins and unknown lipopeptides	*B. cinerea* and *Sphaerotheca fuliginea*	Zhang et al., 2013
	Bacillus subtilis strains S499 and M4 was	Fengycin-type cyclopeptides	Control of grey mould rot in apple fruits caused by Botrytis cinerea (Ongena et al., 2005)	Ongena et al., 2005

TABLE 11.1 (Continued)

Microbial origin	AMP types		Antifungal	Reference
Nonribosomal	*B. subtilis* strains BBG131 and BBG125,		*Bremia lactucae*	Deravel et al., 2014
	Trichoderma pseudokoningii strain SMF2		*Ascochyta citrullina, B. cinerea, F. oxysporum, Phytophthora parasitica,* and *V. dahlia*	Shi et al., 2012
	T. atroviride	Atroviridins A–C Neoatroviridins	Gram-positive bacteria and phytopathogenic fungi	Oh et al., 2002; Neuhof et al., 2007
	T. harzianum	Trichokindins I–X Trichorzin HA I Harzianins	–	Neuhof et al., 2007
	T. saturnisporum	Paracelsin E, Saturnisporins SA II and IV	–	Ritieni et al., 1995; Neuhof et al., 2007
	T. viride	Trichovirins II, Trichodecenin I Suzukacillin A Alamethicin F50	–	Neuhof et al., 2007
	T. pseudokoningii	Trichokonin VI	programmed cell death in *Fusarium oxysporum* plant fungal pathogens	Shi et al., 2012
	T. pseudokoningii	Trichokonins	Induce resistance against Gram-negative *Pectobacterium carotovorum* subsp. *carotovorum* in Chinese cabbage	Li et al., 2014

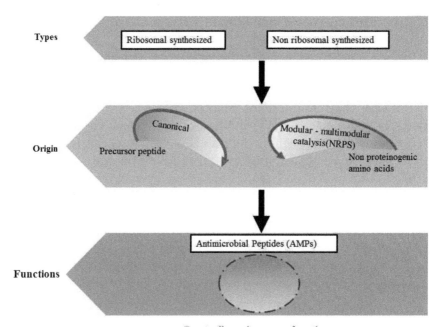

FIGURE 11.1 (See color insert.) Antimicrobial peptides of ribosomal origin are synthesized using canonical pathway, whereas nonribosomal origin in general includes peptaibols which are synthesized through modular pathway. Still irrespective of their origin, both AMPs either target cell membrane or intracellular pathways.

The main structural characteristics of AMPs include their high affinity and selectivity for membranes (Melo et al., 2009) which is determined by the amino acid composition, charge, hydrophobicity, amphipathicity, H-bonding, and flexibility (Findlay and Zhanel, 2010). The net positive charge on peptide is considered essential for initial binding to the negative charged membrane of bacteria, and thus allows selective discrimination between bacterial and host cell membrane, whereas its hydrophobic nature helps in insertion into membrane followed by perturbation (Lohner and Blondelle, 2005; Henderson and Lee, 2013; Malanovic and Lohner, 2016). The AMPs of animal and plant origin are endogenous and are composed of 10–50 amino acids (Malanovic and Lohner, 2016). These are cationic in nature due to excess of lysine and arginine amino acids and can behave as constitutive or inducible after pathogen infection. The structural composition of these peptides can be α-helical or β-sheet elements which is further stabilized by intramolecular disulphide (Malanovic and Lohner, 2016).

11.2.1   BIOGENESIS OF PEPTIDES

The NRPS produce a wide variety of peptides in bacteria and fungi. Although structurally diverse, most of these nonribosomal peptides share a common biosynthetic pathway, that is, presence of multienzyme thio template mechanism. The constituents of the peptides are connected sequentially by the corresponding peptide synthetase system, which is first activated as acyl adenylates through the utilization of ATP as energy source. The activation share its similarity to aminoacyl-tRNA synthetases involved in translation during protein synthesis. However, presence of peptide synthetases instead of tRNA intermediates is involved in covalent linking of the activated amino acid as carboxy thioester (Pavela-Vrancic et al., 1994). This module of peptide synthetase works in semiautonomous fashion. The module activates and modifies a single residue of the final peptide. Each module can be partitioned into three repeating sites: adenylation, thiolation, and condensation (Daniel and Filho, 2007a).

11.2.2   RIBOSOMAL PEPTIDES AND THEIR STRUCTURE

The AMPs of ribosomal origin are produced both by prokaryotes and eukaryotes, and are important component of defense against microbes (Aoki et al., 2012). In general, majority of these peptides are cationic and usually amphiphilic in nature, therefore target the cells by altering cell membrane permeability. Based on the structural organization, the ribosomal peptides are classified into different groups either having a high content of one or two amino acid, usually proline. Structurally, these are predominantly β-sheet structure containing intramolecular disulfide bonds, or α-helical structure with amphiphilic regions (Sarika et al., 2012; Holaskova et al., 2014; Carneiro et al., 2015) or extended, and loop. The α-helix and β-sheet structural groups are more common and among them α-helical peptides are explored in majority (Powers and Hancock, 2003). The best known examples of such AMPs are protegrin, magainin, cyclic indolicin, and coiled indolicin (Huang et al., 2010). β-sheet peptides are composed of minimum of two β-strands and share disulfide bonds between them (Bulet et al., 2004). Ribosomal peptides are difficult to predict in silico from transcriptomic and genomic sequencing projects due to their small size and high diversity and lack of generic cleavage sites that could indicate a potential peptide. The advancement in bioinformatics tools has led to in silico annotation and identification of microbial genome for these peptides (Aleti et al., 2015). The most comprehensive AMPs database to date, named ADAM, is publically available and currently contains 7007 unique peptide sequences and 759 structures (http://bioinformatics.cs. ntou.edu.tw/ADAM/links.html) (Lee et al., 2015; Breen et al., 2015).

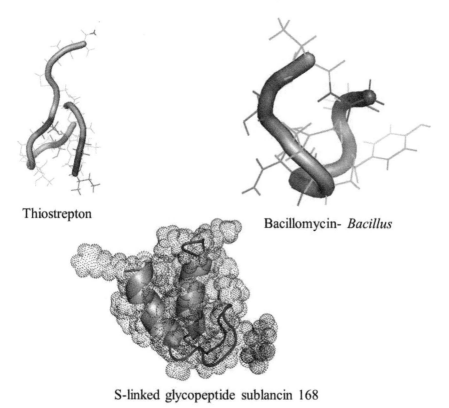

Thiostrepton

Bacillomycin- *Bacillus*

S-linked glycopeptide sublancin 168

FIGURE 11.2a **(See color insert.)** Ribosomal-derived antimicrobial peptides from bacterial source.

Many of biocontrol agents (BCAs) including gram-negative *Pseudomonas fluorescens* and gram-positive *Bacillus subtilis/amyloliquefaciens* are known to produce AMPs like cyclodipopeptides (cLPs) which includes surfactins, iturins, and fengycins, or pseudopeptides (Table 11.2).

Majority of AMPs are encoded by genes, while others are products of secondary metabolism or synthesized by NRPS (Giessen and Marahiel, 2012). Certain AMPs are generated after posttranslational modifications and have cyclic structure or contain unusual amino acids (Laverty et al., 2011; Tables 11.2 and 11.3). Furthermore, independent of their origin, these AMPs can provide resistance against fungal or bacterial pathogens in plant species (Rahnamaeian et al., 2009). Hence, they present innovative approaches for plant protection in agriculture (Fig. 11.2b) (Holaskova et al., 2015).

11.2.3   NON-RIBOSOMAL PEPTIDES AND THEIR STRUCTURE

The nonribosomal-derived peptides are represented by peptaibols which are either linear or rarely cyclic in structure (Fig. 11.2b). These nonribosomal peptides are synthesized by modular peptide synthetases (NRPSs) (Fig. 11.1), and contained 5–20 amino acid residues which are often rich in the nonproteinogenic α-aminoisobutyric acid (Aib) (Szekeres et al., 2005; Lee and Kim, 2015; Patel et al., 2015). Similar to ribosomal-derived AMPs, the peptaibols are amphipathic with hydrophilic and hydrophobic domains. Additionally, a high proportion of nonproteinogenic amino acids such as α-aminoisobutyrate (A, Aib), α,α-dialkylated amino acid residues, and isovaline and 2-amino-6-hydroxy-4-methyl-8- oxodecanoic acid is hallmark of these peptides. The C-terminal is usually amino alcohol phenylalaninol or leucinol or 2-(2-aminopropyl) methylaminoethanol and the N-acyl group is usually acetyl (Daniel and Filho, 2007b). Peptaibols are categorized into three groups depending upon the length of the amino acid sequences. The group I contains 18–20 amino acid residues long and group II are represented by 11–16 amino acid residues. The group III contains 6 or 10 residues (Daniel and Filho, 2007) out of ~1250–1300 peptaibiotics known (Ayers et al. 2012; Carroux et al. 2013; Kimonyo and Brückner 2013; Röhrich et al. 2012; Röhrich et al. 2013a, b; Chen et al. 2013; Panizel et al. 2013; Ren et al. 2013; Stoppacher et al. 2013). So far, over 950 nonribosomal synthesized peptides have been identified from *Trichoderma/Hypocrea* which accounts over 80% from this BCA alone (Brückner et al. 1991; Degenkolb and Brückner et al., 2009). The unique membrane-altering bioactivity due to amphipathic and helical nature is possibly considered to play role in assisting root colonization and defense (Röhrich, et al., 2015).

TABLE 11.2 Classification of Ribosomal-derived Antimicrobial Peptides.

Class	Subclass	Structure	Example	Microbes	References
RiPPs	Lanthipeptides	dehydroalanine/dehydrobutyrine, lanthionine/methyl-lanthionine residues	Subtilin	*B. subtilis*	Kleerebezem 2004; Kleerebezem et al. 2004
	Head to tail cyclized peptides	direct linkage of their N- and C-terminal amino acids	Amylocyclicin	*B. amyloliquefaciens* FZB42	Gonzalez et al., 200; Maqueda et al., 2008; Van Belkum et al., 2011; Scholz et al., 2014
	Sactipeptides	unusual sulfur to α-carbon cross-inks, which are catalyzed by radical S-adenosylmethionine (SAM) enzymes in a leader peptide-dependent manner	Subtilosin A	*B. subtilis*	(Kawulka et al., 2003; Noll et al., 2011; Flühe et al., 2012; Yang and Donk 2013
	Linear azole-containing peptides (LAPs)	Heterocyclic ring of oxazoles and thiazoles derived from serine/threonine and cysteine by enzymatic cyclodehydration and dehydrogenation	Microcin B17 streptolysin S plantazolicin A & B	*E. coli, LAB* *B. amyloliquefaciens, B. methylotrophicus*	Nizet et al., 2000; Heddle et al., 2001; Melby et al., 2011; Scholz et al., 2011; Banala et al., 2013
	Thiopeptides	Nitrogenous macrocycle being central of piperidine/pyridine/dehydropi- peridine and including additional thiazoles and dehydrated amino acid residues	Thiocillins	*B. cereus ATCC 14579*	Bowers et al., 2010; Just-Baringo et al., 2014
	Glycocins	Glycosylated residues.	Sublancin 168	*B. subtilis*	Stepper et al., 2011
	Lasso peptides	N-terminal macrolactam ring that is threaded by the C-terminal tail resulting in a unique lasso structure—the so-called lariat knot			Maksimov et al., 2012; Hegemann et al., 2015

TABLE 11.3   Selected Database of Antimicrobial Peptides of Bacterial and Fungal Origin.

Sr. No.	Database	Antimicrobial peptides	Reference
1.	BAGEL2	Ribosomal	de Jong et al., 2010
2.	Peptaibol	Peptaibol-Nonribsomal	Whitmore and Wallace, 2004
3.	APD	Ribosomal and non ribosomal	Wang et al., 2009; Wang et al., 2016
4.	DAMPD	Updated natural AMPs and contained 1232 AMPs presently	Seshadri et al., 2012
5.	DBAASP	Ribosomal and nonribosomal	Gogoladze et al., 2014; Pirtskhalava et al., 2016
6.	CyBase	Cyclic backbone, in which N- and C-terminals are linked by a peptide bond, including cyclotides and bacteriocins	Wang et al., 2008
7.	BACTIBASE	Bacteriocins, bacterial AMPs that display growth-inhibition activity against other closely related bacteria	Hammami et al., 2010

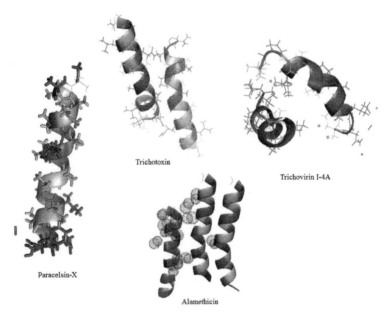

FIGURE 11.2b   **(See color insert.)** Peptaibol of *Trichoderma* representing nonribosomal-derived antimicrobial peptides from *Trichoderma reesei* Aib as light blue and other sky blue.

The α,-dialkyl α-amino acids, or 1-aminocyclopropane-1-carboxylic acid (Acc) family having molecular mass between 0.5 and 2.1 kDa contained 4–21 amino acids with unusual nonproteinogenic amino acids and/or lipoamino acids. The N-terminus is acylated N-terminus, and in linear peptide the C-terminus contains amide bond. The second subfamily of peptaibiotics contained β-amino alcohol and the C-terminus of these peptides contain polyamine, free amino acid, amide, 2,5-diketopiperazine, or a sugar alcohol (Fig. 11.3) (Degenkolb and Brückner, 2008; Stoppacher et al., 2013). The peptides produced by biocontrol fungus such as *Trichoderma* spp. are known to possess broad-spectrum antimicrobial activity (Fig. 11.2b) (Song et al., 2006; Shi et al., 2012). In addition, the biological roles of these peptide includes the elicitation of systemic resistance in tobacco and Chinese cabbage (Luo et al., 2010; Li et al., 2014) root colonization, and programmed cell death induction in tumor cells (Shi et al., 2010; Shi et al., 2016).

```
Trichodecenin_TD_I    -----------------Z--decenoylGlyGlyLeuAibGlyI------------leL
Trichorozin_I         AcAibAsnIleLeuAibProI-----------leLeuAibProValOH----------
Harzianin_HB_I        AcAibAsnLeuIleAibProI-----------ValeuAibProLeuOH----------
Pseudokinin_KLIII     AcAibAsnIleIleAibProL-----------euLeuAibProNH-------------
NA_VII                AcAibAlaAl---aAibIvaGlnAibAibAib---------SerLeuAibOCH------
Paracelsin_A          AcAibAlaAibAlaAibAlaGlnAibValAibGlyAibAibProValAibAibGlnGlnP
Saturnisporin_SA_I    AcAibAlaAibAlaAibAlaGlnAibLeuAibGlyAibAibProValAibAibGlnGlnP
Atroviridin_A         AcAibProAibAlaAibAlaGlnAibValAibGlyLeuAibProValAibAibGlnGlnP
Alamethicin_F-30      AcAibProAibAlaAibAlaGlnAibValAibGlyLeuAibProValAibAibGluGlnP


Trichodecenin_TD_I    euOH
Trichorozin_I         ----
Harzianin_HB_I        ----
Pseudokinin_KLIII     ----
NA_VII                ----
Paracelsin_A          heOH
Saturnisporin_SA_I    heOH
Atroviridin_A         heOH
Alamethicin_F-30      heOH
```

FIGURE 11.3 **(See color insert.)** Multiple alignment of nonribosomal-derived antimicrobial peptides (AMPs) such as peptaibols of *Trichoderma* origin.

11.2.4 MODE OF ACTION

AMPs have varied structural and biological properties. In general, majority of these AMPs share cationic backbone and the amphipathic arrangement target the negatively charged bacterial membrane through electrostatic interaction. Further the segregation of the charged side from a hydrophobic one permits these molecules to enter into the hydrophobic microbial membrane and ultimately leading to membrane disruption and cell death

(Brogden, 2005; Izadpanah and Gallo, 2005). Based on the centric general peptide–membrane interaction, four models: barrel-stave, carpet-like, toroidal-pore, and disordered toroidal pore have been proposed for their function (Melo et al., 2009). So far, it is well accepted that the different physicochemical properties of the phospholipid bilayer membranes allow AMPs to discriminate between bacterial and host cell membranes (Lohner 2001; Lohner and Staudegger, 2001). Cholesterol being a main biomolecule of eukaryotic membrane and its absence in bacterial cell membranes is proposed to play an important role in imparting selectivity (Malanovic and Lohner, 2016). Furthermore, the neutral lipids in model membranes have been demonstrated to inhibit damage to lipid vesicles by different AMPs (Matsuzaki et al., 1995; Lee et al., 2015).

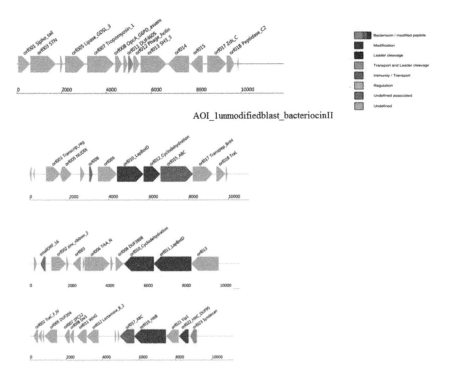

FIGURE 11.4 (See color insert.) Organization of ribosomal-derived antimicrobial peptides (AMPs) in the genome of *Bacillus* species mostly used as biocontrol agents; green color indicates AOI type of antimicrobial peptides.

In the classical membrane disruption model, the APMs incorporate themselves across the membrane and then cause disruption of the outer membrane eventually leading to the death of the target cell. In barrel-stave model, the peptides tend to assemble and form channels in transmembrane in the form of a barrel where hydrophobic parts of peptide aligned with the lipid bilayer and the hydrophilic regions of AMPs constitute the inner part of the pore (Zhang, et al., 2001; Melo et al., 2009). In carpet-like model, AMPs act as detergent and forms micelles or cluster at the membrane surface and leads to cooperative permeabilization of the cytoplasmic membrane. Initially, the AMPs cover the lipid bilayer membrane in a carpet-like fashion and under appropriate conditions, the peptides align themself into the cell membrane due to their detergent-like properties. In toroidal-pore/wormhole model, the hydrophilic part of AMPs remains bounded throughout the polar region of the phospholipid membrane and develop a stable curvature or the membrane phospholipids bend backward upon themselves after magainin monomers which ultimately leads to toroidal pore formation. The disordered toroidal pore represents a modification of toroidal pore model. Here, slightly flexible peptide conformations are formed and the inner side of the pore is also surrounded by hydrophilic region of phospholipid (Li, et al., 2012; Rocha et al., 2012; Bahar and Ren, 2013; Carneiro et al., 2015).

The adsorption of these AMPs such as Nguyen to the membrane can be increased by coupling them with anions across the bilayer using oxidized phospholipids or other substance (Mattila et al., 2008; Nguyen et al., 2011) or alternatively using molecular electroporation techniques which can increase the membrane permeability transiently (Gifford et al., 2005). The specificity of AMPs for microbial cells compared with host cells results from the high amount of anionic lipids in the bacterial cell membrane and electrical-potential gradient. During interaction with bacterial membranes, AMPs usually undergo structural changes in the form of aggregates and hence responsible for antimicrobial activity (Kaiserer et al., 2003).

Amphipathicity resulting from segregation of apolar and polar residues upon secondary structure formation favors the internalization of the peptide and in turn membrane perturbation. In gram-positive bacteria, before interaction with the cytoplasmic membrane, the peptidoglycan layer and teichoic acids of bacteria may help in penetration, interaction with AMPs and acts as a ladder in targeting membrane interaction and then affect domain organization by lipid segregation which affects membrane permeability. Unlike antibiotics which inhibits the peptidoglycan biosynthesis, the AMPs directly target the cell wall precursors containing highly conserved lipid II which can induce membrane pore formation and hence its disruption (Malanovic and Lohner, 2016). In Gram-negative bacteria, LPS acts as a barrier for AMPs interaction by inhibiting the entry of antimicrobial proteins and other antibiotics (Datta et al., 2015).

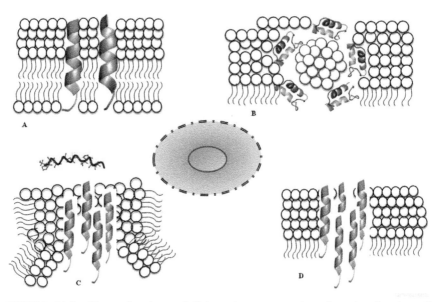

FIGURE 11.5   (See color insert.) Schematic representation of mode of action of antimicrobial peptides A-Torroidal, B-Carpet-like model high concentrations of peptide molecules disrupt the membrane in a detergent-like manner breaking the lipid bilayer into set of separate micelles. C- Barrel stave in which hydrophobic regions of AMPs align with the tails of the lipids and the hydrophilic residues form the inner surface of the forming pore and D- Toroidal pore model in which peptides aggregates and hydrophilic heads of the lipids are electrostatically dragged by charged residues of AMPs.

Similar to AMPs of ribosomal origin, the amphipathic nature of peptaibols of nonribosomal origin is the likely basis of their biological activities in the formation of voltage-dependent ion channels in lipid membranes (Béven et al., 1999; Chugh and Wallace, 2001), which is responsible for their antimicrobial, antitumor activity, and the ability to elicit plant defense (Szekeres et al., 2005). For example, alamethicin, the most extensively studied long- sequence peptaibol, is well known for its antimicrobial activity and the ability to induce plant resistance (Leitgeb et al., 2007; Kredics et al., 2013).

11.2.5   INTRACELLULAR TARGETS

Membrane interactions remain important even for intracellular-targeting peptides because they must have means of translocation. As referred in Figure 11.2, AMPs can also act on multiple intracellular targets, modulating gene expression or inhibiting enzymatic processes important for cell viability maintenance. Some AMPs have the ability to interfere with the metabolism of nucleic acids, as is the case Microcidin B17 targets the DNA gyrase-inhibiting DNA replication (Collin et al., 2013). The peptide MccJ25 of bacterial origin may compromise RNA polymerase activity by preventing the transcription process (Mathavan and Beis, 2012). Other peptides may directly act on proteins synthesis as in the case of apidecins and oncocins, causing a blockage in the formation of new protein products through binding of ribosomal proteins (Krizsan et al., 2014). Changes in cell wall synthesis process can be triggered by peptides with antimicrobial intracellular targets. Mediated receptor binding was observed for Lactococcin G and Enterocin 1071 which binds to UPPP, an enzyme involved in cell wall synthesis. The antibiotic class of antimicrobial peptides produced by bacteria may serve as important precursors for the synthesis of peptidoglycan or by activation of autolysins (Oppegård et al., 2007).

11.3   MINING FOR AMPS IN GENOMIC DATA

The small ORFs hence small size, classical activity, screening, and characterization are often excluded. Thus, there is need for in silico genomic annotations of data (Hancock and Lehrer, 1998; de Jong et al., 2010). Further mining microbial genomes for ribosomal and nonribosomal peptide is a challenging task due to the lack of substantial homology, sequencing, and structural information of these antimicrobial peptides. Still continuous research on AMPs and evolution of molecular tools varying from genomic to metagenomic data have led to the development and improvement of genome mining software for automated screening of AMPs gene clusters.

BAGEL3 (http://bagel.molgenrug.nl/), a follow-up of previous BAGEL and BAGEL2 web-based databases, identifies putative bacteriocins on the basis of three subclasses: Class I contains RiPPs of less than 10 kDa, which is divided into more than 12 supported subclasses; class II contains unmodified peptides not fitting the criteria of the first database; class III contains antimicrobial proteins larger than 10 kDa. BAGEL3 database is a versatile fast tool valid for modified and nonmodified AMPs produced from cultural and nonculturalable bacteria (Table 11.3). BAGEL3 either uses DNA nucleotide sequences in FASTA format as input file or alternatively from drop-down menu bacteria of choice can be selected. The input DNA sequences are analyzed in parallel either for bacteriocin or RiPP or precursor-based mining directly (Zhao and Kuipers, 2016). The peptaibols database (http://peptaibol.cryst.bbk.ac.uk/) store all structural and sequence information of nonribosomal AMPs known as peptaibols, whereas synthetic analogues of peptaibol are excluded from this database. From database, the sequence section allows searches of the peptaibol name, group, motif, and mold, whereas structural section provided information of their structural features.

11.4 ANALYSIS AND PLANT ACTIVITY ASSAY OF AMPS

The AMPs due to their small size are difficult to resolve using Laemmli–SDS-PAGE which is also known as Glycine–SDS-PAGE. Therefore, Tricine–SDS-PAGE is the preferred electrophoretic system for resolving proteins less than 30 kDa (Schägger, 2006). For antimicrobial activity, young plants/seedlings are preferred choice and can be inoculated using clip inoculation-based method. For optimum results, a log phase bacterial culture can be washed with distilled water and then resuspended in phosphate buffer 10 mM, pH 7.4 to a cell density of 10^8 cells/mL. For infection, either sterile scissors dipped into cell suspension or an insulin needle can be used by pricking the veins of the leaves and then bacterial cell suspension can be applied using a cotton swab. Control sets of plants were inoculated with phosphate buffer (pH 7.4) alone and bacterial cell suspension pre incubated for 5 h with AMPs can be used to inoculate the plants. The plants are then observed daily for symptoms after post-infection (dpi). The plants can be uprooted and compared for morphological changes for further analysis (Datta et al., 2015).

11.5 FUTURE AND CONCLUSION

The traditional methods of using hazardous chemicals for agricultural pathogen management has led to development of resistance among pathogens as well as created several environmental and health risks. These AMPs have gained the attention of many researchers due to their efficacy against a broad range of pathogens. The use of AMPs for crop protection can play important role in the management of plant pathogens (Datta et al., 2015) and currently, these AMPs have emerged as alternative for conventional methods of plant disease management. Despite the promise of AMPs against a vast range of pathogens, several successful stories on the use of antimicrobial peptides in food and agriculture sectors indicate the promising future of these peptides. In agricultural sector, AMPs with anti-microbial potential and their advantage to induce plant immune responses are proving useful due to their minimal side effects to the environment.

Many of current AMPs have been used for developing transgenic plants with increased resistance to pathogens under laboratory conditions. The transgenic cotton plants expressing the tobacco peptide NaD1 which can target PIP2 of fungal membrane have been found to enhance the resistance to fungal pathogens in field conditions (Gaspar et al., 2014) and against several other filamentous fungi under in vitro conditions (Lay et al., 2003; Gaspar et al., 2014). So far, a large numbers of peptaibols amino acid sequences have been discovered due to the advances in spectroscopic and MS techniques (Breen et al., 2015). MS techniques have played vital role in the complete sequencing of peptaibols isolated from BCAs *Trichoderma* species. To the best of our knowledge, transgenic plants have only been attempted and developed with ribosomal or artificially designed AMPs. Developing transgenic plants using nonribosomal AMPs is a challenging task as they do not undergo translational synthesis. Additionally, research investment into the biosynthetic mechanism used by the NRPS system and cloning of their peptide synthetase genes and their modulation could be optimized for the production of these peptides for various biotechno-logical and pharmacological applications. The successful stories on the use of antimicrobial peptides in agriculture and food industry indicate a promising future for extensive application of these peptides. Although the genetic manipulation with antimicrobial peptide-encoding genes has been done for plants and animals, still such strategies and products may still have a long way to go before being confirmed by regulatory bodies to surmount technical problems before being accepted as applicable ones.

KEYWORDS

- biocontrol
- antimicrobial peptides
- ribosomal
- nonribosomal
- management

REFERENCES

Aleti, G.; Sessitsch, A.; Brader, G. Genome Mining: Prediction of Lipopeptides and Polyketides from Bacillus and Related Firmicutes. *Comput. Struct. Biotechnol. J.* **2015,** *13,* 192–203, doi: 10.1016/j.csbj.2015.03.003.

Aoki, W.; Kuroda, K.; Ueda, M. Next Generation of Antimicrobial Peptides as Molecular Targeted Medicines. *J. Biosci. Bioeng.* **2012,** *114* (4), 365–370. http://doi.org/10.1016/j.jbiosc.2012.05.001.

Ayers, S.; Ehrmann, B. M.; Adcock, A. F; Kroll, D. J; Carcache de, Blanco. E. J.; Shen, Q; Swanson, S. M.; Falkinham, J. O. III.; Wani, M. C.; Mitchell, S. M.; Pearce, C. J.; Oberlies, N. H. Peptaibols from Two Unidentified Fungi of the Order Hypocreales with Cytotoxic, Antibiotic, and Anthelmintic Activities. *J. Pept. Sci.* **2012,** *18,* 500–510.

Bahar, A. A.; Ren, D. Antimicrobial Peptides. *Pharmaceuticals* **2013,** *6* (12), 1543–1575. http://doi.org/10.3390/ph6121543.

Banala, S.; Ensle, P.; Sussmuth, R. D. Total Synthesis of the Ribosomally Synthesized Linear Azole-containing Peptide Plantazolicin A From Bacillus Amyloliquefaciens. *Angew. Chem. Int. Ed.* **2013,** *52* (36), 9518–9523.

Barbault, F.; Landon, C.; Guenneugues, M.; Meyer, J. P.; Schott, V.; Dimarcq, J. L.; Vovelle, F. Solution Structure of alo-3: A New Knottin-type Antifungal Peptide from the Insect *Acrocinus longimanus*. *Biochemistry* 2003, *42,* 14434–14442.

Bower, A. A.; Walsh, C. T.; Acker, M. G. Genetic Interception and Structural Characterization of Thiopeptide Cyclization Precursors from *Bacillus cereus. J. Am. Chem. Soc.* **2010,** *132* (35), 12182–12184.

Breen, S.; Solomon, P. S.; Bedon, F.; Vincent, D. Surveying the Potential of Secreted Antimicrobial Peptides to Enhance Plant Disease Resistance. *Front. Plant Sci.* **2015,** 900. http://doi.org/10.3389/fpls.2015.00900.

Brogden, K. A. Antimicrobial Peptides: Pore Formers or Metabolic Inhibitors in Bacteria? *Nat. Rev. Microbiol.* **2005,** *3,* 238–250.

Brückner, H.; Graf, H. Paracelsin, a Peptide Antibiotic Containing α-aminoisobutyric Acid, Isolated from *Trichoderma reesei* Simmons. *Part A Experientia.* **1983,** *39,* 528–530.

Brückner, H.; Becker, D.; Gams, W.; Degenkolb, T. Aib and Iva in the Biosphere: Neither Rare Nor Necessarily Extraterrestrial. *Chem. Biodivers.* **2009,** *6,* 38–56.

Brumfitt, W.; Salton, M. R.; Hamilton-Miller, J. M. Nisin, Alone and Combined With Peptidoglycan-Modulating Antibiotics: Activity Against Methicillin-Resistant Staphylococcus Aureus And Vancomycin-Resistant Enterococci. *J. Antimicrob. Chemother.* **2002**, *50*, 731–734.

Bulet, P.; Stocklin, R.; Menin, L. Anti-microbial Peptides: From Invertebrates to Vertebrates. *Immunol. Rev.* **2004**, *198*, 169–184.

Carneiro, V. A.; Duarte, H. S.; Prado, M. G. V.; Silva, M. L.; Teixeira, M. S.; Santos, Y. M. V. Antimicrobial Peptides: From Synthesis To Clinical Perspectives. In *The Battle Against Microbial Pathogens: Basic Science, Technological Advances and Educational Program*; A. Méndez-Vilas, Ed., 2015, pp 81–90.

Carroux, A; van Bohemen, A-I.; Roullier, C.; Robiou du Pont, T.; Vansteelandt, M.; Bondon, A.; Zalouk-Vergnoux, A.; Pouchus, Y. F.; Ruiz N. Unprecedented 17-residue Peptaibiotics Produced by Marine-derived *Trichoderma atroviride*. *Chem Biodivers.* **2013**, *10*, 772–786.

Daniel, J. F.; Filho, E. R. Peptaibols of *Trichoderma*. *Nat. Prod. Reps.* **2007**, *24*, 1128–1141. http://doi.org/10.1039/b618086h.

Datta, A.; Ghosh, A.; Airoldi, C.; Sperandeo, P.; Mroue, K. H.; Jiménez-Barbero, J.; Bhunia, A. Antimicrobial Peptides: Insights Into Membrane Permeabilization, Lipopolysaccharide Fragmentation and Application in Plant Disease Control. *Scientific Reports* **2015**, *5*, 11951. http://doi.org/10.1038/srep11951.

de Jong, A.; van Heel, A. J.; Kok, J.; Kuipers, O. P. BAGEL2: Mining for Bacteriocins in Genomic Data. *Nucleic Acids Res.* **2010**, *38* (2), 647–651. http://doi.org/10.1093/nar/gkq365.

De Lucca, A. J.; Walsh, T. J. Antifungal peptides: Novel Therapeutic Compounds Against Emerging Pathogens. *Antimicrob. Agents Chemother.* **1999**, *43*, 1–11.

De Lucca, A. J.; Bland, J. M.; Jacks, T. J.; Grimm, C.; Walsh, T. J. Fungicidal and Binding Properties of the Natural Peptides Cecropin B and Dermaseptin. *Med. Mycol.* **1998**, *36*, 291–298.

Degenkolb, T.; Brückner, H. Peptaibiomics: Towards a Myriad of Bioactive Peptide Containing Cα-dialkylamino Acids. *Chem. Biodivers.* **2008**, *5*, 1817–1843.

Dubos, R. J. Studies on a Bactericidal Agent Extracted From A Soil Bacillus: I. Preparation of the Agent. Its Activity In Vitro. *J. Exp. Med.* **1939a**, *70*, 1–10.

Dubos, R. J. Studies on a Bactericidal Agent Extracted from a Soil Bacillus: II Protective Effect of the Bactericidal Agent Against Experimental Pneumococcus Infections in Mice. *J. Exp. Med.* **1939b**, *70*, 11–17.

Findlay, B.; Zhanel, G. G.; Schweizer, F. Cationic Amphiphiles, A New Generation of Antimicrobials Inspired by the Natural Antimicrobial Peptide Scaffold. *Antimicrob. Agents Chemother.* **2010**, *54*, 4049–4058.

Flühe, L.; Knappe, T. A.; Gattner, M. J.; Schäfer, A.; Burghaus, O.; Linne, U.; et al. The Radical SAM Enzyme Alba Catalyzes Thioether Bond Formation in Subtilosin A. *Nat. Chem. Biol.* **2012**, *8* (4), 350–357.

Gaspar, Y. M.; McKenna, J. A.; McGinness, B. S.; Hinch, J.; Poon, S.; Connelly, A. A.; et al. Field Resistance to *Fusarium oxysporum* and *Verticillium dahliae* in Transgenic Cotton Expressing the Plant Defensin NaD1. *J. Exp. Bot.* **2014**, *65*, 1541–1550. doi: 10.1093/jxb/eru021.

Giessen, T. W.; Marahiel, M. A. Ribosome-independent Biosynthesis of Biologically Active Peptides: Application of Synthetic Biology to Generate Structural Diversity. *FEBS Lett.* **2012**, *586*, 2065–2075.

Gifford, J. L.; Hunter, H. N.; Vogel, H. J. Lactoferricin: A Lactoferrin-derived Peptide with Antimicrobial, Antiviral, Antitumor and Immunological Properties. *Cell Mol. Life Sci.* **2005**, *62*, 2588–2598.

Gill, E. E.; Franco, O. L.; Hancock, R. E. W. Antibiotic Adjuvants: Diverse Strategies for Controlling Drug-resistant Pathogens. *Chem. Biol. Drug Des.* **2015**, *85* (1), 56–78. http://doi.org/10.1111/cbdd.12478.

Gonzalez, C., Langdon, G. M., Bruix, M., Galvez, A., Valdivia, E., Maqueda, M., et al. Bacteriocin AS-48, A Microbial Cyclic Polypeptide Structurally and Functionally Related to Mammalian Nk-Lysin. *Proc. Natl. Acad. Sci. USA.* **2000**, *97* (21), 11221–11216.

Hancock, R. E. Peptide Antibiotics. *Lancet* **1997**, *349*, 418–422.

Hancock, R. E. W.; Lehrer, R. Cationic Peptides: A New Source of Antibiotics. *Trends Biotechnol.* **1998**, *6*, 10747–10751.

Heddle, J. G.; Blance, S. J.; Zamble, D. B.; Hollfelder, F.; Miller, D. A.; Wentzell, L. M.; et al. The Antibiotic Microcin B17 is a DNA Gyrase Poison: Characterisation of the Mode of Inhibition. *J. Mol. Biol.* **2001**, *307* (5), 1223–1234.

Hegemann, J. D.; Zimmermann, M.; Xie, X.; Marahiel, M. A. Lasso Peptides: An Intriguing Class of Bacterial Natural Products. *Acc. Chem. Res.* **2015**, *48* (7), 1909–1919.

Henderson, J. M.; Lee, K. Y. C. Promising Antimicrobial Agents Designed from Natural Peptide Templates. *Curr. Opin. Solid State Mater. Sci.* **2013**, *17*, 175–192.

Holaskova, E.; Galuszka, P.; Frebort, I.; Oz, M. T. Antimicrobial Peptide Production and Plant-based Expression Systems for Medical and Agricultural Biotechnology. *Biotechnol. Advances* **2014**, *33* (6), 1005–1023. http://doi.org/10.1016/j.biotechadv.2015.03.007.

Hotchkiss, R. D.; Dubos, R. J. Fractionation of the Bactericidal Agent from Cultures of a Soil Bacillus. *J. Biol. Chem.* **1940**, *132*, 791–792.

Huang, Y. B.; Huang, J. F.; Chen, Y. X. Alpha-helical Cationic Antimicrobial Peptides: Relationships of Structure and Function. *Protein Cell* **2010**, *1*, 143–152.

Izadpanah, A.; Gallo, R. L. Antimicrobial Peptides. *J. Am. Acad. Dermatol.* **2005**, *52*, 381–390.

Jenssen, H.; Hamill, P.; Hancock, R. E. W. Peptide Antimicrobial Agents. *Clin. Microbiol. Rev.* **2006**, *19*, 491–511.

Jiang, Z.; Vasil, A. I.; Hale, J. D.; Hancock, R. E.; Vasil, M. L.; Hodges, R. S. Effects of Net Charge and the Number of Positively Charged Residues on the Biological Activity of Amphipathic Alpha-helical Antimicrobial Peptides. *Biopolymers* **2008**, *90*, 369–383.

Just-Baringo, X.; Albericio, F.; Alvarez, M. Thiopeptide Antibiotics: Retrospective and Recent Advances. *Mar Drugs.* **2014**, *12* (1), 317–351.

Kaiserer, L.; Oberparleiter, C.; Weiler-Görz, R.; Burgstaller, W.; Leiter, E.; Marx, F. Characterization of the *Penicillium chrysogenum* Antifungal Protein PAF. *Arch. Microbiol.* **2003**, *180* (3), 204–210.

Kawulka, K.; Sprules, T.; McKay, R. T.; Mercier, P.; Diaper, C. M.; Zuber, P.; et al. Structure of Subtilosin A, an Antimicrobial Peptide from *Bacillus Subtilis* with Unusual Posttranslational Modifications Linking Cysteine Sulfurs to Alpha-carbons of Phenylalanine and Threonine. *J. Am. Chem. Soc.* **2003**, *125* (16), 4726–4727.

Kimonyo, A.; Brückner, H. Sequences of Metanicins, 20-residue Peptaibols from the Ascomycetous Fungus CBS 597.80. *Chem. Biodivers.* **2013**, *10*, 813–826.

Kleerebezem, M. Quorum Sensing Control of Lantibiotic Production; Nisin and Subtilin Autoregulate Their Own Biosynthesis. *Peptides* **2004,** *25* (9), 1405–1414.

Kleerebezem, M.; Bongers, R.; Rutten, G.; de Vos, W. M.; Kuipers, O. P. Autoregulation of Subtilin Biosynthesis in *Bacillus subtilis*: The Role of the Spa-box in Subtilin-responsive Promoters. *Peptides* **2004,** *25* (9), 1415–1424.

Kragol, G.; Lovas, S.; Varadi, G.; Condie, B. A.; Hoffmann, R.; Otvos, L. The Antibacterial Peptide Pyrrhocoricin Inhibits the Atpase Actions if DNAk and Prevents Chaperone-assisted Protein Folding. *Biochemistry* **2001,** *40,* 3016–3026.

Krizsan, A.; Volke, D.; Weinert, S.; Sträter, N.; Knappe, D.; Hoffmann, R. Insect-derived Proline-rich Antimicrobial Peptides Kill Bacteria by Inhibiting Bacterial Protein Translation at the 70S Ribosome. *Angew. Chem. Int. Ed. Engl.* **2014,** *53* (45), 12236–9. doi: 10.1002/anie.201407145. Epub 2014 Sep 12.

Laverty, G.; Gorman, S. P.; Gilmore, B. F. The Potential of Antimicrobial Peptides as Biocides. *Int. J. Mol. Sci.* **2011,** *12,* 6566–6596.

Lay, F. T.; Brugliera, F.; Anderson, M. A. Isolation and Properties of Floral Defensins from Ornamental Tobacco and Petunia. *Plant Physiol.* **2003,** *131,* 1283–1293. doi: 10.1104/pp.102.016626.

Lee, D. G.; Kim, H. K.; Kim, S. A.; Park, Y.; Park, S. C.; Jang, S. H.; Hahm, K. S. Fungicidal Effect of Indolicidin and Its Interaction with Phospholipid Membranes. *Biochem. Bioph. Res. Co.* **2003,** *305,* 305–310.

Lee, D. W.; Kim, B. S. Antimicrobial Cyclic Peptides for Plant Disease Control. *Plant Pathol. J.* **2015,** *31,* 1–11. doi: 10.5423/PPJ.RW.08.2014.0074.

Lee, Y. T.; Kim, D. H.; Suh, J. Y.; Chung, J. H.; Lee, B. L.; Lee, Y.; Choi, S. Structural Characteristics of Tenecin 3, An Insect Antifungal Protein. *Biochem. Mol. Biol. Int.* **1999,** *47,* 369–376.

Lee, D. K.; Bhunia, A.; Kotler, S. A.; Ramamoorthy, A. Detergent-type Membrane Fragmentation by MSI-78, MSI-367, MSI-594, and MSI-843 Antimicrobial Peptides and Inhibition by Cholesterol: A Solid-state Nuclear Magnetic Resonance Study. *Biochemistry* **2015,** *54,* 1897–1907.

Lee, H. T.; Lee, C. C.; Yang, J. R.; Lai, J. Z.; Chang, K. Y. A Large-scale Structural Classification of Antimicrobial Peptides. *Biomed. Res. Int.* **2015,** *475062.* doi: 10.1155/2015/475062.

Li, H. Y.; Luo, Y.; Zhang, X. S.; Shi, W. L.; Gong, Z. T.; Shi, M.; Chen, L. L.; Chen, X. L.; Zhang, Y. Z.; Song, X. Y. Trichokonins from *Trichoderma pseudokoningii* SMF2 Induce Resistance Against Gram-negative *Pectobacterium carotovorum* subsp. carotovorum in Chinese Cabbage. *FEMS Microbiol. Lett.* **2014,** *354,* 75–82.

Lohner, K. *Development of Novel Antimicrobial Agents: Emerging Strategies.* Horizon Scientific Press: Wymondham, UK, 2001; p 270.

Lohner, K.; Staudegger, E. Are we on the Threshold of the Post-antibiotic Era? In *Development of Novel Antimicrobial Agents: Emerging Strategies*; Lohner, K., Ed.; Horizon Scientific Press: Wymondham, UK, 2001, pp 1–15.

Lohner, K.; Blondelle, S. E. Molecular Mechanisms of Membrane Perturbation by Antimicrobial Peptides and the Use of Biophysical Studies in the Design of Novel Peptide Antibiotics. *Comb. Chem. High Throughput Screen* **2005,** *8,* 241–256.

Malanovic, N.; Lohner, K. *Antimicrobial Peptides Targeting Gram-positive Bacteria*; Wang G, Ed.; *Pharmaceuticals* **2016,** p 59. doi:10.3390/ph9030059.

Makovitzki, A.; Viterbo, A.; Brotman, Y.; Chet, I.; Shai, Y. Inhibition of Fungal and Bacterial Plant Pathogens In Vitro and in Planta with Ultrashort Cationic Lipopeptides. *Appl. Environ. Microbiol.* **2007**, *73*, 6629–6636.

Maksimov, M. O.; Pan, S. J.; James, L. A. Lasso Peptides: Structure, Function, Biosynthesis, and Engineering. *Nat. Prod. Rep.* **2012**, *29* (9), 996–1006.

Maqueda, M.; Sanchez-Hidalgo, M.; Fernandez, M.; Montalban-Lopez, M.; Valdivia, E.; Martinez-Bueno, M. Genetic Features of Circular Bacteriocins Produced by Gram-positive Bacteria. *FEMS Microbiol. Rev.* **2008**, *32* (1), 2–22.

Mathavan I., Beis K. The Role of Bacterial Membrane Proteins in the Internalization of Microcin MccJ25 and MccB17. *Biochem. Soc. Trans.* **2012**, *40*, 1539–154310.1042/BST20120176 [Pubmed].

Matsuzaki, K.; Sugishita, K.; Fujii, N.; Miyajima, K. Molecular Basis for Membrane Selectivity of an Antimicrobial Peptide, Magainin 2. *Biochemistry* **1995**, *34*, 3423–3429.

Mattila, J. P.; et al. Oxidized Phospholipids as Potential Molecular Targets for Antimicrobial Peptides. *Biochim. Biophys. Acta* **2008**, *1778*, 2041–2050.

Melby, J. O.; Nard, N. J.; Mitchell, D. A. Thiazole/Oxazole-modified Microcins: Complex Natural Products from Ribosomal Templates. *Curr. Opin. Chem. Biol.* **2011**, *15* (3), 369–378.

Melo, M. N.; Ferre, R.; Castanho, M. A. Antimicrobial Peptides: Linking Partition, Activity and High Membrane-nound Concentrations. *Nat. Rev. Microbiol.* **2009**, *7*, 245–250.

Moerman, L.; Bosteels, S.; Noppe, W.; Willems, J.; Clynen, E.; Schoofs, L.; Thevissen, K.; Tytgat, J.; Van Eldere, J.; van der Walt, J.; et al. Antibacterial and Antifungal Properties of α-helical, Cationic Peptides in the Venom of Scorpions from Southern Africa. *Eur. J. Biochem.* **2002**, *269*, 4799–4810.

Montesinos, E. Antimicrobial Peptides and Plant Disease Control. *FEMS Microbiol. Lett.* **2007**, *270* (1), 1–11.

Neuhof, T.; Dieckmann, R.; Druzhinina, I. S.; Kubicek, C. P.; von Döhren, H. Intact-cell MALDI-TOF Mass Spectrometry Analysis of Peptaibol Formation by the Genus *Trichoderma*/*Hypocrea*: Can Molecular Phylogeny of Species Predict Peptaibol Structures? *Microbiology* **2007**, *153* (10), 3417–3437.

Nguyen, L. T.; Haney, E. F.; Vogel, H. J. The Expanding Scope of Antimicrobial Peptide Structures and Their Modes of Action. *Trends Biotechnol.* **2011**, *29* (9) 464–472.

Nizet, V.; Beall, B.; Bast, D. J.; Datta, V.; Kilburn, L.; Low, D. E.; et al. Genetic Locus for Streptolysin S Production by Group A *Streptococcus*. *Infect Immun.* **2000**, *68* (7), 4245–4254.

Noll, K. S.; Sinko, P. J.; Chikindas, M. L. Elucidation of the Molecular Mechanisms of Action of the Natural Antimicrobial Peptide Subtilosin Against the Bacterial Vaginosis-associated Pathogen *Gardnerella vaginalis*. *Probiotics Antimicrob.* **2011**, *3* (1), 41–47.

Oh, S. U.; Yun, B. S.; Lee, S. J.; Kim, J. H.; Yoo, I. D. Atroviridins A-C and Neoatroviridins A-D, Novel Peptaibol Antibiotics Produced by Trichoderma Atroviride F80317. I. Taxonomy, Fermentation, Isolation and Biological Activities. *J. Antibiot.* **2002**, *55* (6), 557–564.

Ongena, M.; Jacques, P.; Toure, Y.; Destain, J.; Jabrane, A.; Thonart, P. Involvement of Fengycin-type Lipopeptides in the Multifaceted Biocontrol Potential of *Bacillus subtilis*. *Appl. Microbiol. Biotechnol.* **2005**, *69*, 29–38.

Oppegård, C.; Fimland, G.; Thorbæk, L.; Nissen-Meyer, J. Analysis of the Two-peptide Bacteriocins Lactococcin G and Enterocin 1071 by Site-directed Mutagenesis. *Appl. Environ. Microbiol.* **2007,** *73* (9), 2931–2938. doi:10.1128/AEM.02718-06.

Otvos, L.; Rogers, M. E.; Consolvo, P. J.; Condie, B. A.; Lovas, S.; Bulet, P.; Blaszczyk-Thurin, M. Interaction Between Heat Shock Proteins and Antimicrobial Peptides. *Biochemistry* **2000,** *39,* 14150–14159.

Panizel, I.; Yarden, O.; Ilan, M.; Carmeli, S. Eight New Peptaibols from Sponge-associated Trichoderma Atroviride. *Mar. Drugs* **2013,** *11,* 4937–4960.

Patel, S.; Ahmed, S.; Eswari, J. S. Therapeutic Cyclic Lipopeptides Mining from Microbes: Latest Strides and Hurdles. *World J. Microbiol. Biotechnol.* **2015,** *31,* 1177–1193. doi: 10.1007/s11274-015-1880-8.

Pavela-Vrancic, M.; Van Liempt, H.; Pfeifer, E.; Freist, W.; Von Dohre, H. Nucleotide Binding by Multi Enzyme Peptide Synthetases. *Eur. J. Biochem.* **1994,** *220,* 535–542.

Pirtskhalava, M.; Gabrielian, A.; Cruz, P.; Griggs, H. L.; Squires, R. B.; Hurt, D. E.; Grigolava, M.; Chubinidze, M.; Gogoladze, G.; Vishnepolsky, B.; Alekseev, V.; Rosenthal, A.; Tartakovsky M. DBAASP v.2: An Enhanced Database of Structure and Antimicrobial/ Cytotoxic Activity of Natural and Synthetic Peptides. *Nucl. Acids Res.* **2016,** *44* (D1), D1104–D1112.

Powers, J. P.; Hancock, R. E. The Relationship Between Peptide Structure and Antibacterial Activity. *Peptides* **2003,** *24,* 1681–1691.

Rahnamaeian, M.; Langen, G.; Imani, J.; Khalifa, W.; Altincicek, B.; vonWettstein, D.; et al. Insect Peptide Metchni Kowin Confers on Barley a Selective Capacity for Resistance to Fungal Ascomycetes Pathogens. *J. Exp. Bot.* **2009,** *60,* 4105–4114.

Ren, J.; Xue, C.; Tian, L.; Xu, M.; Chen, J.; Deng, Z.; Proksch, P.; Lin, W. Asperelines A-F, Peptaibols from the Marine-derived Fungus *Trichoderma asperellum. J. Nat. Prod.* **2009,** *72,* 1036–1044.

Ritieni, A.; Fogliano, V.; Nanno, D.; Randazzo, G.; Altomare, C.; Perrone, G.; Bottalico, A.; Maddau, L.; Marras, F. Paracelsin E, A New Peptaibol from *Trichoderma saturnisporum. J. Nat. Prod.* **1995,** *58,* 1745–1748.

Röhrich, C. R.; Iversen, A.; Jaklitsch, W. M.; Voglmayr, H.; Berg, A.; Dörfelt, H.; Thrane, U.; Vilcinskas, A.; Nielsen, K. F.; von Döhren, H.; Brückner, H.; Degenkolb, T. Hypopulvins, Novel Peptaibiotics from the Polyporicolous Fungus *Hypocrea pulvinata,* are Produced During Infection of its Natural Hosts. *Fungal Biol.* **2012,** *116,* 1219–1231.

Röhrich, C. R.; Iversen, A.; Jaklitsch, W. M.; Voglmayr, H.; Vilcinskas, A.; Nielsen, K. F.; Thrane, U.; von Döhren, H.; Brückner, H.; Degenkolb, T. Screening the Biosphere: The Fungicolous Fungus *Trichoderma phellinicola,* a Prolific Source of Hypophellins, New 17-, 18-, 19-, and 20-residue Peptaibiotics. *Chem. Biodivers.* **2013a,** *10,* 787–812.

Röhrich, C. R.; Vilcinskas, A.; Brückner, H.; Degenkolb, T. The Sequences of the Eleven-residue Peptaibiotics: Suzukacillins-B. *Chem. Biodivers.* **2013b,** *10,* 827–837.

Röhrich, C. R.; Voglmayr, H.; Iversen, A.; Vilcinskas, A. Europe PMC Funders Group Front Line Defenders of the Ecological Niche! Screening the Structural Diversity of Peptaibiotics from Saprotrophic and Fungicolous *Trichoderma/Hypocrea* Species. *Fungal Diversity* **2015,** *69* (1), 117–146.

Sahl, H. G. Optimizing Antimicrobial Host Defense Peptides. *Chem. Biol.* **2006,** *13,* 1015–1017.

Sarika; Iquebal, M. A.; Rai, A. Biotic Stress Resistance in Agriculture Through Antimicrobial Peptides. *Peptides* **2012**, *36* (2), 322–330.

Schägger, H. Tricine–SDS-PAGE. *Nat. Protocol* **2006**, *1* (1), 16–23. http://doi.org/10.1038/nprot.2006.4.

Scholz, R.; Molohon, K. J.; Nachtigall, J.; Vater, J.; Markley, A. L.; Sussmuth, R. D.; et al. Plantazolicin, A Novel Microcin B17/Streptolysin S-like Natural Product from *Bacillus Amyloliquefaciens* FZB42. *J. Bacteriol.* **2011**, *193* (1), 215–224.

Scholz, R.; Vater, J.; Budiharjo, A.; Wang, Z.; He, Y.; Dietel, K.; et al. Amylocyclicin, A Novel Circular Bacteriocin Produced by *Bacillus amyloliquefaciens* FZB42. *J. Bacteriol.* **2014**, *196* (10), 1842–1852.

Shi, M.; Wang, H. N.; Xie, S. T.; Luo, Y.; Sun, C. Y.; Chen, X. L.; Zhang, Y Z. Antimicrobial Peptaibols, Novel Suppressors of Tumor Cells, Targeted Calcium-mediated Apoptosis and Autophagy in Human Hepatocellular Carcinoma Cells. *Mol. Cancer* **2010**, *9*, 26–41.

Shi, M.; Chen, L.; Wang, X. W.; Zhang, T.; Zhao, P. B.; Song, X. Y.; Sun, C. Y.; Chen, X. L.; Zhou, B. C.; Zhang, Y. Z. Antimicrobial Peptaibols from *Trichoderma Pseudokoningii* Induce Programmed Cell Death in Plant Fungal Pathogens. *Microbiology* **2012**, *158*, 166–175.

Shi, W. L., Chen, X. L., Wang, L. X., Gong, Z. T., Li, S., Li, C. L., et al., Song, X. Y. Cellular and Molecular Insight into the Inhibition of Primary Root Growth of Arabidopsis Induced by Peptaibols, A Class of Linear Peptide Antibiotics Mainly Produced by Trichoderma spp. *J. Exp Bot.* **2016**, *67* (8), 2191–205. http://doi.org/10.1093/jxb/erw023.

Song, X. Y.; Shen, Q. T.; Xie, S. T.; Chen, X. L.; Sun, C. Y.; Zhang, Y. Z. Broad- spectrum Antimicrobial Activity and High Stability of Trichokonins from *Trichoderma koningii* SMF2 Against Plant Pathogens. *FEMS Microbiol. Lett.* **2006**, *260*, 119–125.

Stein, T. *Bacillus subtilis* Antibiotics: Structures, Syntheses and Specific Functions. *Mol. Microbiol.* **2005**, *56*, 845–857.

Stepper, J.; Shastri, S.; Loo, T. S.; Preston, J. C.; Novak, P.; Man, P.; et al. Cysteine S-glycosylation, a New Post-translational Modification Found in Glycopeptide Bacteriocins. *FEBS Lett.* **2011**, *585* (4), 645–650.

Stoppacher, N.; Neumann, N. K.; Burgstaller, L.; Zeilinger, S.; Degenkolb, T.; Brückner, H.; Schuhmacher, R. The Comprehensive Peptaibiotics Database. *Chem. Biodivers.* **2013**, *10*, 734–743.

Strange, R. N.; Scott, P. R. Plant Disease: A Threat to Global Food Security. *Annu. Rev. Phytopathol.* **2005**, *43*, 83–116.

Sutyak, K. E.; Wirawan, R. E.; Aroutcheva, A. A; Chikindas, M. L. Isolation of the Bacillus Subtilis Antimicrobial Peptide Subtilosin from the Dairy Product-derived Bacillus Amyloliquefaciens. *J. Appl. Microbiol.* **2008**, *104* (4), 1067–1074.

Szekeres, A.; Leitgeb, B.; Kredics, L.; Antal, Z.; Hatvani, L.; Manczinger, L.; Vágvölgyi, C. Peptaibols and Related Peptaibiotics of *Trichoderma*. *Acta Microbiologica et Immunologica Hungarica* **2005**, *52*, 137–168.

Terras, F. R.; Schoofs, H. M.; De Bolle, M. F.; Van Leuven, F.; Rees, S. B.; Vanderleyden, J.; Cammue, B. P.; Broekaert, W. F. Analysis of Two Novel Classes of Plant Antifungal Proteins from Radish (*Raphanus sativus* L.) Seeds. *J. Biol. Chem.* **1992**, *267*, 15301–15309.

Van Belkum, M. J.; Martin-Visscher, L. A.; Vederas, J. C. Structure and Genetics of Circular Bacteriocins. *Trends Microbiol.* **2011**, *19* (8), 411–418.

Van der Weerden, N. L.; Hancock, R. E.; Anderson, M. A. Permeabilization of Fungal Hyphae by the Plant Defensin Nad1 Occurs Through a Cell Wall-dependent Process. *J. Biol. Chem.* **2010**, *285*, 37513–37520.

Wang, G.; Li, X.; Wang, Z. APD2: The Updated Antimicrobial Peptide Database and Its Application in Peptide Design. *Nucleic Acids Res.* **2009**, *37*, D933–D937.

Wang J.; Wong E. S.; Whitley J. C.; Li J.; Stringer J. M.; Short K. R.; et al. Ancient Antimicrobial Peptides Kill Antibiotic-resistant Pathogens: Australian Mammals Provide New Options. *PLoS One* **2011**, 6:e24030. 10.1371/journal.pone.0024030.

Wang, G.; Li, X.; Wang, Z. APD3: The Antimicrobial Peptide Database as a Tool for Research and Education. *Nucleic Acids Res.* **2016**, *44*, D1087–D1093.

Whitmore, L.; Wallace, B. A. The Peptaibol Database: a Database for Sequences and Structures of Naturally Occurring Peptaibols. *Nucleic Acids Res.* **2004**, *32*, D593–D594. doi:10.1093/nar/gkh077.

Yang, X., van der Donk, W. A. Ribosomally Synthesized and Post-translationally Modified Peptide Natural Products: New Insights into the Role of Leader and Core Peptides During Biosynthesis. *Chemistry* **2013**, *19* (24), 7662–7677.

Yokoyama, S.; Iida, Y.; Kawasaki, Y.; Minami, Y.; Watanabe, K.; Yagi, F. The Chitin-Binding Capability of Cy-Amp1 From Cycad Is Essential To Antifungal Activity. *J. Pept. Sci.* **2009**, *15*, 492–497.

Zhang, L.; Rozek, A.; Hancock, R. E. Interaction of Cationic Antimicrobial Peptides with Model Membranes. *J. Biol. Chem.* **2001**, *276*, 35714–35722.

PART IV
Biotechnological Approaches and the Impact of Climate Change

CHAPTER 12

Plant Disease Management Using Biotechnology: RNA Interference

NARESH PRATAP SINGH[1,*] and VAISHALI SHAMI[1]

[1]Department of Biotechnology, Sardar Vallabhbhai Patel University of Agriculture and Technology, Meerut (U.P), 250110, India.

*Corresponding author. E-mail: naresh.singh55@yahoo.com

ABSTRACT

World food supply is still threatened by various biotic stresses despite of substantial advances in plant disease management strategies. The situation demands judicious blending of conventional, unconventional and frontier technologies. Biotechnological novel techniques such as tissue culture and genetic engineering help us to achieve this goal by producing new organisms and or products that can be used in variety of ways. In this regard, RNA interference (RNAi) has emerged as a powerful technology for controlling various challenging diseases caused by viruses, fungi and bacteria. RNAi is a mechanism for RNA-guided regulation of gene expression in which double-stranded ribonucleic acid (dsRNA) inhibits the expression of genes with complementary nucleotide sequences. The application of tissue-specific or inducible gene silencing together help in silencing several genes simultaneously will result in protection of crops against destructive pathogens. RNAi application has resulted in successful control of many economically important diseases in plants.

12.1   INTRODUCTION

Currently, science has developed alot in reference to agriculture but still no significant tool developed to control over plant diseases caused by various pathogens such as fungi, viruses etc. Currently, more than 70% major crops yield is lost due to pathogen attack (Wani et al., 2010). With the advent of new technologies in plant genomics, including structural and functional genomics using biotechnological tools for developing improved crops and vegetables such as in rice (Hackauf 2009, Jeon 2008), wheat (Gill 2007), brassica (Wang 2008), maize (Diwedi 2008), soybean (Canon 2008), cotton (Chaudhary 2008) and vegetable crops (Lehtonen 2008). Most of the crops mainly vegetables are susceptible to plethora of biotic stresses like viruses, bacteria, fungal pathogens, insect pests and nematode parasites. Among the different types of biotic stresses, fungal and viral pathogens causes major loss in crop yields and pose huge economic losses globally (Wani et al 2010). The conventional methods to combat plant fungal pathogens includes breeding strategy, limits by the availability of resistant cultivars and evolution of fungal pathogenic races which makes the crop plants susceptible (Gilbert et al 2006) which is major drawback of this method. Plant diseases are usually controlled by the use of chemicals. The overdose use of chemicals may effects the human health, safety and cause environmental risks as well (Gilbert et al., 2006). Today, pathogens like fungi, viruses have evolved the resistant mechanisms through genetic adjustment, by which pathogen becomes less sensitive towards agro-chemicals (Yang 2012). This has fuelled a continual search for novel and alternate strategies for management of fungal pathogens. The silencing of such genes which are required for the invasion, growth of phytopathogenic fungi may be used as an ideal strategy for fungal disease management.

12.2   ABOUT RNAi

RNA interference (RNAi) has turned out to be an effective way to control infection caused by fungal pathogens, through silencing of vital genes associated with pathogens. In other words, crop plants expressing dsRNA targeting essential genes in fungal pathogens i.e., plant-mediated pathogen gene silencing has evolved as an emerging strategy to combat fungal pathogens in crop plants. In 1998, Fire and Mello discovered RNAi in *Caenorhabditis elegans* (Fire 1998). Now-a-days it is considered to be a powerful functional genomics tool to silence any gene of interest with tightly controlled sequence specificity by introducing dsRNA or small interfering RNA (siRNA) containing the target gene sequences into cells or organisms (Agarwal 2003). The loss of functions and altered phenotypes represented by result analysis prove to be the most readily interpretable method for experimentally validating the cellular function of genes (Bhadauria 2009). RNAi pathway can be induced by presence of dsRNA, which can be formed by various ways such as RNA viruses, inverted repeats, *in vitro* transcribed dsRNA, expressed dsRNA or transgenes in genetically modified organisms.

12.3 WORKING OF RNAi

RNA interference refers collectively to diverse RNA based processes that all result in sequence-specific inhibition of gene expression at the transcription, mRNA stability or translational level. The unifying features of this phenomena are the production of small RNAs (21-26 nucleotides (nt) that act as specific determinants for down-regulating gene expression (Issac,1992) and the requirement for one or more members of Argonaute family of protein (Maloy, 2005). RNAi operates by triggering the action of dsRNA intermediates, which are processed into RNA duplexes of 21-24 mucleotides by a ribonuclease III like enzyme called Dicer (Mehrotra and Aggarwal, 2003). Once produced, these small RNA molecules or short interfering RNAs (siRNAs) are incorporated in a multi-subunit complex called RNA induced silencing complex (RISC) (Mehrotra and Aggarwal, 2013): RISC is formed by a siRNA and an endonuclease among other component. The siRNAs within RISC acts as a guide to target the degradation of complementary messenger RNAs (mRNAs). When dsRNA molecules produced during viral replication trigger gene silencing, the process is called virus-induced gene silencing (VGS) (Maloy, 2005). One interesting feature of RNA silencing in plants is that once it is triggered in a certain cell, a mobile signal is produced and spread through the whole plant causing the entire plant to be silenced. This silencing process is also enhanced by the enzymatic activity of the RISC complex, mediating multiple turnover reaction (Broglie et al., 1991). Furthermore, production of the secondary siRNAs leads to enrichment of silencing via its spread from the first activated cell to neighboring cells, and systematically through system (Maloy, 2005). The cell to cell spread can be mediated as passive spread of the small RNAs via plasmodesmata, since it does not spread into meristematic cells. The discovery of RNA binding protein (PSRPI) in the phloem and its stability to build 25 ntssRNA species add further to the argument that siRNAs (24- 26nt) are the key components for systemic silencing signal (Brain and Beathle, 2003).

The first successful demonstration of RNAi like pathway in fungi came in 1992 with the pioneer research of Romano and Machino (Machino 1992).

12.4   INDUCTION METHODS OF RNai IN PLANTS

The first biggest difficulty is the transfer of the such active molecules that will switch on the RNAi pathway in plants. The various methods for the delivery of dsRNA or siRNA into different cells and tissue which include not only the transformation with dsRNA forming vectors for respective gene(s) by an *Agrobacterium* mediated transformations (Wterhouse et al., 2001, Chuang et al., 2000) but also the delivery of (a) dsRNA of *uidA* GUS (β-glucuronidase) and TaGLP2a: GFP (green fluorescent protein) reporter genes into epidermal cells of maize, barley and wheat by particle bombardment (Schweizer et al., 2000), (b) introducing a *Tobacco rattle virus* (TRV)-based vector in tomato plants by infiltration (Liu et al., 2002a), (c) delivery of dsRNA into tobacco suspension cells by cationic oligopeptide polyarginine-siRNA complex, (d) delivery of siRNA into cultured cells of rice, cotton and slash pine for gene silencing by nanosense pulsed laser-induced stress wave (LISW) (Tang et al., 2006). The most reliable and significant approachesd from above said to transfer dsRNA into plants cells are agroinfiltration, micro-bombardment and VIGS which are discussed below.

12.4.1   AGROINFILTRATION METHOD

The transfer of *Agrobacterium* carrying similar DNA constructs into the intracellular spaces of leaves for inducing RNA silencing is known as agro-inoculation or agroinfiltration (Hilly J.M and Liu Z, 2007). Agroinfiltration in mostly cases is used to induce systemic silencing or to know the effect of suppressor genes. Cytoplasmic RNAi can be induced efficiently in plant cells by agroinfiltration, which is similar to the expression of T-DNA vectors after delivery by *Agrobacterium tumefaciens*. The transiently expressed DNA encodes either an ss- or dsRNA, which is typically a hairpin (hp) RNA (Johansen L.K and Carrington J.C, 2001, Voinnet O, 2001).

12.4.2 MICROBOMBARDMENT METHOD

In this method, particles coated with dsRNA, siRNA or DNA that encode hairpin constructs as well as sense or antisense RNA, activate the RNAi pathway are bombarded with cells. Synthetic siRNAs are delivered into plants by biolistic pressure to cause silencing of GFP expression. The silencing effect of RNAi is detected after a day of bombardment, and it continues upto 3 to 4 days of post bombardment. Silencing occurred after 1 or 2 weeks and spread later in the vascular tissues of the non-bombarded leaves that were closest to the bombarded ones. Approximately after one month or above, the loss of GFP expression was seen cells. RNA blot hybridization with systemic leaves indicated delivery of siRNAs which cause systemic silencing by accumulation (Klahre et al., 2002).

12.4.3 VIRUS-INDUCED GENE SILENCING (VIGS) METHOD

Viruses, such as *Tobacco mosaic virus* (TMV), *Potato virus X* (PVX) and TRV, can be used for both protein expression and gene silencing(Kumagai et al., 1995, Mallory et al., 2002). RNA virus-derived expression vectors may have potent anti-silencing proteins so they will not be used as silencing vectors always(Kumagai et al., 1995, Palmer K.E and Rybicki E.P, 2001). Similarly, DNA viruses also have not been used extensively as expression vectors due to their size constraints for movement (Kjemtrup et al., 1998). Firstly, Dallwitz M.J and Zurcher E.J, 1996, demonstrated RNA viruses by inserting sequences into TMV and then for DNA viruses by replacing the coat protein gene with a homologous sequence(Kjemtrup et al., 1998). These reports used for gene silencing phytoene desaturase (*PDS*) and chalcone synthase (*CHS*), which provide a measure of the tissue specificity of silencing. The *PDS* gene protects the chlorophyll from photo oxidation. By silencing this gene, a significant decrease in leaf carotene content that resulted into the appearance of photo bleaching symptom (Liu et al., 2002c, Turnage et al., 2002). In the same way, over expression of *CHS* gene causes an albino phenotype in place of deep orange color(Cogoni et al., 1994). As a result, their action as a phenotypic marker helps in easy understanding of the mechanism of gene silencing. Most viruses are plus-strand RNA viruses or satellites, whereas *Tomato golden mosaic virus* (TGMV) and *Cabbage leaf curl virus* (CaLCuV) are DNA viruses. Both types of viruses induces silencing of endogenous genes but the extent of silencing spread and the severity of viral symptoms vary in different host plants(Teycheney P.Y and Tepfer M, 2001). The continuous development of virus-based silencing vectors can extend VIGS to economically important plants and protection from diseases.

12.5   RNAi DISEASES MANAGEMENT

Our global food supply is still threatened by various types of pathogens and pests .The edvent of new researches/technologies are to develop to respond more efficiently and effectively to this problem. Today, RNAi technology has emerged as one of the most potential and promising strategies for building up the resistance in plants to overcome the loss caused by various fungal, bacteria, viral and nematode diseases (Singh, 2005). Many of the examples listed below illustrate the possibilities for commercial exploitation of this inherent biological mechanism to generate disease resistant plants in the future by taking advantage of this approach e.g. including; *Cladosporium fulvum* (Singh, 2001) *Magnaporthae oryzae*, *Venturia inaequalis* and *Neurospora crassa* (Singh, 2005).

12.5.1   MANAGEMENT OF PLANT FUNGAL INFECTIONS USING RNAi

A potential approach of RNAi-based "host plant mediated pathogen gene silencing" for efficient control of fungal pathogens infecting various agronomical important crop plants found to be the efficient new biotechnological tool. Small non-coding RNAs which play a vital role in the process called RNA silencing. RNAi operates in both plants and animals, and use double stranded RNAi (dsRNA) for degradation or inhibiting transcription and translation in fungi, viruses, bacteria (Maloy, 2005, Mehrotra and Aggarwal, 2003). The concept behind this method is the down-regulation of vital fungal genes required for fungal invasion, normal growth and pathogenesis by the uptake of dsRNA/siRNAs produced by transgenic plants (Rajam et al 1998; Khatri 2007). RNA-mediated gene silencing (RNA silencing) is used as a reverse tool for gene targeting in fungi. The hypermorphic mechanism of RNA interference implies that this technique can also be applicable to all those plant pathogenic fungi, which are polyploid and polykaryotic in nature. Simultaneous silencing of several unrelated genes by introducing a single chimeric construct has been demonstrated in the case of *Venturia inaequalis* (Fitzgerald et al., 2004). *HCf-1*, a gene that codes for a hydrophobin of the tomato *pathogen C.* fulvum (Spanu, 1997), was cosuppressed by ectopic integration of homologous transgenes. Transformation of *Cladosporium fulvum* with DNA containing a truncated copy of the hydrophobin gene *HCf-1* caused co-suppression of hydrophobin synthesis in 30% of the transformants. The co-suppressed isolates had a hydrophilic phenotype, lower levels of *HCf-1* mRNA than wild type and contain multiple copies of the plasmid integrated as tandem repeats at ectopic sites in the genome (Hamada and Spanu, 1998). The transcription rate of *HCf-1* in the co-suppressed isolates was higher than in the untransformed strains, suggesting that silencing acted at the post-transcriptional level (Hamada and Spanu, 1998). Similarly, the silencing of *cgl1* and *cgl2* genes using the *cgl2* hairpin construct in *Cladosporium fulvum* has also been reported (Segers *et al.* 1999), which can be helpful for protecting the consumable products of vegetables and fruit crops from the post harvest diseases caused by different plant pathogens in the future. Fitzgerald et al. (2004), using the hairpin vector technology, have been able to trigger simultaneous high frequency silencing of a green fluorescent protein (*GFP*) transgene and an endogenous trihydroxynaphthalene reductase gene (*THN*) in *V.inaequalis.* High frequency gene silencing was achieved using hairpin constructs for the *GFP* or the *THN* genes transferred by *Agrobacterium* (71 and 61%, respectively). Similarly, multiple gene silencing has been achieved in *Cryptococcus neoformans* using chimeric hairpin constructs (Liu et al., 2002) and in plants using partial sense constructs (Abbott et al., 2002).). Tinoco et al. (2010), reported that the GUS specific siRNAs expressed in the transgenic tobacco could lead to the GUS gene silencing in the GUS transformed Fusarium verticillioides' transformants. They concluded that this could be a result of movement of silencing signal through the germinating spores into the fungal cells. The transgenic lettuce plants expressing a GUS dsRNA could induce specific gene silencing in the parasitizing plant Triphysaria versicolor expressing GUS gene (Tomilov et al., 2008).

12.5.2 MANAGEMENT OF PLANT PATHOGENIC VIRUSES USING RNAi

RNAi technology may be used for viral disease control in human cell lines (Bitko and Barik, 2001; Novina et al., 2002; Jacque et al., 2002) and also helpful in protecting viral infections in plants (Waterhouse et al., 2001; Ullu et al., 2002). Generating virus-resistant plants was first done in potato (Waterhouse et al., 1998; Chapman et al., 2004). DNA viruses like Gemini viruses *Mungbean yellow mosaic India virus* (MYMIV) was expressed as hairpin construct and used as biolistically to inoculate MYMIV-infected black gram plants which showed a complete recovery from infection, which lasted until senescence (Pooggin et al., 2003). RNAi-mediated silencing of Gemini viruses using transient protoplast assay where protoplasts were cotransferred with an siRNA designed to replicase (Rep)-coding sequence of *African cassava mosaic virus* (ACMV) and the genomic DNA of ACMV resulted in 99% reduction in *Rep* transcripts and 66% reduction in viral DNA (Vanitharani et al., 2003; Ruiz-Ferrer and Voinnet, 2007). Multiple suppressors have been reported in *Citrus tristeza virus* (Lu et al., 2003). A 273-bp (base pair) sequence of the *Arabidopsis* miR159 a pre-miRNA transcript expressing amiRNAs can be used against the viral suppressor genes to provide resistance against *Turnip yellow mosaic virus* and *Turnip mosaic virus* infection (Niu et al., 2006). Different amiRNA vector was used to target the 2 b viral suppressor of the *Cucumber mosaic virus* (CMV), which blocked the slicer activity of *AGO*1and confer resistance to CMV infection in transgenic tobacco (Qu et al., 2007). A strong correlation between virus resistance and the expression level of the 2 b-specific amiRNA was shown for individual plant lines. It is significantly proved from abovementioned reports that the RNA components, such as ssRNA, dsRNA, and/or siRNA of the silencing pathways are appropriate targets of most viral suppressors and help in protection of plants from viral infections using RNAi technology.

12.6   CONCLUSION WITH FUTURE PROSPECTS

To feed the expanding human population leads the high stress on agroecosystems to control plants from various biotic stresses. RNAi and miRNA technologies of gene silencing is gaining the novel importance not only in functional genomics but also having great tendency of higher silencing efficiency of vital gene of pathogen to control disease. The RNA silencing is highly sequence specific and it is technologically efficient and economical as well. Therefore, this technique has great potential in agriculture specifically for nutritional improvement of plants and the management of pathogenic plant diseases. Researches on sRNAs have provided lots of studies regarding their types, functions, etc. But still the information is not enough. Various report of researches helped to classified srNAs into mainly three categories, miRNAs, siRNAs, and piRNAs, while many new types of srNAs are under exploration. In plants, both miRNAs and siRNAs are present. They act collectively as well as individually to help the plants with their maintenance, homeostasis, and survival under adverse conditions. Various researches marked RNAi technology as powerful tool to combat plant pathogens in the near future. Development of vectors that can suppress the RNAi pathway but overexpress transgenes in a tissue-specific manner will revolutionize this field in future. Presently, substantial researches are being conducted to find the role of miRNAs and siRNAs in biotic stresses. Regulation by srNAs may be used as a promising tool to improve yields, quality, or resistance to various pathogenic diseases and environmental stresses. Discovery of more srNAs in plant system will help researchers to manipulate these srNAs in favor of plant growth and development. Hence, it will be very appropriate to call srNA "an efficient molecule of the millennium."

KEYWORDS

- RNA interference
- disease management
- crop improvement
- biotic and abiotic stress
- transformation
- genetic engineering

REFERENCES

Abbott, J. C.; Barakate, A.; Pincon, G.; Legrand, M.; Lapierre, C.; Mila, I.; Schuch, W.; Halpin, C. Simultaneous Suppression of Multiple Genes by Single Transgenes. Down-regulation of Three Unrelated Lignin Biosynthetic Genes in Tobacco. *Plant Physiol.* **2002**, *128*, 844–853.

Adenot, X.; Elmayan, T.; Lauressergues, D.; Boutet, S.; Bouche, N.; Gasciolli, V.; Ucheret, H. Uncoupled Production of ta-siRNAs in Hypomorphic *rdr6* and *sgs3* Mutants Uncovers a Role for *TAS3* in AGO7-DCL4-DRB4-mediated Control of Leaf Morphology. *Curr. Biol.* **2006**, *16* (9), 927–932.

Agrawal, N.; Dasaradhi, P. V.; Mohmmed, A.; Malhotra, P.; Bhatnagar, R. K.; et al. RNA Interference: Biology, Mechanism, and Applications. *Microbiol. Mol. Biol. Rev.* **2003**, *67*, 657–685.

Allen, E.; Xie, Z.; Gustafson, A. M.; Carrington, J. C. MicroRNA-directed Phasing During *Trans*-acting siRNA Biogenesis in Plants. *Cell* **2005**, *121*, 207–221.

Ambros, V. The Functions of Animal MicroRNAs. *Nature* **2004**, *431*, 350–355.

Bartel, D. P. MicroRNAs: Genomics, Biogenesis, Mechanism, and Function. *Cell* **2004**, *116*, 281–297.

Baulcombe, D. C. RNA Silencing in Plants. *Nature* **2004**, *431*, 356–363.

Baumberger, N.; Baulcombe, D. C. *Arabidopsis* ARGONAUTE1 is an RNA Slicer that Selectively Recruits Micrornas and Short Interfering RNAs. *Proc. Natl. Acad. Sci.* **2005**, *102*, 11928–11933.

Bhadauria, V.; Banniza, S.; Wei, Y.; Peng, Y. L. Reverse genetics for functional genomics of phytopathogenic fungi and oomycetes. *Comp. Funct. Genomics* **2009**, *11*. doi: 10.1155/2009/380719.

Bitko, V.; Barik, S. Phenotypic Silencing of Cytoplasmic Genes with Sequence Specific Double Stranded Short Interfering RNA and Its Applications in the Reverse Genetics of Wild Type Negative Strand RNA Virus. *BMC Microbiol.* **2001**, *1*, 34–44.

Bohmert, K.; Camus, I.; Bellini, C.; Bouchez, D.; Caboche, M.; Benning, C. AGO1 Defines a Novel Locus of *Arabidopsis* Controlling Leaf Development. *EMBO J.* **1998**, *17*, 170–180.

Borsani, O.; Zhu, J.; Verslues, P. E.; Sunkar, R.; Zhu, J. K. Endogenous siRNAs Derived From a Pair of Natural *Cis*-antisense Transcripts Regulate Salt Tolerance in *Arabidopsis*. *Cell* **2005**, *123*, 1279–1291.

Brain, C. F.; Beathle, G. A. An Overview of Plant Defences Against Pathogen and Herbivores. *Plant Health Instructor* **2003**, *1*. doi:10.1094/PHI-I-2008-0226-01.

Brennecke, J.; Hipfner, D. R.; Stark, A.; Russell, R. B.; Cohen, S. M. Bantam Encodes a Developmentally Regulated MicroRNA that Controls Cell Proliferation and Regulates the Proapoptotic Gene Hid in *Drosophila*. *Cell* **2003**, *113*, 25–36.

Broglie, K. I.; Chet, M.; Holliday, M. N. Transgenic Plants with Enhanced Resistance. *Cell* **1991**, *123*, 1279–1291.

Cakir, C.; Tör, M. Factors Influencing Barley Stripe Mosaic Virus-mediated Gene Silencing in Wheat. *Physiol. Mol. Plant Pathol.* **2010**, *74*, 246–253.

Cannon, S. Legume Comparative Genomics. *Genet. Genomics Soybean* **2008**, 35–54.

Cao, X.; Zhou, P.; Zhang, X.; Zhu, S.; Zhong, X.; Xiao, Q.; Ding, B.; Li, Y. Identification of an RNA Silencing Suppressors from a Plant Double Stranded RNA Virus. *J. Virol.* **2005**, *79*, 13018–13027.

Chapman, E. J.; Prokhnevsky, A. I.; Gopinath, K.; Dolja, V. V.; Carrington, J. C. Viral RNA Silencing Suppressors Inhibit the micro-RNA Pathway at an Interphase Step. *Genes Dev.* **2004**, *18*, 1179–1186.

Chaudhary, B.; Hovav, R.; Rapp, R.; Verma, N.; Udall, J. A.; Wendel, J. F. Global Analysis of Gene Expression in Cotton Fibers from Wild and Domesticated *Gossypium barbadense*. *Evol. Dev.* **2008**, *10*, 567–582.

Chen, J.; Li, W. X.; Xie, D.; Peng, J. R.; Ding, S. W. Viral Virulence Protein Suppresses RNA Silencing-mediated Defense but Upregulates the Role of MicroRNA in Host Gene Regulation. *Plant Cell* **2004**, *16* (5), 1302–1313.

Chen, X. A microRNA as a Translational Repressor of *APETALA2* in *Arabidopsis* Flower Development. *Science* **2004**, *303*, 2022–2025.

Chen, X.; Liu, J.; Cheng, Y.; Jia, D. *HEN1* Functions Pleiotropically in *Arabidopsis* Development and Acts in C Function in the Flower. *Development* **2002**, *129*, 1085–1094.

Chuang, C. F.; Meyerowtiz, E. M. Specific and Heritable Genetic Interference by Double-stranded RNA in *Arabidopsis thaliana*. *Proc. Natl Acad. Sci. USA* **2000**, *97*, 985–4990.

Cogoni, C.; Romano, N.; Macino, G. Suppression of Gene Expression by Homologous Transgenes. *Antonie Leeuwenhoek Int. J. Gen. Mol. Microbiol.* **1994**, *65*, 205–209.

Dallwitz, M. J.; Zurcher, E. J. Plant Viruses Online. In *Descriptions and Lists from the VIDE Database*, Brunt, A. A.; Crabtree, K.; Dallwitz, M. J.; Gibbs, A. J.; Watson, L.; Zurcher, E. J., Eds.; CAB International: UK, 1996; p 1484.

Ding, X. S.; Schneider, W. L.; Chaluvadi, S. R.; Rouf Mian, R. M.; Nelson, R. S. Characterization of a Brome Mosaic Virus Strain and Its Use as a Vector for Gene Silencing in Monocotyledonous Hosts. *Mol. Plant Microbe Interact.* **2006**, *19*, 1229–1239.

Du, T.; Zamore, P. D. MicroPrimer: The Biogenesis and Function of MicroRNA. *Development* **2005**, *132*, 4645–4652.

Dwivedi, S.; Perotti, E.; Ortiz, R. Towards Molecular Breeding of Reproductive Traits in Cereal Crops. *Plant Biotech. J.* **2008**, *6*, 529–559.

Emery, J. F.; Floyd, S. K.; Alvarez, J.; Eshed, Y.; Hawker, N. P.; Izhaki, A.; Baum, S. F.; Bowman, J. L. Radial Patterning of *Arabidopsis* Shoots by Class III HD-ZIP and KANADI Genes. *Curr. Biol.* **2003**, *13*, 1768–1774.

Erdmann, V. A., et al. Regulatory RNAs. *Cell Mol. Life Sci.* **2001**, *58*, 1–18.

Fagwalawa, L. D.; Kutama, A. S.; Yakasai, M. T. Current Issues in Plant Disease Control: Biotechnology and Plant Disease. *Bayero J. Pure Appl. Sci.* **2013**, *6* (2), 121–126.

Finnegan, E. J.; Matzke, M. A. The Small RNA World. *J. Cell Sci.* **2003**, *116*, 4689–4693.

Fire, A.; Xu, S.; Montgomery, M. K.; Kostas, S. A.; Driver, S. E.; et al. Potent and Specific Genetic Interference by Double-stranded RNA in *Caenorhabditis elegans*. *Nature* **1998**, *391*, 806–811.

Fitxgerald, A.; Van Kha, J. A.; Plummer, K. M. Simultaneous Silencing of Multiple Genes in the Apple Scab Fungus *Venturia Inaequalis*, by Expression of RNA With Chimeric Inverted Repeats. *Fungal Genet. Biol.* **2004**, *41*, 963–971.

Fofana, I. B.; Sangare, A.; Collier, R.; Taylor, C.; Fauquet, C. M. A Geminivirus-induced Gene Silencing System for Gene Function Validation in Cassava. *Plant Mol. Biol.* **2004**, *56*, 613–624.

Fujii, H.; Chiou, T. J.; Lin, S. I.; Aung, K.; Zhu, J. K. A miRNA Involved in Phosphate-starvation Response in *Arabidopsis*. *Curr. Biol.* **2005**, *15*, 2038–2043.

Gasciolli, V.; Mallory, A. C.; Bartel, D. P.; Vaucheret, H. Partially Redundant Functions of Arabidopsis DICER like Enzymes and a Role for DCL4 in Producing Trans-acting siRNAs. *Curr. Biol.* **2005**, *15*, 1494–1500.

Gil, S. J.; Candela, H.; Ponce, M. R. Plant microRNAs and Development. *Int. J. Dev. Biol.* **2005**, *49*, 733–744.

Gilbert, J.; Jordan, M.; Somers, D. J.; Xing, T.; Punja, Z. K. Engineering Plants for Durable Disease Resistance. In *Multigenic and Induced Systemic Resistance in Plants*; Springer: New York, 2006.

Gill, B. S.; Li, W.; Sood, S.; Kuraparthy, V.; Friebe, B. R.; Simons, K. J.; Zhang, Z.; Faris, J. D. Genetics and Genomics of Wheat Domestication-driven Evolution. *Israel J. Plant Sci.* **2007**, *55*, 223–229.

Goldoni, M.; Azzalin, G.; Macino, G.; Cogoni, C. Efficient Gene Silencing By Expression of Double Stranded RNA in *Neurospora crassa. Fungal Genet. Biol.* **2004**, *41*, 1016–1024.

Guleria, P.; Mahajan, M.; Bhardwaj, J.; Yadav, S. K. Plant Small Rnas: Biogenesis, Mode of Action and Their Roles in Abiotic Stresses. *Genom. Proteom. Bioinform.* **2011**, *9* (9), 183–199.

Gustafson, A. M.; Allen, E.; Givan, S.; Smith, D.; Carrington, J. C.; Kasschau, K. D. ASRP: The *Arabidopsis* Small RNA Project Database. *Nucleic Acids Res.* **2005**, *33*, D637–D640.

Hackauf, B.; Rudd, S.; van der Voort, J. R.; Miedaner, T.; Wehling, P. Comparative Mapping of DNA Sequences in Rye (*Secale cereale* L.) in Relation to the Rice Genome. *Theor. Appl. Genet.* **2009**, *118*, 371–384.

Hamada, W.; Spanu, P. D. Co-suppression of the Hydrophobin gene *Hcf-1* Is Correlated with Antisense RNA Biosynthesis in *Cladosporium fulvum. Mol. Gen. Genet.* **1998**, *259*, 630–638.

Hammond, T. M.; Keller, N. P. RNA Silencing in *Aspergillus nidulans* is Independent of RNA-dependent RNA Polymerase. *Genetics* **2005**, *169*, 607–617.

Hilly, L. K.; Liu, Z. An Overview of Small RNAs. In *Regulation of Gene Expression in Plants*; Bassett, C. L., Ed.; Springer-Verlag: Berlin, 2007; pp 123–147.

Hiraguri, A.; Itoh, R.; Kondo, N.; Nomura, Y.; Aizawa, D.; Murai, Y.; Koiwa, H.; Seki, M.; Shinozaki, K.; Fukuhara, T. Specific Interactions Between Dicer-like Proteins and HYL1/DRB-family dsRNA-binding Proteins in *Arabidopsis thaliana. Plant Mol. Biol.* **2005**, *57*, 173–188.

Holzberg, S.; Brosio, P.; Gross, C.; Pogue, G. P. Barley Stripe Mosaic Virus-induced Gene Silencing in a Monocot Plant. *Plant J.* **2002**, *30*, 315–327.

Issac, S. *Biotechnology in the Study of Fungalplant Interaction*; Chapman and Mall: London, 1992, pp 327–382.

Jacque, J. M.; Triques, K.; Stevenson, M. Modulation of HIV-1 Replication by RNA Interference. *Nature* **2002**, *418*, 435–438.

Jeon, J.; Choi, J.; Park, J.; Lee, Y. H. Functional Genomics in the Rice Blast Fungus to Unravel the Fungal Pathogenicity. *J. Zhejiang Univ. Sci. B.* **2008**, *9*, 747–752.

Johansen, L. K.; Carrington, J. C. Silencing on the Spot. Induction and Suppression of RNA Silencing in the *Agrobacterium*-mediated Transient Expression System. *Plant Physiol.* **2001**, *126*, 930–938.

Jones-Rhoades, M. W.; Bartel, D. P.; Bartel, B. Micro- RNAs and Their Regulatory Roles in Plants. *Annu. Rev. Plant Biol* **2006**, *57*, 19–53.

Kadotani, N.; Nakayashiki, H.; Tosa, Y.; Mayama, S. RNA Silencing in the Pathogenic Fungus *Magnaporthe oryzae. Mol. Plant-Microb. Interact.* **2003**, *16*, 769–776.

Khatri, M.; Rajam, M. V. Targeting Polyamines of *Aspergillus Nidulans* by Sirna Specific to Fungal Ornithine Decorboxylase Gene. *Med. Mycol.* **2007**, *45*, 211–220.

Khraiwesha, B.; Zhua, J. K.; Zhuc, J. Role of miRNAs and siRNAs in Biotic and Abiotic Stress Responses of Plants. *Biochim. Biophys. Acta.* **2012**, *1819* (2), 137–148.

Kim, V. N. Small RNAs: Classification, Biogenesis, and Function. *Mol. Cells* **2005**, *19*, 1–15.

Kim, V. N., et al. Biogenesis of Small RNAs in Animals. *Nat. Rev. Mol. Cell Biol.* **2009**, *10*, 126–139.

Kjemtrup, S.; Sampson, K. S.; Peele, C. G.; Nguyen, L. V.; Conkling, M. A. Gene Silencing from Plant DNA Carried by a Geminivirus. *Plant J.* **1998**, *14*, 91–100.

Klahre, U.; Crete, P.; Leuenberger, S. A.; Iglesias, V. A.; Meins, F. High Molecular Weight RNAs and Small Interfering RNAs Induce Systemic Post Transcriptional Gene Silencing in Plants. *Proc. Natl Acad. Sci. USA* **2002**, *99*, 11981–11986.

Kumagai, M. H.; Donson, J.; della-Cioppa, G.; Harvey, D.; Hanley, K. Grill, L. K. Cytoplasmic Inhibition of Carotenoid Biosynthesis with Virus-derived RNA. *Proc. Natl Acad. Sci. USA* **1995**, *92*, 1679–1683.

Lee, R. C.; Feinbaum, R. L.; Ambros, V. The *C. elegans* Heterochronic Gene lin-4 Encodes Small RNAs with Antisense Complementarity to lin-14. *Cell* **1993**, *75*, 843–854.

Lehtonen, M. J.; Somervuo, P.; Valkonen, J. P. T. Infection with *Rhizoctonia solani* Induces Defense Genes and Systemic Resistance in Potato Sprouts Grown Without Light. *Phytopathology* **2008**, *98*, 1190–1198.

Li, H. W.; Ding, S. W. Antiviral Silencing in Animals. *FEBS Lett.* **2005**, *579*, 5965–5973.

Li, J.; Yang, Z.; Yu, B.; Liu, J.; Chen, X. Methylation Protects miRNAs and siRNAs from a 3-end Uridylation Activity in *Arabidopsis*. *Curr. Biol.* **2005**, *15*, 1501–1507.

Liu, Y. L.; Schiff, M.; Kumar, D. S. P. Virus Induced Gene Silencing in Tomato. *Plant J.* **2002**, *31*, 777–786.

Liu, Q.; Singh, S. P.; Green, A. G. High-stearic and High-oleic Cottonseed Oils Produced by Hairpin RNA Mediated Post-transcriptional Gene Silencing. *Plant Physiol.* **2002c**, *129*, 1732–1743.

Lu, C.; Fedoroff, N. A mutation in the *Arabidopsis* HYL1 Gene Encoding A dsRNA Binding Protein Affects Responses to Abscisic Acid, Auxin, and Cytokinin. *Plant Cell* **2000**, *12*, 2351–2366.

Lu, C.; Tej, S. S.; Luo, S.; Haudenschild, C. D.; Meyers, B. C.; Green, P. J. Elucidation of the Small RNA Component of the Transcriptome. *Science* **2005**, *309*, 1567–1569.

Lu, R.; Martin-Hernandez, A. M.; Peart, J. R.; Malcuit, I.; Baulcombe, D. C. Virus Induced Gene Silencing in Plants. *Methods* **2003**, *30*, 296–303.

Mallory, A. C.; Parks, G.; Endres, M. W.; Baulcombe, D.; Bowman, L. H. The Ampliconplus System for High-level Expression of Transgenes in Plants. *Nat. Biotechno.* **2002**, *20*, 622–625.

Mallory, A. C.; Reinhart, B. J.; Jones-Rhoades, M. W.; Tang, G.; Zamore, P. D.; Barton, M. K.; Bartel, D. P. MicroRNA Control of *PHABULOSA* in Leaf Development: Importance of Pairing to the MicroRNA 5_ Region. *EMBO J.* **2004**, *23*, 3356–3364.

Maloy, O. C. Plant Disease Management. *Plant Instructor* **2005**, 0202–0201.

Matsukura, C.; Aoki, K.; Fukuda, N.; Mizoguchi, T.; Asamizu, E.; Saito, T.; Shibata, D.; Ezura, H. Comprehensive Resources for Tomato Functional Genomics Based on the Miniature Model Tomato Micro-Tom. *Curr. Genomics* **2008**, *9*, 436–443.

Mehrotra, R. S.; Aggarwal, A. *Plant Pathology*, 2nd ed.; Tata McGraw Hill Publishing Company Limited, 2003, pp. 411–423. ISBN-10: 9383286490.

Nakayashiki, H. RNA Silencing in Fungi: Mechanisms and Applications. *Fed. Eur. Biochem. Soc. Lett.* **2005**, *579*, 5950–5970.

Niu, Q. W.; Lin, S. S.; Reyes, J. L.; Chen, K. C.; Wu, H. W.; Yeh, S. D.; Chua, N. H. Expression of Artificial MicroRNAs in Transgenic *Arabidopsis thaliana* Confers Virus Resistance. *Nat. Biotechnol.* **2006**, *24*, 1420–1428.

Novina, C. D.; Murray, M. F.; Dykxhoorn, D. M.; Beresford, P. J.; Riess, J.; Lee, S. K.; Collman, R. G.; Lieberman, J.; Shanker, P.; Sharp, P. A. siRNA-directed Inhibition of HIV-1 Infection. *Nat. Mediterr.* **2002**, *8*, 681–686.

Palatnik, J. F.; Allen, E.; Wu, X.; Schommer, C.; Schwab, R.; Carrington, J. C.; Weigel, D. Control of Leaf Morphogenesis by MicroRNAs. *Nature* **2003**, *425*, 257–263.

Palmer, K. E.; Rybicki, E. P. Investigation of the Potential of Maize Streak Virus to Act as an Infectious Gene Vector in Maize Plants. *Arch. Virol.* **2001**, *14*, 1089–1104.

Panthee, D. R.; Chen, F. Genomics of Fungal Disease Resistance in Tomato. *Curr. Genomics* **2010**, *11*, 30–39.

Parizotto, E. A.; Dunoyer, P.; Rahm, N.; Himber, C.; Voinnet, O. In Vivo Investigation of the Transcription, Processing, Endonucleolytic Activity, and Functional Relevance of the Spatial Distribution of a Plant miRNA. *Genes Dev.* **2004**, *18*, 2237–2242.

Peragine, A.; Yoshikawa, M.; Wu, G.; Albrecht, H. L.; Poethig, R. S. SGS3 and SGS2/SDE1/RDR6 Are Required for Juvenile Development and the Production of *Trans*-acting siRNAs in *Arabidopsis*. *Genes Dev.* **2004**, *18*, 2368–2379.

Pooggin, M.; Shivaprasad, P. V.; Veluthambi, K.; Hohn, T. RNAi Targetting of DNA Viruses. *Nat. Biotechnol.* **2003**, *21*, 131–132.

Publishing Co. PVT. LTD. New Delhi; pp 178–189

Qi, Y.; Denli, A. M.; Hannon, G. J. Biochemical Specialization Within *Arabidopsis* RNA Silencing Pathways. *Mol. Cell* **2005**, *19*, 421–428.

Qu, J.; Ye, J.; Fang, R. X. Artificial microRNA-mediated Virus Resistance in Plants. *J. Virol.* **2007**, *81*, 6690–6699.

Rajam, M. V. Polyamine Biosynthetic Pathway: A Potential Target for Plant Chemotherapy. *Curr. Sci.* **1998**, *74*, 729–731.

Reinhart, B. J.; Slack, F. J.; Basson, M.; Pasquinelli, A. E.; Bettinger, J. C.; Rougvie, A. E.; Horvitz, H. R.; Ruvkun, G. The 21-nucleotide let-7 RNA Regulates Developmental Timing in *Caenorhabditis elegans*. *Nature* **2000**, *403*, 901–906.

Reinhart, B. J.; Weinstein, E. G.; Rhoades, M. W.; Bartel, B.; Bartel, D. P. MicroRNAs in Plants. *Genes Dev.* **2002**, *16*, 1616–1626.

Rhoades, M.; Reinhart, B.; Lim, L.; Burge, C.; Bartel, B.; Bartel, D. Prediction of Plant MicroRNA Targets. *Cell* **2002**, *110*, 513–520.

Romano, N.; Macino, G. Quelling: Transient Inactivation of Gene Expression in *Neurospora Crassa* by Transformation with Homologous Sequences. *Mol. Microbiol.* **1992**, *6*, 3343–3353.

Rubio-Somoza, I., et al. Regulation and Functional Specialization of Small RNA–target Nodes During Plant Development. *Curr. Opin. Plant Biol.* **2009**, *12*, 622–627.

Ruiz-Ferrer, V.; Voinnet, O. Viral Suppression of RNA Silencing: 2b Wins the Golden Fleece by Defeating Argonaute. *Bioassays* **2007**, *29*, 319–323.

Rymarquis, L. A., et al. Diamonds in the Rough: mRNA-like Non-coding RNAs. *Trends Plant Sci.* **2008**, *13*, 329–334.

Schauer, S. E.; Jacobsen, S. E.; Meinke, D. W.; Ray, A. DICER-LIKE1: Blind Men and Elephants in *Arabidopsis* development. *Trends Plant Sci.* **2002**, *7*, 487–491.

Schweizer, P.; Pokorny, J.; Schulze-Lefert, P.; Dudler, R. Double Stranded RNA Interference with Gene Functions at the Single Cell in Cereals. *Plant J.* **2000**, *24*, 895–903.

Scofield, S. R.; Huang, L.; Brandt, A. S.; Gill, B. S. Development of a Virus-induced Gene Silencing System for Hexaploid Wheat and Its Use in Functional Analysis of the Lr21 Mediated Leaf Rust Resistance Pathway. *Plant Physiol.* **2005**, *138*, 2165–2173.

Segers, G. C.; Hamada, W.; Oliver, R. P.; Spanu, P. D. Isolation and Characteristaion of Five Different Hydrophobin-encoding cDNA from the Fungal Tomato Pathogen *Cladosporium fulvum. Mol. Gen. Genet.* **1999**, *261*, 644–652.

Singh, R. S. *Plant Disease Management*; Oxford and IBN New Delhi, 2001; pp 234–245. ISBN-10: 1578081602.

Singh, R. S. *Introduction to Principles of Plant Pathology*, 4th edn.; Oxford and IBH, 2005. ISBN-9788120415515.

Stricklin, S. L., et al. *C. elegans* Noncoding RNA Genes. In *WormBook*, The *C. elegans* Research Community, Ed.; WormBook, 2005. doi/10.1895/wormbook.1.1.1.

Szymanski, M., et al. Noncoding RNA Transcripts. *J. Appl. Genet.* **2003**, *44*, 1–19.

Talbot, N. J.; Kershaw, M. J.; Wakley, G. E.; de Vries, O. M. H.; Wessels, J. G. H.; Hamer, J. E. MPG1 Encodes a Fungal Hydrophobin Involved in Surface Interactions During Infection-related Development of *Magnaporthe grisea. Plant Cell* **1996**, *8*, 985–999.

Tang, W.; Weidner, D. A.; Hu, B. Y.; Newton, R. J.; Hu, X. Efficient Delivery of Small Interfering RNA to Plant Cells by a Nanosecond Pulsed Laser-induced Wave for Post Transcriptional Gene Silencing. *Plant Sci.* **2006**, *171*, 375–381.

Telfer, A.; Poethig, R. S. Hasty: A Gene That Regulates the Timing of Shoot Maturation in Arabidopsis Thaliana. *Development* **1998**, *125*, 1889–1898.

Teycheney, P. Y.; Tepfer, M. Virus Specific Spatial Differences in the Interference with Silencing of the *chs-A* Gene in Non-transgenic Petunia. *J. Gen. Virol.* **2001**, *82*, 1239–1243.

Tinoco, M. L.; Dias, B. B.; Dall'Astta, R. C.; Pamphile, J. A.; Aragão, F. J. *In vivo* Trans-specific Gene Silencing in Fungal Cells by in Planta Expression of a Double-stranded RNA. *BMC Biol.* **2010**, *8*, 27.

Tomilov, A. A.; Tomilova, N. B.; Wroblewski, T.; Michelmore, R.; Yoder, J. I. Trans-specific Gene Silencing Between Host and Parasitic Plants. *Plant J.* **2008**, *56*, 389–397.

Turnage, M. A.; Muangsan, N.; Peele, C. G.; Robertson, D. Geminivirus-based Vectors for Gene Silencing in *Arabidopsis. Plant J.* **2002**, *30*, 107–117.

Ullu, E.; Djikeng, A.; Shi, H.; Tschudi, C. RNA Interference: Advances and Questions. *Phil. Trans. Royal Soc. London British Biol. Sci.* **2002**, *29*, 65–70.

Vanitharani, R.; Chellappan, P.; Fauquet, C. M. Short Interfering RNA-mediated Interference of Gene Expression and Viral DNA Accumulation in Cultured Plant Cells. *Proc. Natl Acad. Sci. USA* **2003**, *100*, 9632–9636.

Vaucheret, H. Post-transcriptional Small RNA Pathways in Plants: Mechanisms and Regulations. *Genes Dev.* **2006**, *20*, 759–771.

Vaucheret, H.; Vazquez, F.; Crété, P.; Bartel, D. P. The Action of *ARGONAUTE1* in the miRNA Pathway and Its Regulation by the miRNA Pathway are Crucial for Plant Development. *Genes Dev.* **2004**, *18*, 1187–1197.

Vazquez, F.; Gasciolli, V.; Crété, P.; Vaucheret, H. The Nuclear dsRNA Binding Protein Hyl1 is Required for Microrna Accumulation and Plant Development, but Not Posttranscriptional Transgene Silencing. *Curr. Biol.* **2004a**, *14*, 346–351.

Vazquez, F.; Vaucheret, H.; Rajagopalan, R.; Lepers, C.; Gasciolli, V.; Mallory, A. C.; Hilbert, J. L.; Bartel, D. P.; Crete, P. Endogenous *Trans*-acting siRNAs Regulate the Accumulation of *Arabidopsis* mRNAs. *Mol. Cell* **2004b**, *16*, 69–79.

Voinnet, O. RNA Silencing as a Plant Immune System against Viruses. *Trends Genet.* **2001**, *17*, 449–459.

Wang, X. J.; Gaasterland, T.; Chua, N. H. Genomewide Prediction and Identification of *Cis*-natural Antisense Transcripts in *Arabidopsis thaliana*. *Genome Biol.* **2005**, *6*, R30.

Wang, N.; Wang, Y. J.; Tian, F.; King, G. J.; Zhang, C. Y.; Long, Y.; Shi, L.; Meng, J. L. A Functional Genomics Resource for *Brassica Napus*, Development of an EMS Mutagenized Population and Discovery of FAE1 Point Mutations by Tilling. *New Phytologist* **2008**, *180*, 751–765.

Wani, S. H.; Sanghera, G. S. Genetic Engineering for Viral Disease Management in Plants. *Notulae Scientia Biologicae* **2010a**, *2*, 20–28.

Wani, S. H.; Sanghera, G. S.; Singh, N. B. Biotechnology and Plant Disease Control-role of RNA Interference. *Am. J. Plant Sci.* **2010b**, *1*, 55–68.

Waterhouse, P. M.; Graham, M. W.; Wang, M. B. Virus Resistance and Gene Silencing in Plants can be Induced by Simultaneous Expression of Sense and Antisense RNA. *Proc. Natl. Acad. Sci. USA* **1998**, *95*, 13959–13964.

Waterhouse, P. M.; Wang, M. B.; Lough, T. Gene Silencing as an Adaptive Defense against Viruses. *Nature* **2001**, *411*, 834–842.

Xie, Z.; Kasschau, K. D.; Carrington, J. C. Negative Feedback Regulation of *Dicer-Like1* in *Arabidopsis* by MicroRNA-guided mRNA. *Curr. Biol.* **2003**, *13*, 784–789.

Xie, Z.; Johansen, L. K.; Gustafson, A. M.; Kasschau, K. D.; Lellis, A. D.; Zilberman, D.; Jacobsen, S. E.; Carrington, J. C. Genetic and Functional Diversification of Small RNA Pathways in Plants. *PLoS Biol.* **2004**, *2*, 642–652.

Xie, Z.; Allen, E.; Fahlgren, N.; Calamar, A.; Givan, S. A.; Carrington, J. C. Expression of *Arabidopsis* MIRNA genes. *Plant Physiol.* **2005a**, *138*, 2145–2154.

Xie, Z.; Allen, E.; Wilken, A.; Carrington, J. C. DICER-Like 4 Functions in *Trans*-acting Small Interfering RNA Biogenesis and Vegetative Phase Change in *Arabidopsis thaliana*. *Proc. Natl. Acad. Sci.* **2005b**, *102*, 12984–12989.

Yang, J.; Lu, Z.; Luo, Y.; Bi, W.; Zhu, M.; Zhang, K. Q. Involvement of the Putative G-Protein a Subunit Gene PNGPA1 in the Regulation of Growth, Sensitivity to Fungicides, and Osmotic Stress in *Phytophthora nicotianae*. *Afr. J. Microbiol. Res.* **2012**, *6*, 680–689.

Yoshikawa, M.; Peragine, A.; Park, M. Y.; Poethig, R. S. A Pathway for the Biogenesis of *Trans*-acting siRNAs in *Arabidopsis*. *Genes Dev.* **2005**, *19*, 2164–2175.

CHAPTER 13

Engineering Plastid Pathways: An Environment-Friendly Alternative for in Planta Transformation

BHAVIN S. BHATT[1] and ACHUIT K. SINGH[2,*]

[1]Shree Ramkrishna I nstitute of Computer Education and Applied Sciences, Surat, Gujarat, India

[2]Crop Improvement Division, ICAR Indian Institute of Vegetable Research, Varanasi, Uttar Pradesh, India

*Corresponding author. E-mail: achuits@gmail.com

ABSTRACT

The chloroplast, site of photosynthesis in higher plants and algae, fixes atmospheric CO_2 to sugar and thus occupies a central position as the primary source of food. Every living organisms are directly or indirectly associated for their food requirements and thus survival to these green "plants." Plastids of higher plants are semiautonomous and they represent cellular fraction of the cell, transmitted maternally during reproduction. Chloroplast genome is circular, self-replicating, multicopy number, highly polyploid, and has its own transcription–translation machinery. Chloroplast genome, plastome, represents 10–20% of total cellular genome despite of small size and codes for as many 130 genes. Due to their circular genome nature, plastome offers a homely environment to "incoming" transgene embedded in a circular vehicle, provides more chances of site-specific integration through homologous recombination. Transgene restraint due to lack of pollen transmission and maternal inheritance offers a great advantage over facile methods of plant genetic transformation through nuclear genome. Furthermore, high level of transgene expression and lack of gene silencing are an added advantage of plastid transformation. The study of chloroplast genome transformation leads to understanding of biochemistry and physiology of plastid metabolism. This article summarizes plastome organization and regeneration, the transformation process, and highlights selected applications of transplastomic technologies in basic and applied research.

13.1 INTRODUCTION

The current global population of the world is 6.4 billion and is experiencing steady increase by each year and expected to reach 10 billion by the year 2050.[1] On the other hand, the rate of agricultural yield remains stagnant due to either crop loss by various stresses or poor quality yield.[2] Such agricultural food product digression results in widening the existing gap between demand and supply and ultimately leads to malnutrition or starvation, mainly to underdevelop or developing country people.[3] Traditional methods of crop improvement programs often includes selective breeding for desired traits, which is now no more applicable for increasing productivity or quality, where such colossal gap exists between demand and supply. Furthermore, urbanization and marginal or salty acreage are stretching the gap to an extent to be filled by such traditional options. The only hope to this situation seems to be "Green Revolution." As our understanding about system biology and genetic traits contributing to quality and yield has been increased to several fold, it can be combined with traditional breeding programs to produce good amount of "grain" crops. The term Green Revolution has been used by the popular press to describe the spectacular increase in cereal–grain production during the past several years.

Father of green revolution and 1970 Nobel Peace Prize winner, Norman Borlaug also sees hopes in biotechnology to ameliorate environmental concerns, while meeting the rising demand for agricultural production.[4] The main concerns lies in plant transgenic is to combat with various stresses that plant experience in the natural environments. The stresses that plant exhibit can be broadly classified into biotic (fungal, bacterial, viral, etc., infection) and abiotic stress (drought, salinity, etc.). Also, transgenic variety for better or improved yield through the incorporation of novel DNA into the genome possess stress on performance of plant.[5,6] Such transgenics, although often offers better products, are severely criticized and compounded by negative public sentiment through the fear of transgene escape via pollen or seeds. Without the universal public acceptance and clearance from regulatory agencies, such recent advancements are of limited use, up to the bench only. The bench-proven technologies can only be translated to the field once it assures concomitant threat to the environment.

Since much threat to sow transgenic crops centered on its spread and disturbance of native habitat and diversity, solving this problem through restricted or no spreading of such crop or crop parts allows use of transgenic by wider range of people. Crop plants possess two genomes in addition to that of the nucleus, the organelle genomes of mitochondria and chloroplasts. These cytoplasmic organelles are transferred to progeny, upon division, by mother cells. Such maternal inheritance checks any spread of novel genes in the native environment. Chloroplasts genetic engineering of higher plants may offer the potential to mitigate certain limitations of agricultural productivity. Technological advances, most notably the invention of the particle accelerator[7] and the ability to express foreign genes in plastids,[8,9] have provided the opportunity to explore the chloroplast genome as a new platform to address current and future demands for improved food production. Such novel idea has already been explored for number of traits conferring to plants like insect resistance, herbicide resistance, salt tolerance, drought tolerance, phytoremediation, etc. Results of such experiments are very promising and will cast plastid transformation as most widely accepted technology in future for in planta transformation/s.

Green plants are autotrophic in nature and hence they directly or indirectly are important for all forms of present day living organisms for their food requirement. The central mechanisms, exclusively found in green plants is photosynthesis and is maintained in chloroplasts which are present in all plants, with the exception of the few parasitic plants which have lost autotrophy. To maintain a high photosynthetic capacity, the number of chloroplasts per cell has tremendously increased during evolution.

Endosymbiotic theory suggests that chloroplasts have originated from cyanobacteria. Mereschkowsky in 1905 first suggested this hypothesis after an observation by Schimper in 1883.[10] Such first-generation endosymbionts gave birth to second- and third-generation endosymbionts by engulfing the previous organism. Such series of endosymbionts shows degree of changes in structure and appearance of chloroplast. For example, Protists, Euglenozoa, etc. are second-generation endosymbionts containing chloroplasts, forming chloroplasts with three or four membrane layers. In the alga *Chlorella*, there is only one chloroplast, which is bell shaped.

In higher plants, chloroplasts are derived from proplastids, from pre-existing chloroplasts or from other forms of plastids. Many a times, chloroplasts are referred as different names depending upon their tissue existence. There are several types of plastid including: (1) chlorophyll containing chloroplasts; (2) yellow, orange, or red carotenoid-containing chromoplasts; (3) starch-storing amyloplasts; (4) oil-containing elaio-plasts; (5) proplastids (plastid precursors found in most plant cells); and (6) etioplasts (partially developed chloroplasts that form in dark-grown seedlings). The conversion of photosynthetic chloroplasts into yellow carotenoid-rich chromoplasts is seen in the ripening of bananas; the conversion of chloroplasts to lycopene-containing red chromoplasts is seen in the ripening of tomatoes. Each compartment of the eukaryotic cell is unique. A particular biochemistry can be favored in one compartment (e.g., chloroplasts or chromoplasts), while the environment in another compartment (e.g., the cytoplasm) is unfavorable.

Protoplastids are originated from meristematic cells and developed according to tissue type to which they are located. Seed germination in soil is light independent process. Etioplasts, present in cotyledon cells, are marked with pseudocrystalline structure, the prolamellar body, and many ribosomes to support active "compartmental" translation and setting up the photosynthetic apparatus.[11] In the presence of light, thylakoids are formed, emerging from the crystalline body. The determination of the components of the chloroplast genetic system and the analysis of their regulation is essential in at least two ways: (1) Determining pathways for early chloroplast division and differentiation and (2) to understand spatial and temporal expression of chloroplast genome.

The chloroplast is surrounded by a double-layered composite membrane, which is analogous to the outer and inner membranes of the ancestral cyanobacteria, with an intermembrane space; further, it has reticulations, or many infoldings, filling the inner spaces. The chloroplast has its own DNA, which codes for redox proteins involved in electron transport in photosynthesis; this is termed as the plastome. Plastome is a kind of ances-tral genome, mostly similar in organization and structure to prokaryotic genome. Presence of such exclusive genome, chloroplasts are emerging as new sight for novel gene transformation.

13.2 CHLOROPLAST GENOME ORGANIZATION

The chloroplast genomes of vascular plants and most algae are quite similar. The structure and organization of chloroplast genomes deduce greatest diversity and differences with respect to nuclear genome counterparts. With one possible exception of Acetabularia, all known chloroplast genomes are circular DNA molecules. Size variation is greatest among green algae in which most chloroplast genomes range between about 85 and 300 kb, while that of angiosperm is 120–180 kb in range, with majority is 135–160 kb.

Pioneering works of Kowallik and Herrmann (1970) identified a series of discrete areas spread throughout the plastid, which supports the hypothesis of nucleoid organization of plastid genome.[12] These nucleoids are readily observed in chloroplasts stained by DAPI (4', 6- diamidino-2-phenylindole) using fluorescent microscopy.[13] The plastid DNA, also named plastid chromosome or plastome, is a circular double-stranded, negatively supercoiled molecule with 85% single copy sequences and multicopy number per plastid, organized into several nucleoids. Nucleoids appear interconnected in young and mature chloroplasts. A small number of nucleoids are present in proplastids but this number is readily increased as plastid matures. During active plastid development and division, nucleoids are attached with inner membrane of plastid through protein named PEND (plastid envelope DNA binding) and might be involved in DNA replication. Such PEND association is diminished upon maturation and sets nucleotides free to attach with formed thylekoid membrane. The plastid chromosome exists as a negatively supercoiled molecule.[14] The analysis of DNA conformation by pulse-field electrophoresis showed that molecules are present as monomers, dimers, trimers, and tetramers in a relative amount of 1, 1/3, 1/9, and 1/27, respectively.

A number of genes have been located on the circle and one of the important features is the presence of two copies of the ribosomal DNA sequences. These sequences are often but not always present on a large inverted repeat. Other genes mapped include those for the large subunit of RuBP-Case (ribulose-1, 5-bisphosphate oxygenase/carboxylase), tRNAs, subunits of ATP synthase, and cytochrome oxidase.[15] Plastid genome organization and structural features are conserved during the path of evolution. The circular molecule can be divided into three distinct domains: large single copy (LSC), small single copy (SSC), and the inverted repeat (IR) which is present in exact duplicate separated by the two single copy regions. Restriction fragment length polymorphism (RFLP) analysis indicates that the molecule exists in two orientations present in equimolar proportions within a single plant.[16] The circular molecule undergoes interconversion to a dumbbell-shaped conformation that is believed to be facilitated by the presence of the IR. Concerted evolution within the IR suggests intramolecular recombination between the repeats is a possible mechanism. The plastid RecA homolog is thought to be responsible for the site-specific integration of foreign DNA sequences in the plastid genome by homologous recombination. The A (adenine)–T (thymine) content is not evenly distributed in the plastome. It is higher in noncoding regions and is lower in regions coding for tRNA and for the rRNAs. The plastome of higher plants contains four ribosomal RNA genes, 30 tRNA genes, more than 72 genes encoding polypeptides, and several conserved reading frames (ycf) coding for proteins of yet unknown function (Fig. 13.1).[17,18] Transcriptome of chloroplast represents total pool of RNA required for polypeptide synthesis required for chloroplast functioning, and no RNA is transported from outside chloroplast. The plastid genes coding for polypeptides can be classified into several categories: genes coding for prokaryotic RNA polymerase core enzyme; genes coding for proteins of the translational apparatus; for the photosynthetic apparatus and genes coding for subunits of the NADH-dehydrogenase (NDH).

13.3   CHLOROPLAST TRANSFORMATION

13.3.1   HISTORY AND DEVELOPMENT

In plant cell, there are three gene factories, which are working independently and are available for in planta transformations are nucleus, chloroplast, and mitochondria. For very obvious reason, expression of foreign genes by nuclear transformation is most adapted route since long. When concept was emerged, chloroplast genetic engineering is limited to transform foreign gene/s to propoplasts followed by in vitro propagation. Such procedure is very tedious and requires up to date and accurate methodology for trans-formation and plant regeneration.[9,19] Later, invention of newer advanced methods for transformation have eroded out such protracted procedures and allowed direct introduction of foreign genes into chloroplast embedded in plant cell. Furthermore, availability of newer marker genes for selection of true transformants has made plastid transformation as one of the good alternative for sustainable agriculture. Boynton et al. (1988) first reported stable transgene integration in intact chloroplast in *Chlamydomonas.*[7]

After the first in situ chloroplast transformation in *Chlamydomonas reinhardtii,* notion had been extended to higher plant, *Nicotiana tabacum.*[9,20] Up to date, plastid transformation has extended to many other higher plants, such as *Arabidopsis,*[21] grape,[22] potato,[23] lettuce,[24] soybean,[25] cotton,[26] carrot,[27] and tomato.[28] However, plastid transformation is routine only in tobacco and with no known reason, the efficiency of transformation is much higher in tobacco than any other plants.[29] With technical developments of transformation system, plastid gene expression and plastome gene availability, "transplastomic" studies is widely getting attention of plant biologists.

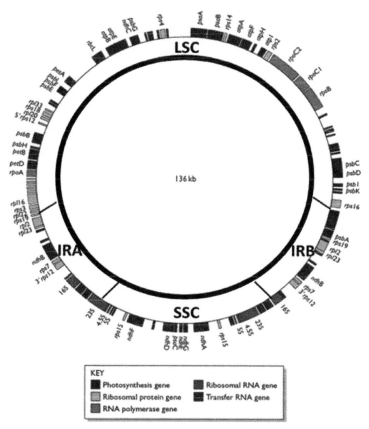

FIGURE 13.1 General structure of chloroplast genome. IR, inverted repeat region; LSC, large single copy subunit; SSC, small single copy subunit.

13.3.2   *TRANSFORMATION METHOD DEVELOPMENT*

There are apparent differences between transgenomic and transplastomic procedures (Table 13.1). Unlike nuclear transformation, genetic transformation in plastids follows integration of transgene through homologous recombination. Transgene should be flanked by plastid DNA sequences for site-specific insertion (Fig. 13.2).[30] Foreign gene transplantation to plastid, in general, involves four distinct steps.

1.   Construction and delivery of foreign gene cassette into chloroplast.
2.   Integration of foreign gene with chloroplast geneome through homologus recombination.
3.   Selection of transformants.
4.   Regeneration of true and stable transformants expressing desired trait/s of foreign gene/s transplanted.

TABLE 13.1   Apparent Differences of Nuclear and Chloroplast Genome of Angiosperms.

	Nuclear genome	Chloroplast genome
Chromosomes	Present in duplicate copies, one of which serves as a dominant allele of gene while other is recessive. Linear form	Multiple copies. ~60–100 copies per plastid and ~50–60 plastids per cell. Circular form
Genetic organization	Genes are varied in a range of thousand per chromosome	~120–150 genes per plastome
Gene arrangements	Monocistronic. One gene is under the control of single promoter	Polycistronic. In an operon cluster, as in prokaryotic genome. Many genes are under the control of single promoter c and be transcribed together

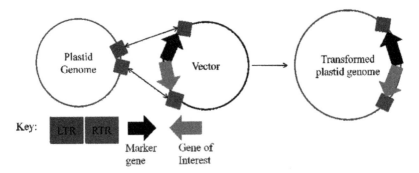

Key: LTR   RTR   █▶ Marker gene   ◀█ Gene of Interest

FIGURE 13.2   (See color insert.) Foreign gene integration in chloroplast genome through homologous recombination.

Biolistic or particle bombardment method is generally employed for chloroplast transformation, in which the plasmid containing a marker gene and the gene of interest were introduced into chloroplasts or plastids. The foreign genes were inserted into plasmid DNA by homologous recombination via the flanking sequences at the insertion site. Also, *Agrobacterium*[31] and polyethylene glycol (PEG)[32]-mediated transformation procedures were adopted in earlier days; they were quickly diminished due to lower transformation/regeneration capacity. The first successful chloroplast transformation in *Chlamydomonas reinhardtii* was employed by particle bombardment. The availability of standardized protocols for efficient transformation and regeneration, make biolistic is a choice of procedure (Table 13.2).

TABLE 13.2 Chloroplast Transformation Methods and Expressed Foreign Gene for Selected Plant Species.

Species	Methods	Selection of transformants	Expressed genes	Reference
Chlamydomonas reinhardtii	Particle bombardment	Photosynthetic proficiency	*atp*B (*ATP Synthase β Subunit*)	[7]
Nicotiana tabaccum	PEG	Spectinomycin	*Rrn*16 (16S rRNA)	[20]
Nicotiana tabaccum	Particle bombardment	Kanamycin	*Npt II* (Neomycin Phosphotransferase)	[33]
Arabidopsis thaliana	Particle bombardment	Spectinomycin	*aadA* (Aminoglycoside adenyl transferase)	[21]
Daucus carota (carrot)	Particle bombardment	Spectinomycin	*BADH* (betaine aldehyde dehydrogenase)	[27]
Rice	Particle bombardment	Spectinomycin	*aadA* and GFP	[34]

13.3.3 SELECTABLE/SCREENABLE MARKER GENE/S

Plastid DNA is present in multiple copies, hence choice of selectable marker genes are critically important to achieve uniform transformation of all genome copies during an enrichment process that involves gradual sorting out of nontransformed plastids on a selective medium.[29] The first selection marker gene used in chloroplast transformation was plastid 16S rRNA (*rrn16*) gene.[20] Transgenic lines were selected by spectinomycin resistance and the efficiency was low. The alternate is *aadA* gene encoding aminoglycoside 3'-adenylyltransferase was used as a selection marker gene[35,36] which increases recovery of plastid transformants. Kanamycin-resistant gene *npt II* was used as a selectable marker for plastid transformation in tobacco, but the transformation efficiency was low. A dramatic improvement in plastid transformation efficiency was obtained by a highly expressed *neo* gene, confers resistance to kanamycin. The bacterial *bar* gene, encoding phosphinothricin acetyltransferase, has also been tested as a marker gene, but it was not good enough.[37] Another marker gene is the betaine aldehyde dehydrogenase (BADH) gene which confers resistance to betaine aldehyde. Chloroplast transformation efficiency was 25-fold higher with betaine aldehyde (BA) selection than with spectinomycin in tobacco.[19] Transgenic carrot plants expressing BADH could be grown in the presence of high concentrations of NaCl (up to 400 mM).[27] But there is no additional report about the use of BA selection.

13.3.4   INSERTION SITES

As discussed earlier, plastid transformation is always accomplished by homologous recombination and hence insertion cassette should possess left and right flanking sequences each with 1–2 kb in size from the host plastid genome.[38] The site of insertion in the plastid genome is determined by the choice of ptDNA (plastid DNA) segment flanking the marker gene and the gene of interest. Insertion of foreign DNA in intergenic regions of the plastid genome had been accomplished at 16 sites, of which three are most commonly used. Two of three insertion sites are located in IR, insertion in which results into rapid doubling of transgene, while the rest one is in the LSC region of the ptDNA, and the gene inserted should have only one copy per ptDNA.[39]

13.3.5   REGULATORY SEQUENCES

The gene expression level in plastids is predominately determined by promoter sequence and 5'-untranslated region (UTR) elements.[40] Therefore, plastid expression vectors should have apposite 5'-UTRd including a ribosomal-binding site. The foremost aim of plastid transformation is expression of transgene at a higher level for protein production. Such high level protein production and accumulation from expression of the transgene is achieved by strong promoter upstream of transgene, which supports multicopy transcription of gene. Protein accumulation from the transgene depends on the 5'-UTR inserted upstream of the open reading frame encoding the genes of interest. Plastid rRNA operon (*rrn*) promoter (*Prrn*) promoter provides such higher level of inducible gene expression and may provide amplification up to 10,000-fold. Stability of the transgenic mRNA is ensured by the 5'-UTR and 3'-UTR sequences flanking the transgenes.

13.3.6 CONTROLLED EXPRESSION OF PLASTID TRANSGENE IN PLANTS

Although plastid gene expression provides numbers of advantages over nuclear transformation, it mainly lacks tightly controllable systems for transgene expression and tissue-specific developmentally regulated control mechanisms. Deleterious phenotypic effect and significant metabolic burden due to higher level of transgene expression calls for tissue or organ or stage definite transgene expression. Deleterious effects are often results of constitutive transgene expression which could be accomplished by making transgene expression dependent on an inducer. Such inducible systems can be constructed by expression of marker gene under the pressure of external stimuli. Examples are β-glucuronidase (GUS) reporter gene under the control of phage T7 promoter was introduced into the plastid genome of plants. GUS expression was dependent on nuclear-encoded plastid targeted T7 RNA polymerase (T7 RNAP) activity.[41] More recently, a Lac repressor-based IPTG-inducible expression system for plastids has been reported for external control of plastid gene expression which is based entirely on plastid components and can therefore be established in a single transformation step.[42]

13.3.7 CHLOROPLAST TRANSFORMATION: ENVIRONMENT FRIENDLY AND ADVANTAGEOUS IN PLANTA TRANSFORMATION ALTERNATIVE

Chloroplasts are main source of photosynthesis in plants and green algae. They are the sites which fix atmospheric CO_2 to organic carbon and thus harboring autotrophy to plants and are the primary source of the world's food productivity and they sustain life on this planet. Apart from working as "food factories," plastids are also nature's cleaning agents which are actively involved in evolution of oxygen, sequestration of carbon, production of starch, synthesis of amino acids, fatty acids, and pigments, and key aspects of sulfur and nitrogen metabolism. The main advantage with plastid transformation is lower environmental risks through biological containment.[19,43] In most angiosperm plant species, plastid genes are maternally inherited[44,45] and therefore transgenes in these plastids are not disseminated by pollen, thus abolishing any chances of transgenes spread through breeding. This makes plastid transformation a valuable tool for the creation and cultivation of genetically modified plants without unnecessary escape of transgene to nearby habitats and allows coexistence of conventional and genetically modified crop.[19,46] Cytoplasmic male sterility presents a further genetic engineering approach for transgene containment.[47] Stable integration of transgene through site-specific homologous recombination results into accumulation of large amounts of foreign protein (up to 46% of total leaf protein) due to the polyploidy of the plastid genetic system with up to 10,000 copies of the chloroplast genome in each plant cell. Such site-specific integration into the chloroplast genome by homologous recombination of flanking chloroplast DNA sequences present in the chloroplast vector eliminates the concerns of position effect which is frequently observed in nuclear transformations.[19] Other advantages seen in chloroplast transgenic plants include the lack of transgene silencing and transgene stacking, that is, simultaneous expression of multiple transgenes, creating an opportunity to produce multivalent vaccines in a single transformation step. Moreover, foreign proteins synthesized in chloroplasts are properly folded with appropriate posttranscriptional modifications, including disulfide bonds[48,49] and lipid modifications[50] Furthermore, plant-derived therapeutic proteins are free of human pathogens and mammalian viral vectors. Therefore, plastids provide a viable alternative to conventional production systems such as microbial fermentation or mammalian cell culture. Comparison between transgenomic and transplastomic technologies is summarized in Table 13.3.

13.4   APPLICATIONS OF PLASTID TRANSFORMATION

13.4.1   ENGINEERING THE CHLOROPLAST GENOME FOR HERBICIDE RESISTANCE

A herbicide, commonly known as a weed killer, is a type of pesticide used to kill unwanted plants. Glyphosate is a potent, broad-spectrum herbicide that is highly effective against grasses and broad-leaf weeds. Glyphosate works by competitive inhibition of an enzyme in the aromatic amino acid biosynthetic pathway, 5-enol-pyruvyl shikimate-3-phosphate synthase (EPSPS).[51] Unfortunately, like most commonly used herbicides, glyphosate does not distinguish crops from weeds, thereby restricting its use. The apparent solution to this problem is engineering of desired crop for herbicide resistance. However, this approach raises the concern that if the engineered resistance gene escapes via pollen dispersal, it might result in resistant weeds or might cause genetic pollution among other crops.[52] Since chloroplasts are maternally inherited, they offer a solution to this problem. The chloroplast of pollen is metabolically active but the plastid DNA is lost during pollen maturation and hence is not transmitted to the next generation.[53,54] In addition, the target proteins for many herbicides are compartmentalized within the chloroplast. Petunia EPSPS nuclear gene is expressed in chloroplast and resultant transgenic plants are resistant to 10-fold higher levels of glyphosate than the lethal dosage, and the transgene is maternally inherited. Recently, the *Agrobacterium* EPSPS gene (*C4*) was expressed in tobacco plastids and resulted in 250-fold higher levels of the glyphosate-resistant C4 protein than were achieved via nuclear transformation. Even though C4 expression in plastids was enhanced more than nuclear expression levels, field tolerance to glyphosate remained the same, showing that higher levels of expression do not always proportionately increase herbicide tolerance.[55] Similarly, expression of *bar* gene in the plastid genome provides herbicide resistance in an environmentally proscribed manner.[37]

TABLE 13.3 Comparison Between Transgenomic and Transplastomic Technologies.

Property	Transgenomic technologies	Transplastomic technologies
Biological containment	Nuclear genes are inherited equally from both parents. There will be high risk of transgene transfer into nearby nontransgenic plants.	Chloroplasts are maternally inherited in most of angiosperms. There is no risk of transgene transmission through pollen to undesired non transgenic plants.
Level of expression	Level of gene expression is limited as there is only one gene copy number per cell.	Level of gene expression is high as number of chloroplasts per cell is high. It has been estimated that chloroplast transformation will result in 40% of rise in transgenic protein production.
Gene silencing	There will be chance of no expression of transgene through posttranscriptional gene silencing or through RNA interference.	The transgene expression in chloroplast is more stable. Transgene expression is independent of effect of nuclear gene silencing.

TABLE 13.3   *(Continued)*

Property	Transgenomic technologies	Transplastomic technologies
Elimination of toxic effects of transgenes	Nuclear genes are universally expressed in all parts of the plant.	Chloroplast gene is downregulated in non-green parts of the plants (*viz.* flowers, fruits etc.). The effect of transgene products in these parts is minimum.
Multiple gene transfer	Expression of polycistronic mRNA, under single promoter, is not successful in nuclear transformation.	Chloroplast can able to express polycistronic mRNA, under single promoter. Thus, a bacterial operon or complete biosynthetic pathway genes can be transformed to chloroplast genome.
Protein synthesis	Nuclear genes are translated through eukaryotic 80S ribosomes, which is not suitable for microbial gene translation.	Chloroplast possess 70S ribosome, which is more suitable for expression of bacterial genes.
Frequency	Frequency of nuclear transformation is higher.	Frequency of chloroplast transformation is lower.
Section procedures	Mendelian inheritance of transgenes. Selection can be done in F_1 hybrids.	Maternal inheritance of transgenes. To obtain pure line of hybrids, it require 2-4 generations and selection pressure.
Method of transformation	Many biological (*Agrobacterium* mediated) and non-biological (Biolistic) methods are standardized and available in literature	The transgene transfer into chloroplast is very peculiar process and require stringent regeneration protocol from chloroplast.
Site of accumulation	Transgenomic transgenes will show their expression uniformly in any parts of the plants.	Transplastomic transgenes show product accumulation in green parts of the plants only.

13.4.2   *ENGINEERING BACTERIAL OPERONS VIA CHLOROPLAST GENOMES IN POLYCISTRONIC MANNER*

Typical eukaryotic and hence plant nuclear mRNAs are monocistronic. This poses a serious drawback when engineering multiple genes or metabolic engineering, end product of either of which is resultant of cascade of pathway.[56,57] For such nuclear transformations, single genes were first introduced into individual plants, which were then back-crossed to reconstitute the entire pathway or the complete protein.[58] The striking example is "Golden" rice expressing a biosynthetic pathway for β-carotene expression.[59] By contrast, most chloroplast genes are arranged in an operon structure under the action of single promoter and cotranscribed as polycistronic RNAs, which are subsequently processed to form translatable transcripts. Therefore, introduction of multiple chloroplast transgenes arranged in an operon should allow expression of entire pathways in a single transformation event. Recently, the *Bt cry*2Aa2 three gene operon was used as a model system to test the feasibility of multigene operon expression in engineered chloroplasts.[60] Operon-derived Cry2Aa2 protein accumulates in transgenic chloroplasts as cuboidal crystals, to a level of 45.3% of the total soluble protein (tsp) and remains stable even in senescing leaves (46.1%). This is the highest level of foreign gene expression ever reported in transgenic plants, killing insects that are exceedingly difficult to control.[61] Importantly, pollens are free of such insecticidal proteins, thus eliminating potential harm to nontarget insects. This first demonstration of bacterial operon expression in transgenic plants opens the door to engineer novel pathways in a single transformation event.

13.4.3 ENGINEERING THE CHLOROPLAST GENOME FOR PATHOGEN RESISTANCE

Plant pathogens possess serious threat to crop yields and quality, sometimes up to 100% depending upon infection severity. Hence, continuous efforts are made to engineer plants that are resistant to pathogenic bacteria and fungi. Amphipathic peptides are possible hope for fighting a battle against such pathogens. Such amphipathic peptides like MSI- 99 forms α-helical molecule with affinity for the negatively charged phospholipids found in the outer membrane of bacteria and fungi. Upon contact with these membranes, aggregation of individual peptides forms barrel-like structure embedding the plasma membrane of bacteria or fungi, resulting in lysis. Because of the concentration-dependent action of antimicrobial peptides, MSI- 99 was expressed via the chloroplast genome to accomplish high-dose release at the point of infection. In vitro and in planta assays confirmed that the peptide was expressed at high levels (up to 21.5% tsp) and retained biological activity against *Pseudomonas syringae*, a major plant pathogen.[61] Importantly, growth and development of the transgenic plants were unaffected by hyper expression of MSI-99 within chloroplasts. Because the outer membrane is an essential and highly conserved part of all microbial cells, microorganisms are unlikely to develop resistance against these peptides. Therefore, these results give a new option in the combat against phytopathogens.

13.4.4   ENGINEERING THE CHLOROPLAST GENOME FOR DROUGHT TOLERANCE

Apart from phytopathogens, abiotic stress like water stress caused by drought, salinity, or freezing also possesses great risk to the plant growth and development.[62] Trehalose is a non-reducing disaccharide of glucose which is synthesized by the trehalose-6-phosphate (T6P) synthase and trehalose-6-phosphate phosphatase complex in *Saccharomyces cerevisiae*. Trehalose protects against damage imposed by these stresses.[63,64] Therefore, engineering high levels of trehalose in plants might confer drought tolerance.[65] Again, gene containment in transgenic plants is a serious concern when plants are genetically engineered for drought tolerance because of the possibility of creating drought-tolerant weeds and passing on undesired pleiotropic traits to related crops. On the other hand, it is always desirable to have high level of expression of transgene in plant. Both of these two opposite consequences can be solved via the chloroplast genome instead of the nuclear genome. Recently, the yeast *trehalose phosphate synthase* (*TPS1*) gene was introduced into the tobacco chloroplast and nuclear genomes to study the resultant phenotypes and chloroplast transgenic plants showed up to 25-fold higher accumulation of trehalose than nuclear transgenic plants. Also, nuclear transgenic plants with significant amounts of trehalose accumulation exhibited a stunted phenotype, sterility, and other pleiotropic effects, whereas chloroplast transgenic plants grew normally and had no visible pleiotropic effects. Investigations have confirmed that trehalose functions by protecting the integrity of biological membranes rather than regulating water potential.[66] Therefore, this study shows that compartmentalization of trehalose within chloroplasts confers drought tolerance without undesirable phenotypes.

13.4.5   ENGINEERING THE CHLOROPLAST GENOME TO OBTAIN TRANSGENIC PLANT LACKING ANTIBIOTIC RESISTANCE GENE

Every genetic transformation procedures require stringent selection procedures for true transformation. Traditional vectors include one or more antibiotic resistance gene/s, which confer resistance to respective antibiotics.[67] Such antibiotic resistance passes to human or other microflora by horizontal gene transfer, when consumed. There is concern that their overuse might lead to the development of resistant bacteria.[68] Therefore, several studies have explored strategies for engineering chloroplasts that are free of antibiotic-resistance markers. One strategy includes use of native chloroplast marker system from distinct species for selection. Such strategy, in addition to gene containment, should ease public concerns over genetically modified crops. The spinach BADH gene has been developed as a plant-derived selectable marker to transform chloroplast genomes.[69] The selection process involves conversion of toxic betaine aldehyde (BA) by the chloroplast-localized BADH enzyme to nontoxic glycine betaine, which also serves as an osmo-protectant.[70,71] Because the BADH enzyme is present only in chloroplasts of a few plant species adapted to dry and saline environments, it is suitable as a selectable marker in many crop plants. The transformation study showed that BA selection was 25-fold more efficient than spectinomycin, exhibiting rapid regeneration of transgenic shoots within 2 weeks. Another approach to develop marker-free transgenic plants is to eliminate the antibiotic resistance gene after transformation using endogenous chloroplast recombinases that delete the marker genes via engineered direct repeats. Recently, another strategy to eliminate selectable marker genes has been developed, using the P1 bacteriophage CRE-lox site-specific recombination system. Altogether, these reports show that efficient removal of selectable marker(s) from chloroplast genomes is feasible.[72,73]

13.4.6 RESEARCH ON RNA EDITING

Plastid transformation played an important role in understanding the RNA editing process by mainly three approaches, namely, minigenes, translational fusion with a reporter gene, and incorporation of an editing segment in the 3' UTR.[74] The most complete information is available for the *psbL* editing site. *psbL* is a plastid photosynthetic gene, in which the translation initiation codon is created by conversion of an ACG codon to an AUG codon at the mRNA level.

13.4.7 "PHARMING" THROUGH "FARMING": CHLOROPLAST AS A BIOREACTOR FOR PRODUCTION OF PHARMACEUTICAL PRODUCTS

Expression and production of human proteins/therapeutics in chloroplast has an added advantage over the usage of prokaryotic cellular machineries that they stably express "eukaryotic" form of protein. Protein function, especially of enzymes and hormones', is largely dependent upon their three-dimensional spatial structure, which are the results of cascade of several posttranslational mechanisms. Such mechanism is either absent or in trivial stage to correctly fold human originated proteins. Hence, the use of microbial cells as bioreactors limits at this stage. Being a part of eukaryotic cell, chloroplast offers such correct folding to be used directly as therapeutics over other added advantages.[75,76] Stable expression of a pharmaceutical protein in chloroplasts was first reported for GVGVP, a protein-based polymer with medical uses such as wound coverings, artificial pericardia, and programmed drug delivery.[77] Human ST (hST) is a multimeric soluble protein which was expressed inside chloroplasts in a soluble, biologically active and disulfide-bonded form.[78] The type I IFNs (interferons) are cytokines that are produced and evoke immune response against range of human pathogens, parasites, tumor cells, and allogeneic cells from graft. IFNa2b ranks third in world market use for a biopharmaceutical, behind only insulin and erythropoietin. IFNa2b was expressed in tobacco chloroplasts with levels of up to 20% of tsp or 3 mg/g of leaf (fresh weight) and facilitated the first field production of a plant-derived human blood protein.[79,80]

13.4.8 PLASTIDS AS VACCINE BIOREACTORS

As opposed to injected subunit vaccines, oral delivery and low-cost purification make plastid-derived subunit production quite plausible.[81] Such plastid-derived vaccines produced very hopeful results when tested in animal models. They are capable of inducing correct line of immune defense when given orally and also withstand a pathogen challenge. The only drawback of plant-derived vaccines is their bioavailability and controlled release at the site of action. However, bioencapsulation can protect the vaccine in the stomach and gradually releases the antigen in the gut. Vaccine antigens against cholera,[82] tetanus,[83] anthrax,[84] and plague[85] have been expressed in transgenic chloroplasts. Bioterrorism is an increased threat in the post 9/11 world. Anthrax is always fatal if not treated immediately. Weapon grade spores can be produced and stored for decades and can be spread by missiles, bombs, or even through the mail. Because of this, it is an ideal biological warfare agent.[86] Plastid produced anthrax vaccine is an immediate workable option in such case. Recently, malaria vaccine has been produced by engineering chloroplast of *Chlamydomonas*.[87]

13.4.9 PLASTIDS AS BIOMATERIAL BIOREACTORS

Besides vaccine antigens, biomaterial and amino acids have also been expressed in chloroplasts. Normally, p-hydroxybenzoic acid (phBA) is produced in small quantities in all plants by series of 10 consecutive reactions from pyruvate, while in *E. coli*, ubiC-encoded chorismate pyruvate lyase catalyzes the direct conversion of chorismate to pyruvate and phBA. Stable integration of the ubiC gene into the tobacco chloroplast resulted in hyperexpression of the enzyme and accumulation of this polymer up to 25% of dry weight.[88] In another study, the gene for thermostable xylanase was expressed in the chloroplasts of tobacco plants.[89] Xylanase accumulated in the cells to approximately 6% of tsp. Zymography assay demonstrated that the estimated activity was 140,755 units per kg fresh leaf tissue.

The use of chloroplast for molecular pharming suggests that chloroplast contain mechanism for correct folding and stable accumulation of foreign protein. Despite of such proceedings of chloroplast molecular biology, expression of many important sugar conjugated proteins, glycoproteins, are not expressed due to the fact that N- or O-glycosylation is required for stability and functionality of many proteins.

13.4.10 ENGINEERING THE CHLOROPLAST GENOME OF EDIBLE CROP PLANTS

Apart from leaves, plastid are also present, in other forms, to other parts of plant, mainly in rudimentary or developing form, many of which are edible. Pharming in such "edible" compartment for the production of orally delivered pharmaceuticals is very propitious approach for delivery. Chromoplast in tomato and amyloplast in potato has recently been explored for the feasibility of an approach. Western blot analysis revealed more promising result in tomato that protein expression was almost doubled in fruits than that of the leaves while in potato, results are not very promising since accumulation is much lower in microtubers. This study predicts the feasibility of expressing high-levels of foreign proteins in the plastids of edible plant organs.[28]

13.5   CONCLUSION

Chloroplast genetic engineering is an exciting technology that has the tremendous competence for gaining "Green Revolution" in a true sense. The plastid transformation offers gene amplification along with gene stacking due to their multicopy number and operon organized genome. Also, maternal inheritance restricts spread of transgene is an added advantage. Plastid genetic engineering has become a powerful tool for basic research in plastid biogenesis and function. Recent advances in plastid engineering provide an efficient platform for the production of therapeutic proteins, vaccines, and biomaterials using an environmentally friendly approach. Although concept of environmentally sustained plant transformation through chloroplast compartmentalization is older now, there are many challenges for successive use of this technique for diverse range of plant materials. A main lacunae lie is the unavailability of chloroplast genome sequences and species specific transformation vectors, to which transformation efficiency depends greatly. Plastid transformation inexorably followed by plant regeneration in vitro. Hence, accurate in vitro regeneration protocol should be available in hand for successful regeneration of transformed plants. Overall, plastid transformation and related technologies are now on the horizon. In spite of being overlooked by the transgenomic technologies earlier days, noticeable environmental constrains put forward plastid transformation again in the light. Such initiatives, surely in future will give new way of understanding molecular processes and thus provide newer route of societal upliftment through better and healthier plants and plant products.

KEYWORDS

- plastome
- chloroplast transformation
- transplastomic technologies
- transgene
- in planta

REFERENCES

1. O' Neill, B. C.; Balk, D.; Brickman, M.; Ezra, M. A Guide to Global Population Projection. *Demogr. Res.* **2011,** *4* (8), 203–288.
2. Oerke, E. -C.; Dehen, H. -W. Safeguarding Production: Losses in Major Crops and the Role of Crop Protection. *Crop Protect* **2004,** *23,* 275–285.
3. Rice, A. L.; Sacco, L.; Hyder, A.; Black, R. E. Malnutrition as an Underlying Cause of Childhood Deaths Associated with Infectious Diseases in Developing Countries. *Bull. World Health Organization* **2000,** *78* (10), 1207–1221.
4. Borlaug, N. *The Green Revolution, Peace and Humanity.* "Nobel Lecture", Nobel Peace Prize, 1970.
5. Goto, F.; Yoshihara, T.; Shigemoto, N.; Toki, S.; Takaiwa, S. Iron Fortification of Rice Seed by the Soybean Feitin Gene. *Nat. Biotechnol.* **1999,** *17,* 282.
6. Chin, H. G.; Kim, G. D.; Martin, I.; Mersha, F.; Evans, T. C. Jr.; Chen, L.; Xu, Mq.; Pradhan, S. Protein Trans Splicing in Transgenic Plant Chloroplast: Reconstruction of Herbicide Resistance from Split Genes. *Proc. Natl. Acad. Sci. USA* **2003,** *100* (8), 4510–4515.
7. Boynton, J. E., et al. *Chloroplast Transformation in Chlamydomonas with High Velocity Micropojectiles. Science* **1988,** *240* (4858), 1534–1538.
8. Daniell, H.; McFadden, B. A. Uptake and Expression of Bacterial and Cyanobacterial Genes by Isolated Cucumber Etioplasts. *Proc. Natl. Acad. Sci. USA* **1987,** *84* (18), 6349–6353.
9. Daniell, H., et al. Transient Foreign Gene Expression in Chloroplasts of Cultured Tobacco Cells After Bolistic Delivery of Chloroplast Vectors. *Proc. Natl. Acad. Sci. USA* **1990,** *87* (1), 88–92.
10. Martin, W.; Kowallik, K. V. On the Nature and Origin of Chromatophores (Plastids) in the Plant Kingdom: Annotated English Translation of Mereschkowsky's 1905 Paper. *Eur. J. Phycol.* **1999,** *34,* 287–295.
11. Mache, R.; Mache, S. L. Chloroplast Genetic System of Higher Plants: Chromosome Replication, Chloroplast Division and Elements of the Transcriptional Apparatus. *Curr. Sci.* **2001,** *80* (2), 217–224.
12. Hermann, R. G.; Kowallik, K. V. Selective Presentation of DNA–Region and Membranes in Chloroplast and Mitochondria. *J. Cell Biol.* **1970,** *45,* 198–202.
13. Kubista, M.; Aakerman, B.; Norden, B. Characterization of Interaction Between DNA and 4', 6-Diamidino-2-Phenylindole by Optical Spectroscopy. *Biochemistry* **1987,** *26* (14), 4545–4553.
14. Sato, N. et al. Molecular Characterization of the PEND Protein, a Novel BZIP Protein Present in the Envelope Membrane That Is the Site of Nucleoid Replication in Developing Plastids. *Plant Cell* **1998,** *10,* 859–872.
15. Barbrook, A. C.; Howe, C. J.; Kurniawan, D. P.; Tarr, S. J. Organization and Expression of Organelle Genome. *Phil. Trans. Royal Soc. B.* **2010,** *365,* 785–797.
16. Palmer, J. D. Chloroplast DNA Exists in Two Orientations. *Nature* **1983,** *92,* 301.
17. Sugiura, M. The Chloroplast Genome. *Plant Mol. Biol.* **1992,** *19,* 149–168.
18. Regert, B. J.; Fairfieldt, S. A.; Epler, J. L.; Barnett, W. E. Identification and Origin of Some Chloroplast Aminoacyl-tRNA Synthetases and tRNAs. *Proc. Natl. Acad. Sci. USA* **1970,** *67* (3), 1207–1213.

19. Daniell, H. Molecular Strategies for Gene Containment in Transgenic Crops. *Nat. Biotechnol.* **2002**, *20*, 581–586.

20. Svab, Z.; Hajdukiewicz, P.; Maliga, P. Stable Transformation of Plastids in Higher Plants. *Proc. Natl. Acad. Sci. USA* **1990**, *87*, 8526–8530.

21. Sikdar, S. R.; Serino, G.; Chaudhuri, S.; Maliga, P. Plastid Transformation in Arabidopsis Thaliana. *Plant Cell Reports* **1998**, *18*, 20–24.

22. Hou, B. K.; Zhou, Y. H.; Wan, L. H.; Zhang, Z. L.; Shen, G. F.; Chen, Z. H.; Hu, Z. M. Chloroplast Transformation in Oilseed Rape. *Transgenic Res.* **2003**, *12*, 111–114.

23. Sidorov, V. A.; Kasten, D.; Pang, S. Z.; Hajdukiewicz, P. T. J.; Staub, J. M.; Nehra, N. S. Stable Chloroplast Transformation in Potato: Use of Green Fluorescent Protein as a Plastid Marker. *Plant J.* **1999**, *19*, 209–216.

24. Lelivelt, C.; McCabe, M.; Newell, C.; DeSnoo, C.; Dun, K.; Birch-Machin, I.; Gray, J.; Mills, K.; Nugent, J. Stable Plastid Transformation in Lettuce (*Lactuca sativa* L.). *Plant Mol. Biol.* **2005**, *58*, 763–774.

25. Dufourmantel, N.; Pelissier, B.; Garcon, F.; Peltier, G.; Ferullo, J. M.; Tissot, G. Generation of Fertile Transplastomic Soybean. *Plant Mol. Biol.* **2004**, *55*, 479–489.

26. Kumar, S.; Dhingra, A.; Daniell, H. Stable Transformation of the Cotton Plastid Genome and Maternal Inheritance of Transgenes. *Plant Mol. Biol.* **2004**, *56*, 203–216.

27. Kumar, S.; Dhingra, A.; Daniell, H. Plastid-expressed Betaine Aldehyde Dehydrogenase Gene in Carrot Cultured Cells, Roots, and Leaves Confer Enhanced Salt Tolerance. *Plant Physiol.* **2004**, *136*, 2843–2854.

28. Ruf, S.; Hermann, M.; Berger, I. J.; Carrer, H.; Bock, R. Stable Genetic Transformation of Tomato Plastids and Expression of a Foreign Protein in Fruit. *Nat. Biotechnol.* **2001**, *19*, 870–875.

29. Maliga, P. Plastid Transformation in Higher Plants. *Annu. Rev. Plant Biol.* **2004**, *55*, 289–313.

30. Maliga, P. Engineering the Plastid Genome of the Higher Plants. *Curr. Opin. Plant Biol.* **2002**, *5*, 164–172.

31. Block, M. D.; Schell, J.; Montagu, M. V. Chloroplast Transformation by *Agrobacterium tumefaciens*. *EMBO J.* **1985**, *4*, 1367–1372.

32. Golds, T.; Maliga, P.; Koop, H. U. Stable Plastid Transformation in Peg-treated Protoplasts of *Nicotiana tabacum*. *Nat. Biotechnol.* **1993**, *11*, 95–97.

33. Carrer, H.; Hockenberry, T. N.; Svab, Z;, Maliga, P. Kanamycin Resistance as a Selectable Marker for Plastid Transformation in Tobacco. *Mol. Gen. Genet.* **1993**, *241*, 49–56.

34. Lee, S. M., et al. Plastid Transformation in the Monocotyledonous Cereal Crop, Rice (*Oryza sativa*) and Transmission of Transgenes to Their Progeny. *Mol. Cells* **2006**, *21*, 401–410.

35. Goldschmidt-Clermont, M. Transgenic Expression of Aminoglycoside Adenine Transferase in the Chloroplast: A Selectable Marker for Site-directed Transformation of Chlamydomonas. *Nucleic Acids Res.* **1991**, *19*, 4083–4089.

36. Svab, Z.; Maliga, P. High-frequency Plastid Transformation in Tobacco by Selection for a Chimeric aadA Gene. *Proc. Natl. Acad. Sci. USA* **1993**, *90*, 913–917.

37. Lutz, K. A. Expression of Bar in Plastid Genome Confers Herbicide Resistance. *Plant Physiol.* **2001**, *125*, 1585–1590.

38. Zoubenko, O. V.; Allison, L. A.; Svab, Z.; Maliga, P. Efficient Targeting of Foreign Genes into the Tobacco Plastid Genome. *Nucleic Acids Res.* **1994**, *22*, 3819–3824.

39. Staub, J. M.; Maliga, P. Long Regions of Homologous DNA Are Incorporated into the Tobacco Plastid Genome by Transformation. *Plant Cell* **1992**, *4*, 39–45.

40. Deng, X. –W.; Gruissem, W. Control of Plastid Gene Expression During Development: The Limited Role of Transcriptional Regulation. *Cell* **1987**, *49* (3), 379–387.

41. McBride, K. E.; Schaaf, D. J.; Daley, M.; Stalker, D. M. Controlled Expression of Plastid Transgenes in Plants Based on a Nuclear DNA-encoded and Plastid-targeted T7 RNA Polymerase. *Proc. Natl. Acad. Sci. USA* **1994**, *91*, 7301–7305.

42. Muhlbauer, S. K.; Koop, H. U. External Control of Transgene Expression in Tobacco Plastids Using the Bacterial Lac Repressor. *Plant J.* **2005**, *43*, 941–946.

43. Khan, M. S.; Khalid, A. M.; Malik, K. Intein-mediated Protein Trans-splicing and Transgene Containment in Plastids. *Trends Biotechnol.* **2005**, *23* (5), 217–221.

44. Reboud, X.; Zeyl, C. Organelle Inheritance in Plants. *Heredity* **1994**, *72*, 132–140.

45. Connett, M. B. Mechanisms of Maternal Inheritance of Plastids and Mitochondria: Developmental and Ultrastructural Evidence. *Plant Mol. Biol. Reporter* **1987**, *4* (4), 193–205.

46. Daniell, H. Transgene Containment by Maternal Inheritance: Effective or Elusive? *Proc. Natl. Acad. Sci. USA* **2007**, *104*, 6879–6880.

47. Ruiz, O. N.; Daniell, H. Engineering Cytoplasmic Male Sterility Via the Chloroplast Genome by Expression of β-Ketothiolase. *Plant Physiol.* **2005**, *138*, 1232–1246.

48. Verma, D., Daniell, H. Chloroplast Vector Systems for Biotechnology Applications. *Plant Physiol.* **2007**, *145*, 1129–1143.

49. Ruhlman, T.; Ahangari, R.; Devine, A.; Samsam, M.; Daniell, H. Expression of Cholera Toxin B-Proinsulin Fusion Protein in Lettuce and Tobacco Chloroplasts-Oral Administration ProtectsAgainst Development of Insulitis on Non-obese Diabetic Mice. *Plant Biotechnol. J.* **2007**, *5*, 495–510.

50. Glenz, K.; Bouchon, B.; Stehle, T.; Wallich, R.; Simon, M. M.; Warzecha, H. Production of a Recombinant Bacterial Lipoprotein in Higher Plant Chloroplasts. *Nat. Biotechnol.* **2006**, *24*, 76–77.

51. Shah, D. M., et al. Engineering Herbicide Tolerance in Transgenic Plants. *Science* **1986**, *233*, 478–481.

52. Ruf, S.; Karcher, D.; Bock, R. Determining the Transgene Containment Level Provided by Chloroplast Transformation. *Proc. Natl. Acad. Sci. USA* **2007**, *104* (17), 6998–7002.

53. Scott, S. E.; Wilkinson, M. J. Low Probability of Chloroplast Movement from Oilseed Rape (*Brassica napus*) Into Wild Brassica Rapa. *Nat. Biotechnol.* **1999**, *17*, 390–392.

54. Daniell, H.; Muthukumar, B.; Lee, S. B. Marker Free Transgenic Plants: Engineering the Chloroplast Genome Without the Use of Antibiotic Selection. *Curr. Genet.* **2001**, *39*, 109–116.

55. Ye, G. -N.; Colburn, S. M.; Xu, C. W.; Hajdukiewicz, P. T. J.; Staub, J. M. Persistence of Unselected Transgenic DNA During a Plastid Transformation and Segregation Approach to Herbicide Resistance. *Plant Physiol.* **2003**, *133*, 402–410.

56. Quesada-Vargas, T.; Ruiz, O. N.; Daniell, H. Characterization of Heterologous Multigene Operons in Transgenic Chloroplasts: Transcription, Processing, and Translation. *Plant Physiol.* **2005**, *138*, 1746–1762.

57. Zhang, J. - Y.; Zhang, Y.; Song, Y. -R. Chloroplast Genetic Engineering in Higher Plants. *Acta Botanica Sinica* **2003**, *45* (5), 509–516.

58. Nawrath, C., et al. Targeting of the Polyhydroxybutyrate Biosynthetic Pathway to the Plastids of Arabidopsis Thaliana Results in High Levels of Polymer Accumulation. *Proc. Natl. Acad. Sci. USA* **1994**, *91*, 12760–12764.

59. Ye, X., et al. Engineering the Provitamin A (B-Carotene) Biosynthetic Pathway into (Carotenoid) Free Rice Endosperm. *Science* **2000**, *287*, 303–305.

60. De Cosa, B. Overexpression of yhe Bt Cry2aa2 Operon in Chloroplast Leads to Formation of Insecticidal Crystals. *Nat. Biotechnol.* **2001**, *19*, 71–74.

61. DeGray, G.; Rajasekaran, K.; Smith, F.; Sanford, J.; Daniell, H. Expression of an Antimicrobial Peptide Via the Chloroplast Genome to Control Phytopathogenic Bacteria and Fungi. *Plant Physiol.* **2001**, *127*, 852–862.

62. Grover, A. et al. Understanding Molecular Alphabets of the Plant Abiotic Stress Responses. *Curr. Sci.* **2001**, *80* (2), 206–216.

63. ByKylie, F.; Mackenzie, T. K.; Singh, K. K.; Brown, A. D. Water Stress Plating Hypersensitivity of Yeasts: Protective Role of Trehalose *in Sacharomyces cerevisiae*. *J. Gen. Microbiol.* **1998**, *134*, 1661–1666.

64. Oscar, J. M. G.; Kees, V. D. Trehalose Metabolism in Plants. *Trends Plant Sci.* **1999**, *4* (8), 315–319.

65. Pilon-Smits, et al. Trehalose Producing Transgenic Tobacco Plants Show Improved Growth Performance Under Drought Stress. *J. Plant Physiol.* **1998**, *152*, 525–532.

66. Romero, C., et al. Expression of Yeast Trehalose-6 -Phosphate Synthase Gene in Transgenic Tobacco Plants: Pleiotropic Phenotypes Include Drought Tolerance. *Planta* **1997**, *201*, 293–297.

67. Iamtham, S.; Day, A. Removal of Antibiotic Resistance Genes from Transgenic Tobacco Plastids. *Nat. Biotechnol.* **2000**, *18*, 1172–1176.

68. Chambers, P. A.; Duggan, P. S.; Heritage, J.; Forbes, J. M. The Fate of Antibiotic Resistance Marker Genes in Transgenic Plant Feed Material Fed to Chicken. *J. Antimicrob. Chemother.* **2002**, *49*, 161–164.

69. Rathinasabapathy, B.; McCue, K. F.; Gage, D. A.; Hanson, A. D. Metabolic Engineering of Glycine Betaine Synthesis: Plant Betaine Aldehyde Dehydrogenases Lacking Typical Transit Peptides are Targeted to Tobacco Chloroplast Where They Confer Aldehyde Resistance. *Planta* **1994**, *193*, 155–162.

70. Rontein, D.; Basse, G.; Hanson, A. D. Metabolic Engineering of Osmoprotectant Accumulation in Plants. *Metabolic Eng.* **2002**, *4* (1), 49–56.

71. Ashraf, M.; Foolad, M. R. Roles of Glycine Betaine and Proline in Improving Plant Abiotic Stress Resistance. *Environ. Exp. Botany* **2007**, *59* (2), 206–216.

72. Corneille, S.; Lutz, K.; Svab, Z.; Maliga, P. Efficient Elimination of Selectable Marker Genes from the Plastid Genome by the CRE-lox Site-specific Recombination System. *Plant J.* **2001**, *27* (2), 171–178.

73. Hajdukiewicz, P. T. J.; Gilbertson, L.; Staub, J. M. Multiple Pathways for Cre/ Lox-mediated Recombination in Plastids. *Plant J.* **2001**, *27* (2), 161–170.

74. Shikanai, T. RNA Editing in Plant Organelles: Machinery, Physiological Function and Evolution. *Cellular Mol. Life Sci.* **2006**, *63*, 698–708.

75. Ma, J. K.; Drake, P. M.; Christou, P. The Production of Recombinant Pharmaceutical Proteins in Plants. *Nat. Rev. Genet.* **2003**, *4*, 794–805.

76. Miao, Y.; Ding, Y.; Sun, Q. -Y.; Xu, Z. –F.; Jiang, L. Plant Bioreactors for Pharmaceuticals. *Biotechnol. Genetic Eng. Rev.* **2008**, *25*, 363–380.

77. Guda, C., et al. Stable Expression of Biodegradable Protein Based Polymer in Tobacco Chloroplasts. *Plant Cell Reports* **2000**, *19*, 257–262.

78. Staub, J. M., et al. High-yield Production of a Human Therapeutic Protein in Tobacco Chloroplasts. *Nat. Biotechnol.* **2000**, *18*, 333–338.

79. Arlen, P. A., et al. Field Production and Functional Evaluation of Chloroplast-derived Interferon-A2b. *Plant Biotechnol. J.* **2007**, *5* (4), 511–525.

80. Daniell, H. Production of Biopharmaceuticals and Vaccines in Plants via the Chloroplast Genome. *Biotechnol. J.* **2006**, *1* (10), 1071–1079.

81. Sala, F., et al. Vaccine Antigen Production in Transgenic Plants: Strategies, Gene Constructs and Perspectives. *Vaccine* **2003**, *21*, 803–808.

82. Daniell, H. Expression of Cholera Toxin B Subunit Gene and Assembly as Functional Oligomers in Transgenic Tobacco Chloroplasts. *J. Mol. Biol.* **2001**, *311*, 1001–1009.

83. Tregoning, J. S., et al. Expression of Tetanus Toxin Fragment C in Tobacco Chloroplasts. *Nucleic Acids Res.* **2003**, *31*, 1174–1179.

84. Watson, J.; Koya, V.; Leppla, S. H.; Daniell, H. Expression of *Bacillus Anthracis* Protective Antigen in Transgenic Chloroplasts of Tobacco, a Nonfood/Feed Crop. *Vaccine* **2004**, *22*, 4374–4384.

85. Daniell, H.; Chebolu, S.; Kumar, S.; Singleton, M.; Falconer, R. Chloroplast-derived Vaccine Antigens and Other Therapeutic Proteins. *Vaccine* **2005**, *23*, 1779–1783.

86. Jernigan, J. A., et al. Bioterrorism-related inhalational anthrax: The First 10 Cases Reported in the United States. *Emerging Infectious Dis.* **2001**, *7* (6), 933–944.

87. Dauvillee, D., et al. Engineering the Chloroplast Targeted Malarial Vaccine Antigens in Chlamydomonas Starch Granules. *PLoS One* **2010**, *5* (12), e15424.

88. Viitanen, P. V.; Devine, A. L.; Khan, M. S.; Deuel, D. L.; Van Dyk, D. E.; Daniell, H. Metabolic Engineering of the Chloroplast Genome Using *Escherichia Coli* Ubic Gene Reveals That Chorismate is a Readily Abundant Plant Precursor for P-Hydroxybenzoic Acid Biosynthesis. *Plant Physiol.* **2004**, *136*, 4048–4060.

89. Daniell, H.; Khan, M. S.; Allison, L. Milestones in Chloroplast Genetic Engineering: An Environmentally Friendly Era in Biotechnology. *Trends Plant Sci.* **2004**, *7* (2), 84–91.

CHAPTER 14

Impact of Meteorological Variables and Climate Change on Plant Diseases

A. K. MISRA,[1*] S. B. YADAV,[2] S. K. MISHRA,[3] and M. K. TRIPATHI[4]

[1]*Monsoon Mission Division, Indian Institute of Tropical Meteorology, Dr. Homi Bhabha Road, Pashan, Pune 411008, Maharashtra, India*

[2]*Punjab Agricultural University, Regional Research Station, Faridkot 151203, Punjab, India*

[3]*Department of Agrometeorology, Punjab Agricultural University, Regional Research Station, Faridkot 151203, Punjab, India*

[4]*Department of Natural Resource Management, College of Horticulture, Rajmata Vijayaraje Scindia Krishi Vishwa Vidyalaya, Mandsaur 458001, Madhya Pradesh, India*

*Corresponding author. E-mail: ashueinstein@gmail.com

ABSTRACT

Plant disease is an output of abnormal changes in the physiological processes resulting from biotic and abiotic factors. The individual weather elements as well as their combination play an important role in the disease occurrence and their infestations. Therefore, agrometeorological information becomes pivotal for prediction of disease outbreaks for effective and judicious use of control measures, the prediction of crop yields and of the market potential for the crop. The major meteorological factors responsible for the plant disease outbreaks are temperature (both air and soil), precipitation (rainfall and dew), moisture (relative humidity, soil moisture), solar radiation (intensity and cloudiness), wind, etc. Among these variables, temperature and moisture are considered as the most important factors since all the pathogen have an optimum temperature requirement range for their growth and disease development becomes accelerated in this range. In the similar manner, pathogen replicates with a very high rate under favorable moisture conditions, which enhances the severity of disease incidence and intensity. Apart from this, soil moisture content plays its dominant role on the severity of soilborne diseases, while solar radiation affects the epidemiology and has a profound influence on the developmental cycle of the parasite. Wind speed influences dispersal of the pathogen, disease spread, and epidemic development. Apart from all these meteorological variables, climate change has emerged as another major threat in recent times which may bring new diseases and challenges ahead. It is expected that climate change may affect plant–pathogen interactions as well as disease epidemiology, hence an effective planning and management will be required to overcome this challenge to achieve the food security for all.

14.1 INTRODUCTION

Globally, agricultural activities are highly sensitive to weather aberrations. Plant diseases are one of the most significant factors that affect the global food production and their severity varies with crops and regions. The Great Bengal Famine in India during 1943 is a classic example which was triggered by a simple fungus and resulted in the deaths of about 3 million peoples. Total estimated losses for major food and cash crops in various regions of the world have been brilliantly described by Oerke et al. (1994) and it has been found that due to disease alone, there are 16% yield losses in eight of the most important food and cash crops including rice, wheat, maize, barley, cotton, coffee, etc. Hardwick (2002) has summarized that combined damage due to pests and diseases reduces about 30% global food production, whereas Strange and Scott (2005) estimated minimum 10% food production losses only due to plant diseases. Similarly, Savary et al. (2012) summarized that direct yield losses between 20% and 40% of global agricultural productivity are caused by pathogens, animals, and weeds. Gautam et al. (2013) reported total loss of attainable yield of cotton may be as high as 82% after including losses occurred due to postharvest wastage and quality deterioration.

Plant disease is mainly caused due to the alteration in their physiological process caused by biotic or abiotic factors. Biotic factors include living organisms, for example, fungi, bacteria, viruses, nematodes, insects, and animals while the abiotic factors include weather elements (e.g., heat, cold and drought, freezing and wind injury, excessive precipitation, etc.), pH, nutrition, chemical injury, nutrient deficiency, and inappropriate cultural practices. There are three crucial factors often referred as "the plant disease triangle" which are responsible for an infectious plant disease to occur:

1. A susceptible host plant in vulnerable state.
2. Presence of the pathogen/parasite to cause the disease.
3. Environmental conditions favorable for disease development.

14.2 EFFECT OF WEATHER ELEMENTS ON PLANT DISEASES

Significant role of weather for the commencement and development of plant disease has been established by several researchers (Miller, 1953; Colhoun, 1973; Hardwick, 2002; Te Beest, et al., 2008; Das et al., 2011). Pathogens have their own environmental requirements for infection; therefore, agrometeorological information becomes crucial for protecting the crop through optimal use of available resources. If plant losses through diseases and pests can be reduced to zero and weather information can be efficiently exploited along with the introduction of high yielding varieties, the food production may be enhanced to significant level.

Different diseases occur at different seasons based on their climatic requirement. For example, most of the powdery mildew diseases are observed in late summer. Weather affects growth and development of plants in several ways. Every plant disease requires specific temperature, humidity, wind, radiation, soil quality, and nutrition for their growth. If these conditions are unfavorable for them, there may be a high probability for the plants to be affected by diseases. The pathogens also require certain set of optimum conditions that must continue for a critical period for infection to occur. When environmental conditions (both in air and soils) are favorable, disease development accelerate causing rapid colonization of host tissues and disease spread. It has been observed that specific temperature ranges along with high humidity enhances the possibility for many fungal diseases (Hardwick, 2002). For instance, high humidity, rainfall, or dew with a combination of 10–15°C ambient air temperature have been found to be suitable for yellow rust disease of wheat in Punjab (Gill et al., 2012). The progress and development of powdery mildew disease in mustard was at peak when maximum temperature ranged between 27.2°C and 28.9°C and afternoon relative humidity ranged between 27 and 42% in the Saurashtra region of Gujarat, India (Kanzaria et al., 2013). Kumar and Chakravarty (2008) developed a weather based forewarning model for predicting the white rust incidence for Brassica using hourly weather observations for temperature, relative humidity, and sunshine duration which resulted in the development of a thumb rule for forewarning of white rust.

The impact of individual weather events on plant epidemiology has been summarized below.

14.2.1   TEMPERATURE

Temperature is considered as one of the most significant weather factor that affect host, pathogen, and disease development together. The heat stress in plants reduces photosynthetic and transpiration efficiencies and adversely affects root development with negative impact on crop yields. It also influences all the three sections of disease epidemiology, namely, the incubation period, the generation time, and the infectious period. All the disease pathogen has a specific optimum temperature range for their growth and activities and under favorable conditions, disease development continues till the healthy plant tissue is alive. As the air temperature gets closer to the optimum for the host development, the likelihood of getting infestation decreases and vice-versa. For example, majority of rice varieties are vulnerable for rice blast disease if the night temperature is less than 26°C. When the temperature goes beyond this limit, the symptoms of rice blast are rarely observed. In general, extreme temperatures are harmful for the pathogens but incremental temperature changes can lead to dissimilar effects. For a vector living in such environment where the mean temperature approaches to the extremes of physiological tolerance limit for the pathogen, a minor change in temperature may also have detrimental effect for the pathogen. On the other hand, a vector that lives in low temperature environment, a minute rise in temperature may enhance the development, incubation, and replication of the pathogens. The rate of the disease cycle has a direct relationship with temperatures; hence, it increases with increase in temperature often resulting in rapid epidemics development. In a study conducted by Jhorar et al. (1997) for Indian Punjab, a linear relationship between maximum temperature and ascochyta blight disease of chickpea was obtained. Coakley et al. (1999) reported that the host plants, namely, wheat and oats became extremely susceptible to rust diseases with increase in temperature while some other forage species became resistant to fungi.

Similar to air temperature, soil temperature is another important factor that affects the physiology of a plant. Du and Tachibana (1994) observed that high root temperature may enhance root respiration in cucumber plants. When the roots temperature increases, its dry weight and pectin content along with leaf area reduces while root sugars, predominantly raffinose, increases to a significant instant. Leaf blight disease becomes severe in warm soils. However, moist soils with low temperature have been found as ideal for fungal root diseases. Reddick (1917) has concluded that changes in soil temperature have more profound effect on the host physiology as compared with air temperatures. Likewise, Fir et al. (1983) found soil temperature to be the critical limiting factor for the root rot disease which determines the timing and severity of this disease. Arora and Pareek (2013) suggested that high soil temperature coupled with low moisture content favors charcoal rot disease of Sorghum at Rajasthan in India. Pivonia et al. (2002) concluded that soil temperature has its profound effect on the incidence of melon collapse which was resulted due to *Monosporascus cannonballus*. High correlation between soil temperatures (above 20°C) and *Monosporascus cannonballus* in the first month after planting was observed. Artificial heating of plots till 35°C during the winter season boosted the disease effects up to 85%.

14.2.2 PRECIPITATION

There are various forms of precipitation but rain and dew are most significant in plant disease epidemics. The intensity of rain which is a function of the velocity, size, and number of the water droplets are of critical importance for the determination of wetness of the plant surface and pathogen dispersal in plant communities. Rain removes spores and pollen from the surface of crop by washing them or by shaking impact. Raindrops act as a transport medium for spores which carry away spores with them from one place to another, thus helping in the spread and transfer of disease to new places (Van der Wal, 1978). Moreover, raindrops also help in inoculum dispersal into areas where the pathogen is not present through rainsplash (Huber et al., 1998; Geagea et al., 2000).

Among the several attributes of rain, the time, frequency, and duration are crucial factors that determine the plant wetness as well as pathogen's dispersal by trickling and splashing of rain water. The combined effect of these factors affects the plant disease epidemic outbreaks. High rainfall has been found to have a positive correlation with chickpea blind incidence during the winter season. On the other hand, summer rains have been reported to have a negative stimulus on the disease. When the spring chickpeas in the Mediterranean region get regular rains during the harvest season, the impact of ascochyta blight of chickpea becomes severe in North-West India and Pakistan (Malhotra et al., 1996).

Schwartz et al. (2003) developed a multiple regression model to find the relationship of rainfall and temperature with bacterial leaf blight disease caused by *Xanthomonas campestris* and *Pantoea ananatis* for onion in Colorado, USA. They found that these parameters, that is, rainfall and temperature have significant influence on disease symptoms during late vegetative and early reproductive stages. Singh et al. (2010) reported that rainfall during the 3rd week of January was having favorable role in the formation and further multiplication of secondary spordia of karnal bunt disease in wheat in Karnal region of Haryana, India. A recent study conducted by Pal et al. (2017) reported that heavy rainfall was found to be conducive for initiation of the sheath blight disease of rice, while low and intermittent rainfall of 13–38 mm was found to have a favorable effect for progression of the disease.

Dew is another important form of precipitation and a vital source of moisture in certain arid regions. Dew is the moisture which condenses from the atmosphere on surfaces near the ground including plants, soils, leaves, etc. (Leopold, 1951). The dew is a major source of leaf wetness or free moisture which is a prime requirement for disease infection in several plants such as leaf blight on sweet corn (Levy and Pataky, 1992) or foliar infection of tomato (Byrne et al., 1998). The presence of dew on plant leaf surface significantly reduces the transpiration from the plant.

14.2.3 MOISTURE

Both air or soil moisture plays a pivotal role in the incidence of pest and diseases. All fungal pathogens affecting plants are strongly influenced by the moisture in different forms. In case, pathogen moisture requirements are fulfilled under favorable conditions, it replicates with the maximum possible rate which enhances the severity of disease incidence and intensity. Water movement in a crop canopy is generally associated with rain. Moisture content of air and soil at any place are highly dependent on precipitation. It is a well-known fact that precipitation enhances the moisture content of air, namely, relative humidity of any place due to evaporation. Evaporation results in the cooling and increases the absolute moisture content of the air at a small scale. Similarly, rainfall also increases the soil moisture content of a place and has a major role in soil water dynamics (Xu et al., 2012; Li et al., 2016). Furthermore, moisture is also known to be a major determining factor for growth and development of various microorganisms, for example, a fungi. Soil moisture content plays its dominant role on the severity of soil-borne diseases. Moderate temperature and moderate humidity are favorable for most of the pests and diseases.

As the individual effect of soil moisture on plant disease epidemiology for field crops are difficult to predict, limited research findings are available for this. In general, soil moisture is known to have its impact on plant water potentials to which pathogens in leaves, stems, and fruits are subjected (Van der Wal, 1978). Soil water content influences the microbial activity through its influence on under surface water movement. It is the most critical factor for the determination of *Macrophomina phaseolina* infection to the host plant (Dhingra and Sinclair, 1975; Short et al., 1980). Soil moisture is also a major contributor for root rot disease in several important crops including chickpea (Bhatti and Kraft, 1992), wheat (Gill et al., 2001), and navy bean (Tu and Tan, 2003).

Atmospheric moisture also termed as humidity (or relative humidity) is another form of free water which is a significant factor for plant disease development. Sometimes, high relative humidity in the absence of free water could be sufficient enough for spore germination. In normal scenario, humidity is an efficient indicator of wetness and dryness of the plant surface especially for leaves (Jhorar et al., 1998; Sentelhas et al., 2008). High relative humidity (80–90%) near the leaf and other plant surfaces is sufficient enough to bring infection with spores of several fungi.

Ambient air humidity affects the host without directly influencing the pathogen. Humidity requirement for the germination of various fungus spores for a number of categories such as fungi, downy mildews, powdery mildews, and rusts have been summarized by Yarwood (1978). He found that among the various fungi, *Aspergillus niger* is having the lowest humidity requirement of 76%; however, for occurrence of *Monilinia fructicola* and *Venturia inaequalis,* more than 95% humidity is favorable. High air moisture level favors bacterial infection, while low humidity prevents it. For occurrence of all three types of rust diseases for wheat in India, higher than 70% relative humidity is a necessity (Mavi, 1994). The combined effect of air temperature and relative humidity on the pathogen intensity for gray leaf spot on maize leaves has been found to be significant. Similarly, the maximum rate of spore production has been reported in the temperature range of 25–30°C when RH was nearly 100% (Paul and Munkvold, 2005).

14.2.4 RADIATION

Solar radiation affects the epidemiology of pathogens in two ways, that is, directly as well as indirectly. Direct effect of radiation includes its influence on the developmental cycle of the parasite due to diurnal and seasonal changes in radiation. Certain disease only develop when there is absence of solar exposure to the plants, for example, shade has been found to be beneficial for the coffee rust as compared with direct sunlight in standard fruit load conditions (López-Bravo et al., 2012). Indirect effect includes effect of radiation intensity at different wavelengths on the parasite and host plant (Friesland and Schroedter, 1988). It has been reported by Kirkham et al. (1974) that erratic and short reductions in sunlight intensity showed a marked lack of reproducibility in scab disease of apple plants. Recently David et al. (2016) found that the solar radiation and relative humidity were most effective predictors of ascospore release because these are most influential meteorological factors for the release of ascospores of *Fusarium graminearum.*

14.2.5 WIND

Wind also influences the crop plants in several ways including plant growth, reproduction, distribution, death, as well plant evolution (Nobel, 1981; Ennos, 1997; de Langre, 2008). However, complete understanding of plant response to wind is a complex phenomenon which has not been fully understood (Onoda and Anten, 2011). Wind effects on plants may be categorized as physiological or mechanical. The physiological impact of wind may include effects on transpiration (Dixon and Grace, 1984), photosynthesis (Sinoquet et al., 2001; Smith and Ennos, 2003), and insect communication (Cocroft and Rodríguez, 2005; de Langre, 2008). Cold wind may result in chilling injury, while hot winds may result in plant sunburn. In a longer time series, wind influences the plant development and alters their morphology (Smith and Ennos, 2003). The mechanical impacts of wind include uprooting, lodging, flower and fruit shedding, as well as soil erosion (Cleugh et al., 1998; Onoda and Anten, 2011). Wind adversely affects the dew formation and helps raindrops and dew to dry quickly. Therefore, it reduces the likelihood of disease infection in this context. However, it also helps in dispersal of several organisms including pollen, plant propagules, and disease organisms at distant places. McCartney (1994) and Bock et al. (2011) found that wind speed and rainfall are the major contributors for dispersal of spores and pollen from crops. In this context Bock et al. (2010) observed a direct linear relationship between bacterial densities with wind speed. Wind speed reduction also reduces dispersal of the pathogen, resulting in the lesser disease spread and epidemic development. Hence, wind suppression techniques, for example, windbreaks help in minimizing the numbers of bacteria dispersed in the orchards.

14.2.6 CLIMATE CHANGE

Climate change has become a reality now and the whole world is witnessing it as one of the biggest threat in near future. It is expected that there will be increase in temperature and changes in the rainfall or precipitation patterns in addition to increase in the severity and frequency of hazardous extreme weather events. During recent decades, climate changes have resulted in major impacts on natural and anthropogenic systems throughout the globe including oceans. Emission of greenhouse gases from anthropogenic sources is considered as the most significant driver of climate change. There are very high chances that till 2100, the global mean temperature may rise between 1.8°C and 4.0°C. However, this heating of earth may not be uniform across the globe and their impact will be higher on glaciers and land areas as compared with ocean.

Agriculture is considered as one of the most climate vulnerable sector and a minor alteration in climate may affect the crop production to a significant level. Climate change affects all the four pillars of food security, namely, food availability, access to food, utilization, and stability (Wheeler and von Braun, 2013). As per an estimate, average temperature increase of 2°C may shrink the world GDP by 1% and also reduce the per capita income of Africa and South Asia in the range of 4–5% (World Bank, 2010). An increase in the temperature in the range of 2.5–4.9°C may reduce the rice yields by 32–40% and wheat yields by 41–52 % which may reduce the Indian GDP by 1.8–3.4% (Kalisch et al., 2011).

14.3   IMPACT OF CLIMATE CHANGE ON PLANT DISEASES

It is expected that climate change will directly influence the occurrence of various plant diseases and their severity. Carbon dioxide (CO_2) is the main greenhouse gas responsible for climate change. However, it has a beneficial impact over several plant growth and developmental processes. In numerous studies conducted worldwide, it has been reported that with the increased concentration of the CO_2 in atmosphere, there will be significant improvement in the plant biomass and yield. Higher CO_2 concentrations enhance the plant photosynthetic activities which can favor the better productivity and improved water and nutrient cycles. Manning and Tiedemann (1995) reported that increased CO_2 concentrations is helpful for several plant diseases, for example, leaf spots, rusts, powdery mildew, and blights as it increases the size and density of the plant canopy combined with a higher microclimate relative humidity. Another important greenhouse gas, namely, ozone (O_3) helps to enhance the senescence processes and necrosis and also promote attacks on plants by necrotrophic fungi. Furthermore, higher ozone concentrations modify the structure and properties of leaf surfaces in such a way that it affects the inoculation and infection process.

Since climate change may result in higher temperature and increased carbon dioxide concentrations, which may result in to the spreading of pathogen and vector distributions to new geographical locations. It will provide newer opportunities for the pathogen to hybridize and spread of disease epidemic. Temperature is a major limiting factor for determining the period for reproduction of several pathogens. Longer seasons resulted from increased temperatures will provide additional time for the evolution of pathogens. In addition to this, the pathogen evolution may become faster due to the presence of large pathogen populations (Gautam et al.,

2013). Susceptibility of cereal crops increases due to enhancement in temperature. High night temperatures, especially during winter seasons help in the pathogens survival. It also enhances the life cycles of vectors and fungi with increased sporulation and aerial fungal infection (Yáñez-López et al., 2012). Due to climate change led global warming, disease primarily caused by fungi are expected to experience elongated periods of temperatures which is considered to be optimum for pathogen growth and reproduction activities. However, the effects of higher temperature on plants will not be uniform during all the seasons. During winter season, warming may help to relieve the plant stress, while it may cause heat stress during summers (Garrett et al., 2006).

14.4 CONCLUSION

Weather and its associated variables have a predominant role in the disease infestations. Weather elements influence several biological aspects of the host plants, namely, phenology, sugar and starch contents, root and shoot biomass, etc. Diseases severity also depends on weather changes. Therefore, it is advisable to understand the combination of weather elements that may result in to disease outbreak so that risk of plant diseases can be minimized or avoided. Climate change has become another major threat in recent times which may bring new diseases and challenges ahead. And proper planning and execution is a necessity to overcome with the challenges ahead.

KEYWORDS

- weather
- plant disease
- infection
- abiotic factors
- climate change
- environmental conditions

REFERENCES

Arora, M.; Pareek, S. Effect of Soil Moisture and Temperature on the Severity of Charcoal Rot of Sorghum Macrophomina Charcoal Rot of Sorghum. *Ind. J. Scientific Res.* **2013,** *4* (1), 155–158.

Bhatti, M. A.; Kraft, J. M. Influence of Soil Moisture on Root Rot and Wilt of Chickpea. *Plant Dis.* **1992,** *76* (12), 1259–1262.

Bock, C. H.; Graham, J. H.; Gottwald, T. R.; Cook, A. Z.; Parker, P. E. Wind Speed Effects on the Quantity of Xanthomonas Citri Subsp. Citri Dispersed Downwind ffrom Canopies of Grapefruit Trees Infected with Citrus Canker. *Plant Dis.* **2010,** *94* (6), 725–736. https://doi.org/10.1094/PDIS-94-6-0725.

Bock, C. H.; Gottwald, T. R.; Cook, A. Z.; Parker, P. E.; Graham, J. H. The Effect of Wind Speed on Dispersal of Splash-borne Xanthomonas Citri Subsp. Citri at Different Heights and Distances Downwind of Canker-infected Grapefruit Trees. *J. Plant Pathol.* **2011,** *93* (3), 667–677. https://doi.org/10.4454/jpp.v93i3.1234

Byrne, J. M.; Hausbeck, M. K.; Meloche, C.; Jarosz, A. M. Influence of Dew Period and Temperature on Foliar Infection of Greenhouse-grown Tomato By Colletotrichum Coccodes. *Plant Dis.* **1998,** *82* (6), 639–641.

Cleugh, H. A.; Miller, J. M.; Böhm, M. Direct mechanical Effects of Wind on Crops. *Agroforestry Sys.* **1998,** *41* (1), 85–112. https://doi.org/10.1023/A:1006067721039.

Coakley, S. M.; Scherm, H.; Chakraborty, S. Climate Change and Plant Disease Management. *Annu. Rev. Phytopathol.* **1999,** *37* (1), 399–426. https://doi.org/10.1146/annurev. phyto.37.1.399.

Cocroft, R. B.; Rodríguez, R. L. The Behavioral Ecology of Insect Vibrational Communication. *BioScience* **2005,** *55* (4), 323–334.

Colhoun, J. Effects of Environmental Factors on Plant Disease. *Annu. Rev. Phytopathol.* **1973,** *11* (1), 343–364. https://doi.org/10.1146/annurev.py.11.090173.002015.

Das, S.; Pandey, V.; Patel, H. R.; Patel, K. I. Effect of Weather Parameters on Pest-disease of Okra During Summer Season in Middle Gujarat. *J. Agrometeorol.* **2011,** *13* (1), 38–42.

David, R. F.; BozorgMagham, A. E.; Schmale, D. G.; Ross, S. D.; Marr, L. C. Identification of Meteorological Predictors of Fusarium Graminearum Ascospore Release Using Correlation and Causality Analyses. *Eur. J. Plant Pathol.* **2016,** *145* (2), 483–492. https:// doi.org/10.1007/s10658-015-0832-3.

de Langre, E. Effects of Wind on Plants. *Ann. Rev. Fluid Mechanics* **2008,** *40* (1), 141–168. https://doi.org/10.1146/annurev.fluid.40.111406.102135.

Dhingra, O. D.; Sinclair, J. B. Survival of Macrophomina Phaseolina Sclerotia in Soil: Effects of Soil Moisture, Carbon, Nitrogen Ratios, Carbon Sources, and Nitrogen Concentrations. *Phytopathology* **1975,** *65* (3), 236–240. https://doi.org/10.1094/Phyto-65-236.

Dixon, M.; Grace, J. Effect of Wind on the Transpiration of Young Trees. *Ann. Botany* **1984,** *53* (6), 811–819.

Du, Y. C.; Tachibana, S. Effect of Supraoptimal Root Temperature on the Growth, Root Respiration and Sugar Content of Cucumber Plants. *Scientia Horticulturae* **1994,** *58* (4), 289–301. https://doi.org/10.1016/0304-4238(94)90099-X.

Ennos, A. R. Wind as an Ecological Factor. *Trends Ecol. Evol.* **1997,** *12* (3), 108–111. https://doi.org/10.1016/S0169-5347(96)10066-5.

Friesland, H.; Schroedter, H. The Analysis of Weather Factors in Epidemiology. In *Experimental Techniques in Plant Disease Epidemiology*; Kranz, J.; Rotem, H., Eds; Springer: Berlin, Heidelberg, 1988, pp 115–134. https://doi.org/10.1007/978-3-642-95534-1

Garrett, K. A.; Dendy, S. P.; Frank, E. E.; Rouse, M. N.; Travers, S. E. Climate Change Effects on Plant Disease: Genomes to Ecosystems. *Annu. Rev. Phytopathol.* **2006,** *44* (1), 489–509. https://doi.org/10.1146/annurev.phyto.44.070505.143420.

Gautam, H. R.; Bhardwaj, M. L.; Kumar, R. Climate Change and Its Impact on Plant Diseases. *Curr. Sci.* **2013,** *105* (12), 1685–1691. https://doi.org/10.1016/j.eswa.2010.09.036.

Geagea, L.; Huber, L.; Sache, I.; Flura, D.; McCartney, H. A.; Fitt, B. D. L. Influence of Simulated Rain on Dispersal of Rust Spores from Infected Wheat Seedlings. *Agri. Forest Meteorol.* **2000,** *101* (1), 53–66. https://doi.org/10.1016/S0168-1923(99)00155-0.

Gill, J. S.; Sivasithamparam, K.; Smettem, K. R. J. Effect of Soil Moisture at Different Temperatures on Rhizoctonia Root Rot of Wheat Seedlings. *Plant Soil* **2001,** *231* (1), 91–96. https://doi.org/10.1023/A:1010394119522.

Gill, K. K.; Sharma, I.; Jindal, M. M. Effect of Weather Parameters on the Incidence of Stripe Rust in Punjab. *J. Agrometeorol* **2012,** *14*(2), 167–169.

Hardwick, N. V. Weather and Plant Diseases. *Weather* **2002,** *57* (5), 184–190. https://doi.org/10.1002/wea.6080570507.

Huber, L.; Madden, L. V.; Fitt, B. D. L. Rain-splash and Spore Dispersal: A physical Perspective. In *The Epidemiology of Plant Diseases*; Gareth Jones, D., Ed.; Springer: Dordrecht, Netherlands, 1998. https://doi.org/10.1007/978-94-017-3302-1_17.

Jhorar, O. P.; Mathauda, S. S.; Singh, G.; Butler, D. R.; Mavi, H. S. Relationships Between Climatic Variables and Ascochyta Blight of chickpea in Punjab, India. *Agri. Forest Meteorol.* **1997,** *87* (2–3), 171–177. https://doi.org/10.1016/S0168-1923(97)00014-2.

Jhorar, O. P.; Butler, D. R.; Mathauda, S. S. Effects of Leaf Wetness Duration, Relative Humidity, Light and Dark on Infection and Sporulation by Didymella Rabiei on Chickpea. *Plant Pathol.* **1998,** *47* (5), 586–594. https://doi.org/10.1046/j.1365-3059.1998.0280a.x.

Kalisch, A.; Zemek, O.; Schellhardt, S. Adaptation in Agriculture. In *Adaptation to Climate Change with a Focus on Rural Areas and India;* Ilona Porsché; Anna Kalisch; Rosie Füglein Ed.: Deutsche Gesellschaft für Internationale Zusammenarbeit (GIZ) GmbH, New Delhi, India, **2011,** 44–82.

Kalisch, A.; Zemek, O.; Schellhardt, S. Adaptation in Agriculture. *Adaptation to Climate Change with a Focus on Rural Areas and India* **2011,** 44–82.

Kanzaria, K. K.; Dhruj, I. U.; Sahu, D. D. Influence of Weather Parameters on Powdery Mildew Disease of Mustard Under North Saurashtra Agroclimatic Zone. *J. Agrometeorol.* **2013,** *15* (1), 86–88.

Kirkham, D. S.; Hignett, R. C.; Ormerod, P. J. Effects of Interrupted Light on Plant Disease. *Nature* **1974,** *247* (5437), 158–160. https://doi.org/10.1038/247158a0.

Kumar, G.; Chakravarty, N. V. K. A Simple Weather Based Forewarning Model for White Rust in Brassica. *J. Agrometeorol.* **2008,** *10* (1), 75–80.

Leopold, L. B. Dew as a Source of Plant Moisture. *Pacific Sci.* **1951,** *6*, 259–261.

Levy, Y.; Pataky, J. K. Epidemiology of Northern Leaf Blight on Sweet Corn. *Phytoparasitica* **1992,** *20* (1), 53–66. https://doi.org/10.1007/BF02995636.

Li, B.; Wang, L.; Kaseke, K. F.; Li, L.; Seely, M. K. The Impact of Rainfall on Soil Moisture Dynamics in a Foggy Desert. *PLoS One* **2016,** *11* (10). https://doi.org/10.1371/journal.pone.0164982.

López-Bravo, D. F.; Virginio-Filho, E. de M.; Avelino, J. Shade is Conducive to Coffee Rust as Compared to Full Sun Exposure Under Standardized Fruit Load Conditions. *Crop Protection* **2012**, *38*, 21–29. https://doi.org/10.1016/j.cropro.2012.03.011.

Malhotra, R. S.; Singh, K. B.; Rheenen, H. A. van; Pala, M. Genetic Improvement and Agronomic Management of Chickpea with Emphasis on the Mediterranean Region. In *Adaptation of Chickpea in the West Asia and North Africa Region*; Saxena, N. P.; Saxena, M. C.; Johansen, C.; Virmani, S. M.; Harris, H., Eds; ICARDA: Aleppo, Syria, 1996, pp 217–232.

Manning, W. J.; V. Tiedemann, A. Climate Change: Potential Effects of Increased Atmospheric Carbon Dioxide (CO2), Ozone (O3), and Ultraviolet-B (UV-B) Radiation on Plant Diseases. *Environ. Pollution* **1995**. https://doi.org/10.1016/0269-7491(95)91446-R

Mavi, H. S. *Introduction to Agrometeorology*, 2nd ed.; Oxford & IBH Publishing Co. Pvt. Ltd.: New Delhi, 1994.

McCartney, H. A. Dispersal of Spores and Pollen from Crops. *Grana* **1994**, *33* (2), 76–80. https://doi.org/10.1080/00173139409427835.

Miller, P. R. The Effect of Weather on Diseases. *Yearbook Agriculture* **1953**, 83–93. https://doi.org/10.1002/j.1477-8696.1977.tb04568.x.

Nobel, P. S. Wind as an Ecological Factor. In *Physiological Plant Ecology I: Responses to the Physical Environment*; O. L. Lange; P. S. Nobel; C. B. Osmond; H. Ziegler, Eds; Springer: Berlin, Heidelberg, 1981. https://doi.org/10.1007/978-3-642-68090-8_16.

Oerke, E. C.; Dehne, H. W.; Schönbeck, F.;Weber, A. *Crop Production and Crop Protection: Estimated Losses in Major Food and Cash Crops*. Elsevier Science: Amsterdam, The Netherlands, 1994.

Onoda, Y.; Anten, N. P. R. Challenges to Understand Plant Responses to Wind. *Plant Signal. Behav.* **2011**, *6* (7), 1057–1059. https://doi.org/10.4161/psb.6.7.15635.

Pal, R.; Mandal, D.; Kumar, M.; Panja, B. N. Effect of Weather Parameters on the Initiation and Progression of Sheath Blight of Rice. *J. Agrometeorol.* **2017**, *19* (1), 39–43.

Paul, P. A.; Munkvold, G. P. Influence of Temperature and Relative Humidity on Sporulation of Cercosporazeae-maydis and Expansion of Gray Leaf Spot Lesions on Maize Leaves. *Plant Dis.* **2005**, *89* (6), 624–630.

Pivonia, S.; Cohen, R.; Kigel, J.; Katan, J. Effect of Soil Temperature on Disease Development in Melon Plants Infected by *Monosporascus Cannonballus*. *Plant Pathol.* **2002**, *51* (4), 472–479. https://doi.org/10.1046/j.1365-3059.2002.00731.x.

Reddick, D. Effect of Soil Temperature on the Growth of Bean Plants and on Their Susceptibility to a Root Parasite. *Am. J. Botany* **1917**, *4* (9), 513–519.

Savary, S.; Ficke, A.; Aubertot, J. N.; Hollier, C. Crop Losses Due to Diseases and their Implications For Global Food Production Losses and Food Security. *Food Security* **2012**, *4* (4), 519–537. https://doi.org/10.1007/s12571-012-0200-5.

Schwartz, H. F.; Otto, K. L.; Gent, D. H. Relation of Temperature and Rainfall to Development of Xanthomonas and Pantoea Leaf Blights of Onion in Colorado. *Plant Dis.* **2003**, *87* (1), 11–14. https://doi.org/10.1094/PDIS.2003.87.1.11.

Sentelhas, P. C.; Dalla Marta, A.; Orlandini, S.; Santos, E. A.; Gillespie, T. J.; Gleason, M. L. Suitability of Relative Humidity as an Estimator of Leaf Wetness Duration. *Agri. Forest Meteorol.* 148 (3), 392–400. https://doi.org/10.1016/j.agrformet.2007.09.011.

Shew, H. D.; Benson, D. M. Influence of Soil Temperature and Inoculum Density of Phytophthora Cinnamomi on Root Rot of Fraser Fir. *Plant Dis.* **1983**, *67* (5), 522–524. https://doi.org/10.1094/PD-67-522.

Short, G. E.; Wyllie, T. D.; Bristow, P. R. Survival of Macrophomina Phaseolina in Soal and in Residue of Soybean. *Phytopathology* **1980,** *70* (1), 13–17.

Singh, R.; Singh, R.; Singh, D.; Mani, J. K.; Karwasra, S. S.; Beniwal, M. S. Effect of Weather Parameters on Karnal Bunt Disease in Wheat in Karnal Region of Haryana. *J. Agrometeorol.* **2010,** *12* (1), 99–101.

Sinoquet, H.; Le Roux, X.; Adam, B.; Ameglio, T.; Daudet, F. A. RATP: A model for Simulating the Spatial Distribution of Radiation Absorption, Transpiration and Photosynthesis Within Canopies: Application to an Isolated Tree Crown. *Plant Cell Environ.* **2001,** *24* (4), 395–406. https://doi.org/10.1046/j.1365-3040.2001.00694.x.

Smith, V. C.; Ennos, A. R. The Effects of Air Flow and Stem Flexure on the Mechanical and Hydraulic Properties of the Stems of Sunflowers Helianthus Annuus l. *J. Exp. Botany* **2003,** *54* (383), 845–849. https://doi.org/10.1093/jxb/erg068.

Strange, R. N.; Scott, P. R. Plant Disease: A Threat to Global Food Security. *Ann. Rev. Phytopathol.* **2005,** *43* (1), 83–116. https://doi.org/10.1146/annurev.phyto.43.113004.133839.

Te Beest, D. E.; Paveley, N. D.; Shaw, M. W.; van den Bosch, F. Disease–weather Relationships for Powdery Mildew and Yellow Rust on Winter Wheat. *Phytopathology* **2008,** *98* (5), 609–617. https://doi.org/10.1094/PHYTO-98-5-0609.

Tu, J. C.; Tan, C. S. Effect of Soil Moisture on Seed Germination, Plant Growth and Root Rot Severity of Navy Bean in Fusarium Solani Infested Soil. *Comm. Agri. Appl. Biol. Sci.* **2003,** *68* (4B), 609–612.

Van der Wal, A. F. Moisture as a Factor in Epidemiology and Forecasting. In *Water Deficits and Plant Growth*; Kozlowski, T. T., Ed.; Academic Press, 1978, pp 253–295. https://doi.org/https://doi.org/10.1016/B978-0-12-424155-8.50015-8

Wheeler, T.; von Braun, J. Climate Change Impacts on Global Food Security. *Science* **2013,** *341* (508), 508–513. https://doi.org/10.1126/science.1239402.

World Bank. *World Development Report 2010*. The World Bank: Washington DC, 2010.

Xu, Q.; Liu, S.; Wan, X.; Jiang, C.; Song, X.; Wang, J. Effects of Rainfall on Soil Moisture and Water Movement in a Subalpine Dark Coniferous Forest in Southwestern China. *Hydrol. Proc.* **2012,** *26* (25), 3800–3809. https://doi.org/10.1002/hyp.8400.

Yáñez-López, R.; Torres-Pacheco, I.; Guevara-González, R. G.; Hernández-Zul, M. I.; Quijano-Carranza, J. A.; Rico-García, E. The Effect of Climate Change on Plant Diseases. *Afr. J. Biotechnol.* **2012,** *11* (10), 2417–2428. https://doi.org/10.5897/AJB10.2442.

Yarwood, C. E. Water and the Infection Process. In *Water Deficits and Plant Growth*; Kozlowski, T. T., Ed.; Academic Press Inc.; 1978, pp 141–173.

Index

Printed and bound by CPI Group (UK) Ltd, Croydon, CR0 4YY
01777703-0003